Rudolf Bock, Reinhard Niessner

Separation Techniques in Analytical Chemistry

Also of interest

Food Analysis.
Using Ion Chromatography
Edward Muntean, 2022
ISBN 978-3-11-064438-8, e-ISBN 978-3-11-064440-1

Analytical Chemistry.
Principles and Practice
Victor Angelo Soffiantini, 2021
ISBN 978-3-11-072119-5, e-ISBN 978-3-11-072120-1

Atomic Emission Spectrometry.
AES – Spark, Arc, Laser Excitation
Heinz-Gerd Joosten, Alfred Golloch, Jörg Flock and Susan Killewald, 2020
ISBN 978-3-11-052768-1, e-ISBN 978-3-11-052969-2

Chemical Analysis in Cultural Heritage
Luigia Sabbatini, Inez Dorothé van der Werf (Eds.), 2020
ISBN 978-3-11-045641-7, e-ISBN 978-3-11-045753-7

Rudolf Bock, Reinhard Niessner

Separation Techniques in Analytical Chemistry

Distribution in Non-Miscible Phases or by Different Migration Rates in One Phase

DE GRUYTER

Authors
Prof. Dr. Rudolf Bock †

Prof. Dr. Reinhard Niessner
Alter Berg 10A
82319 Starnberg
reinhard.niessner@mytum.de

ISBN 978-3-11-117979-7
e-ISBN (PDF) 978-3-11-118141-7
e-ISBN (EPUB) 978-3-11-118206-3

Library of Congress Control Number: 2023936600

Bibliographic information published by the Deutsche Nationalbibliothek
The Deutsche Nationalbibliothek lists this publication in the Deutsche Nationalbibliografie; detailed bibliographic data are available on the internet at http://dnb.dnb.de.

Cover image: blueringmedia/iStock/Getty Images Plus
Typesetting: Integra Software Services Pvt. Ltd.
Printing and binding: CPI books GmbH, Leck

www.degruyter.com

Preface

Separation techniques are needed everywhere. Even the latest generations of mass spectrometry instruments will thank the user for having subjected a sample mixture to intensive preparation beforehand, whether for reasons of the necessary detection strength or the longevity of the hardware. Countless questions from the fields of life sciences, materials sciences and environmental sciences would be doomed to failure if an optimal *separation of wheat and chaff* did not help to achieve a breakthrough. Similarly, permanently increasing purity requirements are forcing the use of the best and most efficient separation techniques.

The present monograph *Separation Methods in Analytical Chemistry* represents an attempt to provide systematic and comprehensive information on the current state of the art in separation techniques for analytical problems. The first author, Prof. Dr. Rudolf Bock †, published the first edition in 1974. His classification and evaluation of the methods in use at that time was groundbreaking. Not only at my university was this monograph, which was actually only a part of his more comprehensive work, *Methods of Analytical Chemistry,* an essential building block for the in-depth teaching of prospective analytical chemists. To this day, this work is used intensively, as no similarly structured book exists even in the international book market.

Fortunately, it did not take much persuasion to get the publisher de Gruyter to publish a new edition. I was very happy to take on the task of a comprehensive revision, since I myself have benefited considerably from this work in research and teaching. The systematic overall view of separation techniques *à la Bock* also allows a first approach to new tasks in metrology. The classification chosen by Bock according to distribution in two immiscible phases or separation by different migration velocities in one phase is classical. However, he then logically subdivided the various separation methods in such a way that the techniques that had been added in the last 50 years could be incorporated without constraint. This speaks for the division he chose. The reader is also not deprived of the historical roots of a separation technique in the respective preface to a chapter. This seems important to me, as it encourages reading in long-forgotten journals. Doctoral and post-doctoral students in particular should know that the ancients were indeed extremely inventive in dealing with separation problems. And as is well known, every separation process has the option of substance enrichment.

The revised work has been considerably extended in various areas. Especially techniques based on charged particles or Brownian molecular motion have experienced a strong increase. Each chapter concludes with important literature citations to further work, making it easy to get started on a topic. In addition, and for me this is the greatest charm of this book, techniques are also presented which at first glance appear to be exotic and insignificant. In my own working groups, various new developments based on such *forgotten* technologies have been achieved.

https://doi.org/10.1515/9783111181417-202

It is therefore my firm hope that the revised edition (now in English language) will contribute significantly to the comprehensive education of scientists interested in metrology in science & technology, and perhaps also whets the appetite for the further development of tempting separation methods that have not yet been tested with new materials and molecules.

Unfortunately, I myself never had the honor of meeting Rudolf Bock in person. He passed away in September 2012 at the age of 97. His contributions to German analytical chemistry were fundamental and should be preserved. May this book with his thoughts live on long and find numerous uses.

I would like to take this opportunity to thank Karin Sora and her publishing team for their highly efficient and patient cooperation.

Reinhard Niessner
Munich, January 2023

Contents

Preface — V

Part I: **Introduction**

1 Evaluation of separation processes — 3
1.1 Ideal and real separations — **3**
1.2 Separation factor – enrichment factor – depletion factor — **3**
1.3 Separation factors required for analytical separations — **4**
1.4 Limits of applicability of the separation factor – selectivity and specificity of separations — **4**
General literature — **6**

2 Classification of separation processes — 7
2.1 Separations due to different distribution between two immiscible phases — **7**
2.2 Separations due to different migration rates in one phase — **7**
General literature — **8**

3 Chemical reactions during separations — 9
3.1 Reactions of the substance to be separated — **9**
3.2 Masking of disturbances — **9**
3.3 Separation by transport reaction in the gas phase — **9**
3.4 Destruction of contaminants — **11**
General literature — **12**

4 Analytical application of incomplete separations — 14
4.1 General — **14**
4.2 Empirical yield determination — **14**
4.3 Determination of yield with the aid of partition coefficients — **14**
4.4 Isotope dilution method — **15**
4.1.1 Definitions — **15**
4.4.2 Simple isotope dilution with labeled substance (so called *direct* isotope dilution) — **15**
4.4.3 Simple isotope dilution with unlabeled substance — **16**
4.4.4 Double isotope dilution with unlabeled substance — **17**
4.5 Separations with substoichiometric reagent addition — **18**
4.5.1 Principle — **18**
4.5.2 Saturation analysis — **18**
4.5.3 Double isotope dilution with labeled substance — **19**

4.5.4 Method according to Růžička & Starý — 19
4.5.5 Scope of application and effectiveness of the method — 20
4.6 Standard addition method — 21
Literature to the text — 21

5 Concentration data — 23
Literature to the text — 24

Part II: Separations due to different distribution between two immiscible phases

6 Introduction — 27
6.1 Features of the single step – auxiliary phases – partition coefficient – partition isotherm — 27
6.2 Effectiveness of separations by distribution – separation factor – graphical representation of effectiveness — 29
6.3 Practically achievable separation factors — 32
6.4 Improving separations through optimized choice of conditions — 33
6.5 Improving separations by inserting auxiliary substances — 33
6.6 Improve separations by exploiting different rates in setting the distribution equilibria or different reaction rates — 33
6.7 Improving separations by repeating the single step — 34
6.7.1 General – discontinuous and continuous mode of operation – closed-loop process — 34
6.7.2 One-sided repetition — 34
6.7.3 Systematic repetition (cascade) — 36
6.7.4 Systematic repetition (triangular scheme – separation series) — 37
6.7.5 Systematic repetition (separation columns) — 41
6.7.6 Systematic repetition (thin film technology) — 56
6.7.7 Countercurrent method — 57
6.7.8 Cross-flow method — 58
6.8 Overview — 58
Literature to the text — 59

7 Distribution between two liquids — 61
7.1 General — 61
7.1.1 Historical development — 61
7.1.2 Auxiliary phases – distribution isotherm – rate of equilibration–separation factors — 61
7.1.3 Solvents – solvent mixtures – synergistic effect – extracted compounds – chemical reactions during distribution — 63

7.1.4 pH distribution curves — 68
7.1.5 Influence of equionic additives — 73
7.1.6 Influence of complexing agents — 74
7.1.7 Incomplete separations by distribution between two liquids (cf. 1st part, Chap. 4) — 75
7.1.8 Disturbances — 75
7.2 Separations by one-time equilibrium adjustment — 76
7.2.1 Working method and devices — 76
7.2.2 Applications — 76
7.3 Separations by one-sided repetition — 78
7.3.1 Discontinuous mode of operation — 78
7.3.2 Continuous mode of operation — 79
7.4 Separations by systematic repetition: Separation series — 82
7.4.1 Discontinuous transport of the moving phase — 82
7.4.2 Continuous advancement of the moving phase — 84
7.4.3 Circulation method — 85
7.5 Separations by systematic repetition: Column method (distribution chromatography) — 85
7.5.1 Principle — 85
7.5.2 Support material of the stationary phase — 85
7.5.3 Column material and dimensions — 86
7.5.4 Substance introduction — 86
7.5.5 Stationary and mobile phase — 87
7.5.6 Flow velocity of the mobile phase – pressure columns — 88
7.5.7 Influence of the temperature — 89
7.5.8 Stepwise elution – gradient elution — 90
7.5.9 Isolation of components – elution curves – detectors — 90
7.5.10 Efficacy of partition chromatography — 91
7.6 Separations by systematic repetition: Thin-layer technique (thin-layer distribution chromatography) — 92
7.7 Countercurrent distribution — 92
Literature to the text — 93

8 Solubility of gases in liquids — 95
8.1 General — 95
8.1.1 Historical development — 95
8.1.2 Auxiliary phases – distribution isotherm – equilibration rate – separation factors — 95
8.1.3 Solvents — 97
8.2 Separations by one-time equilibrium adjustment — 97
8.2.1 Discontinuous mode of operation — 97
8.2.2 Continuous mode of operation — 100

8.2.3 Circulation method — 101
8.3 Separations by one-sided repetition — 101
8.4 Separations by systematic repetition: Column method (gas chromatography) — 103
8.4.1 Principle — 103
8.4.2 Inert support material for the stationary phase — 103
8.4.3 Column material and dimensions — 104
8.4.4 Sample transfer — 104
8.4.5 Stationary phase — 106
8.4.6 Mobile phase – circulation method — 108
8.4.7 Flow rate of the mobile phase — 109
8.4.8 Influence of temperature on the separations — 109
8.4.9 Gradient elution and reverse gas chromatography — 111
8.4.10 Isolation of individual components from the eluate — 112
8.4.11 Elution curves and detectors — 113
8.4.12 Identifying the bands of an elution diagram — 113
8.4.13 Efficacy and scope of the method — 117
8.5 Countercurrent method — 118
8.6 Cross flow method — 118
General literature — 118
Literature to the text — 119

9 Adsorption and absorption of gases on solids — 120
9.1 General — 120
9.1.1 Historical development — 120
9.1.2 Definitions — 120
9.1.3 Auxiliary phases – distribution isotherms – speed of equilibrium adjustment — 120
9.1.4 Desorption — 122
9.1.5 Adsorbents — 122
9.1.6 Absorbents — 127
9.2 Separations by one-time equilibrium adjustment — 127
9.3 Separations by one-sided repetition — 129
9.3.1 Adsorption process — 129
9.3.2 Absorption method — 129
9.3.3 Circulation method — 129
9.4 Separations by systematic repetition: Column method (gas adsorption chromatography) — 130
9.4.1 Principle — 130
9.4.2 Column material and dimensions — 130
9.4.3 Stationary and mobile phase — 131
9.4.4 Temperature influence — 132

9.4.5 Gradient elution – reverse gas chromatography – displacement technique — 132
9.4.6 Sampling of components – elution curves – detectors — 133
9.4.7 Efficacy and scope of the method — 133
9.5 Countercurrent method — 134
General literature — 134
Literature to the text — 134

10 Adsorption of dissolved substances on solids — 137
10.1 General — 137
10.1.1 Historical development — 137
10.1.2 Auxiliary phases – distribution isotherms – speed of equilibration – irreversible adsorption — 137
10.1.3 Adsorbents — 138
10.1.4 Solvents in adsorption from solutions — 143
10.1.5 Distribution coefficients and separation factors — 144
10.2 Separations by one-time equilibrium adjustment — 145
10.3 Separations by one-sided repetition — 146
10.4 Separations by systematic repetition: Separation series — 148
10.5 Separations by systematic repetition: Column methods (adsorption chromatography) — 149
10.5.1 Principle — 149
10.5.2 Column material and dimensions – filling the columns — 149
10.5.3 Stationary and mobile phase — 150
10.5.4 Substance task and loading of the columns — 150
10.5.5 Influence of the flow velocity of the mobile phase — 151
10.5.6 Influence of the temperature — 151
10.5.7 Gradient elution — 152
10.5.8 Closed loop recirculation procedure — 153
10.5.9 Displacement technique – interposition of auxiliary substances — 154
10.5.10 Detection — 154
10.5.11 Efficacy and scope of the method — 155
10.6 Separations by systematic repetition: Planar methods (thin-layer chromatography, paper chromatography) — 157
10.6.1 Principle and adsorbents — 157
10.6.2 Performing thin-layer chromatography — 158
10.6.3 Performing paper chromatography — 159
10.6.4 Detection and identification of the separated substances — 160
10.6.5 Efficacy and scope of thin film methods — 160
10.7 Cross-flow method — 162
General literature — 162

11 Ion exchange — 167
11.1 General — 167
11.1.1 Historical development — 167
11.1.2 Definitions – functional groups – exchange reactions – regeneration – titration curves – exchange capacity — 167
11.1.3 Exchanger types – porosity and swelling — 170
11.1.4 Auxiliary phases – exchange equilibria – distribution coefficients – exchange isotherm – selectivity and separation factors – exchange of ions of unequal valences – order of binding strengths of different ions — 173
11.1.5 Influence of the distribution coefficients — 179
11.1.6 Rate of equilibration — 181
11.1.7 Side effects of ion exchange: neutral salt adsorption – uptake of nonelectrolytes – sieve effect – irreversible adsorption — 182
11.2 Separations by one-time equilibration — 183
11.3 Separations by one-sided repetition — 184
11.4 Separations by systematic repetition: Column methods (ion exchange chromatography) — 186
11.4.1 Principle — 186
11.4.2 Column material and dimensions — 187
11.4.3 Stationary and mobile phase – loading — 187
11.4.4 Influence of the flow rate of the mobile phase — 188
11.4.5 Influence of the temperature — 189
11.4.6 Stepwise elution – gradient elution — 189
11.4.7 Displacement technique — 189
11.4.8 Isolation of components – detectors — 190
11.4.9 Efficacy and scope of the method — 190
11.5 Separations by systematic repetition: Planar technique (thin-layer ion exchange chromatography) — 193
11.6 Countercurrent method — 194
General literature — 194

12 Solubility: Precipitation methods — 196
12.1 General — 196
12.1.1 Historical development — 196
12.1.2 Definitions — 196
12.1.3 Auxiliary phases – precipitation reactions – precipitation pH values – speed of precipitation reactions — 196
12.1.4 Solubility – influencing the solubility — 199
12.1.5 Supersaturation – failure to precipitate at low concentrations — 202
12.1.6 Entrainment effect – reducing the carry away — 203
12.1.7 Separation of precipitate and solution: Filtration – centrifugation — 206

12.2 Separations by one-time equilibration: Working with an auxiliary phase — **209**
12.2.1 Inorganic precipitation by double conversion — **209**
12.2.2 Organic precipitation reagents for inorganic ions — **210**
12.2.3 Precipitation reactions for organic substances: Double reaction – addition – condensation – inclusion compounds – antigen-antibody reaction – protein precipitation for structure elucidation — **212**
12.2.4 Precipitation by reduction: Gaseous and dissolved reducing agents — **216**
12.2.5 Precipitation by reduction or oxidation: Electrolysis — **216**
12.3 Separations by one-time equilibration: Working with two auxiliary phases — **230**
12.3.1 Precipitation — **230**
12.3.2 Precipitation with solid reagents – precipitation exchange — **234**
12.3.3 Two-phase precipitation — **235**
12.3.4 Cementation — **235**
12.4 Separations by one-sided repetition — **236**
12.4.1 Re-precipitation — **236**
12.4.2 Fractionated precipitation — **237**
12.4.3 Precipitation exchange – precipitation with ion exchangers — **237**
12.5 Separations by systematic repetition: Precipitation chromatography — **238**
12.6 Separations by systematic repetition: Thin-layer technique (precipitation paper chromatography) — **238**
General literature — **239**
Literature to the text — **241**

13 Solubility: Extraction and phase analysis — 243
13.1 General – definitions – auxiliary phases — **243**
13.2 Separations by one-time equilibrium adjustment — **243**
13.3 Separations by one-sided repetition — **243**
13.3.1 Repeated extraction — **243**
13.3.2 Circulation method — **245**
13.3.3 Fractionated extraction – gradient extraction — **246**
13.3.4 Scope of application and effectiveness of the method — **248**
13.4 Countercurrent method — **250**
General literature — **250**
Literature to the text — **251**

14 Solubility: Crystallization —— 253
14.1 General (definitions – auxiliary phases – melting and solubility diagrams) —— **253**
14.2 Separations by one-time crystallization —— **255**
14.3 Separations by one-sided repetition —— **256**
14.3.1 Repeated crystallization from melts (zone melts) —— **256**
14.3.2 Repeated crystallization from solutions —— **258**
14.4 Systematic repetition (crystallization in the triangular scheme – separation series – column method) —— **258**
14.5 Countercurrent method —— **261**
General literature —— **261**
Literature to the text —— **261**

15 Volatilization: Distillation and related processes —— 263
15.1 General —— **263**
15.1.1 Historical development —— **263**
15.1.2 Definitions – auxiliary phases (auxiliary substances) —— **263**
15.1.3 Vapor pressure curves of pure substances – graphical and mathematical representation – Clausius-Clapeyron equation – Cox Diagrams —— **263**
15.1.4 Vapor pressures of binary liquid mixtures – Raoult's law – vapor pressure diagrams – boiling diagrams – equilibrium diagrams —— **266**
15.1.5 Volatility – relative volatility (separation factor) – presentation of the result of distillations (distillation curves) —— **275**
15.2 Separations by one-time equilibrium adjustment —— **278**
15.2.1 Head space analysis – extraction of dissolved gases (gases in metals and solutions) —— **278**
15.2.2 Suction of dissolved gases (gases in metals and solutions) —— **280**
15.3 Separations by one-sided repetition without auxiliary substance —— **284**
15.3.1 Drying —— **284**
15.3.2 Solvent evaporation – overhead distillation —— **285**
15.3.3 Short path distillation —— **287**
15.3.4 Microdiffusion —— **288**
15.4 Separations by one-sided repetition with auxiliary substance —— **290**
15.4.1 Distillation in the stream of a non-condensed gas: Gases in metals – expulsion of gases from solutions with an auxiliary gas – vacuum distillation —— **290**
15.4.2 Distillation in the stream of a subsequently condensed gas (*codistillation*): Steam distillation – distillation from aqueous solutions – distillation in the vapor stream of organic auxiliary liquids —— **292**
15.4.3 Azeotropic distillation —— **296**
15.4.4 Closed loop procedure —— **296**

15.5 Countercurrent method without auxiliary substance — 298
15.5.1 General – use of boiling and equilibrium diagrams – theoretical bottoms (separation stages) – reflux ratio – operating contents — 298
15.5.2 Determining the number of theoretical plates of a distillation column — 302
15.5.3 Column types – return control – vacuum distillation – cryogenic distillation — 302
15.6 Countercurrent method with auxiliary substance — 305
15.6.1 Extractive distillation — 305
15.6.2 Azeotropic distillation — 307
15.6.3 Inserting auxiliary substances — 307
15.7 Efficacy and scope of the method — 307
General literature — 309
More specific introductions — 309
Literature to the text — 309

16 Volatilization: Sublimation — 312
16.1 General — 312
16.1.1 Historical — 312
16.1.2 Definitions – p, T diagrams – auxiliary gases — 312
16.2 Separations by one-sided repetition — 313
16.2.1 Sublimation under normal pressure – vacuum sublimation – freeze drying — 313
16.2.2 Fractionated sublimation – resublimation — 314
16.2.3 Sublimation in the stream of an auxiliary gas – chlorination – evaporation analysis – pyrohydrolysis — 315
16.2.4 Efficacy and scope of the method — 317
General literature — 318
Literature to the text — 318

17 Condensation — 320
17.1 General — 320
17.2 Separations by one-time equilibrium adjustment — 320
17.2.1 Freezing out of a component from a static gas volume – fractional condensation — 320
17.2.2 Condensation in the temperature gradient — 320
17.2.3 Freezing out components from a gas stream — 322
17.3 Systematic repetition of condensations — 322
Literature to the text — 323

Part III: Separations due to different migration rates in one phase

18 Introduction — 327

19 Migration of charge carriers in electric and magnetic fields (mass spectrometry) — 329

19.1 Historical development — **329**
19.2 General — **329**
19.2.1 Definitions — **329**
19.2.2 Principle structure of a mass spectrometer — **330**
19.2.3 Ionization process — **330**
19.2.4 Acceleration of ions in the electric field – energy dispersion — **334**
19.3 One-dimensional mass spectrometric separations (time-of-flight mass spectrometers) — **335**
19.4 Two-dimensional mass spectrometric separations — **337**
19.4.1 Separation of ion species in the magnetic sector field – single focusing mass spectrometers — **337**
19.4.2 Behavior of ion species in the electric sector field – double focusing mass spectrometers — **338**
19.4.3 Quadrupole mass spectrometer — **340**
19.4.4 Ion traps – mass spectrometry — **340**
19.4.5 Continuous mode of operation – differential mobility analyzer — **342**
19.4.6 Preparative mass spectrometry — **344**
19.5 Efficacy of the method – resolving power — **344**
General literature — **346**
Literature to the text — **347**

20 Migration of dissolved charge carriers in an electric field (electrophoresis; electrodialysis) — 350

20.1 Historical development — **350**
20.2 General — **350**
20.2.1 Definitions — **350**
20.2.2 Appearances during current passage through electrolytes (formation of interfaces – heat of current) — **350**
20.2.3 Ion migration in liquid phase – ion mobilities — **352**
20.3 One-dimensional electrophoretic separations without carriers (Tiselius method) — **357**
20.4 One-dimensional electrophoretic separations with carrier (carrier electrophoresis) — **359**

20.4.1 Carrier materials – special features of the use of carriers (adsorption – electroosmosis – suction flow – change in ion migration velocities – relative mobilities) — **359**
20.4.2 Column electrophoresis — **360**
20.4.3 Thin-layer methods – paper electrophoresis – high-voltage electrophoresis – micro- and ultramicro methods — **361**
20.5 Special effects with inhomogeneous separating zones — **362**
20.5.1 Interface stabilization according to Kendall — **363**
20.5.2 Zone sharpening – disc electrophoresis — **364**
20.5.3 Ion focusing — **369**
20.6 One-dimensional separations with semipermeable membrane (electrodialysis) — **370**
20.7 Countercurrent electrophoresis — **371**
20.8 Two-dimensional mode of operation — **371**
20.8.1 Overview — **371**
20.8.2 Procedure with uniform buffer solution — **372**
20.8.3 Immuno-electrophoresis — **373**
20.9 Efficacy and scope of the method — **374**
General literature — **376**
Literature to the text — **376**

21 Migration of particles in the concentration gradient (diffusion) — 378
21.1 Historical development — **378**
21.2 General — **378**
21.2.1 Definitions — **378**
21.2.2 Fick's law — **378**
21.2.3 Membranes — **379**
21.3 One-dimensional separations by diffusion — **381**
21.3.1 Separations in the gas phase — **381**
21.3.2 Separations in liquid phase (dialysis) — **382**
21.4 Pervaporation — **383**
21.5 Countercurrent method — **384**
21.6 Two-dimensional separations by diffusion — **384**
21.7 Efficacy and scope of the method — **390**
General literature — **390**
Literature to the text — **391**

22 Migration of particles in the gravitational field (sedimentation – flotation) — 393
22.1 General – definitions – centrifugal force — **393**
22.2 Separations using heavy liquids — **394**

22.3 Separations by sedimentation in the gravitational field of the earth — **394**
22.4 Separations with the aid of the ultracentrifuge — **394**
22.5 Efficacy and scope of the method — **395**
General literature — **396**
Literature to the text — **397**

23 Separation of particles in the crossed force field — 399
23.1 General – definitions – asymmetric field-flow fractionation – applications — **399**

Index — 403

Part I: **Introduction**

1 Evaluation of separation processes

1.1 Ideal and real separations

The aim of a separation operation is to separate a mixture of – in the simplest case – two substances A and B as completely as possible. If the separation is complete, the mixture is divided into two parts, one of which contains exclusively substance A, the other exclusively B.

However, such a complete partitioning is not achievable in practice; there will always still be some B as impurity in the part with substance A, and correspondingly some A in the part with B (cf. Fig 1.1).

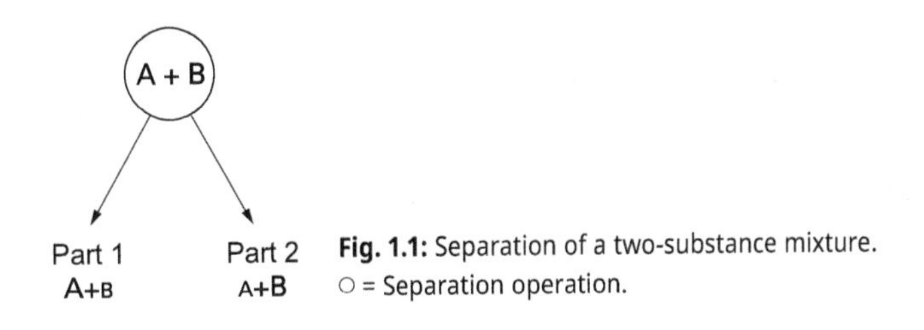

Fig. 1.1: Separation of a two-substance mixture. ○ = Separation operation.

1.2 Separation factor – enrichment factor – depletion factor

The result of a separation operation can be expressed by specifying the impurities in A and in B or by specifying the yields of A and B; thus, to unambiguously characterize the efficiency, two numbers are required for the separation of two substances (correspondingly more for systems with more than two components). To arrive at a simpler description, one uses the *separation factor* β (probably first given by Chlopin), which is defined as follows:

$$\beta = \frac{(\text{concentration of A}/\text{concentration of B})\text{ in part 1}}{(\text{concentration of A}/\text{concentration of B})\text{ in part 2}}. \tag{1}$$

The separation factor is a measure of the effectiveness of the separation of two substances. If, after separation, the concentration ratio in both parts is the same, i.e. [A]/[B] in part 1 = [A]/[B] in part 2, then $\beta = 1$; no separation has taken place.

For an (ideal) complete separation, either the numerator or the denominator of the double fraction would have to become zero, so that β would take the values 0 or ∞. β and $\frac{1}{\beta}$ thus denote the same separation effect, since the choice of numerator and denominator is arbitrary. However, it is common to write the fraction so that $\beta \geq 1$, so that the separation factor increases with an improvement in separation.

Instead of the separation factor, the term *enrichment factor* is sometimes used. For example, the enrichment factor f for substance A (cf. Fig 1.1) is defined as the

https://doi.org/10.1515/9783111181417-001

concentration ratio in Part 1 after separation divided by the concentration ratio in the starting material before separation:

$$f = \frac{[A]/[B]\ \text{in part 1}}{[A]/[B]\ \text{after separation}}$$

Accordingly, one obtains the *depletion factor* f' for A:

$$f' = \frac{[A]/[B]\ \text{in part 2}}{[A]/[B]\ \text{before separation}} \tag{2}$$

The depletion factor becomes smaller as the separation improves (decreasing concentration of A in part 2); more illustrative is the reciprocal value d of this factor, which increases as the efficiency of the separation increases. Referring again to the depletion of A (Fig. 1.1):

$$d = \frac{[A]/[B]\ \text{before separation}}{[A]/[B]\ \text{in part 2}} \tag{3}$$

The quantity d is used above all in radiochemistry when the effectiveness of the removal of interfering radioactive substances is to be described; *d is* then referred to as the *decontamination factor*.

1.3 Separation factors required for analytical separations

In analytical chemistry, the term *quantitative separation* is generally used. This is usually understood to mean the 100 percent separation of the substance sought. According to what has been said so far, complete separations cannot be achieved in practice, and one must therefore define the term *quantitative* arbitrarily.

If a separation is to be sufficient for analytical purposes, it will normally be required that at least 99.9% of substance A is in part 1 and at least 99.9% of substance B is in part 2 after the separation operation (cf. Fig 1.1). The *separation factor* β is then

$$\beta = \frac{99{,}9:0{,}1}{0{,}1:99{,}9} \simeq 10^6.$$

Such a high separation factor is a requirement that can be modified depending on the needs present in the specific case, but on the whole it should be justifiable.

1.4 Limits of applicability of the separation factor – selectivity and specificity of separations

A value of about 10^6 for the separation factor is a necessary but not sufficient condition for analytical applications. If one component of the mixture to be separated is present to an extreme extent in one part after separation, a very large separation factor can be achieved without the other component having to be sufficiently separated.

If, for example, the iron is precipitated with ammonia from a solution containing 100 mg Fe^{3+} and 100 mg Zn^{2+} ions, 10 mg Zn^{2+} ions may be entrained by the iron precipitate and, on the other hand, 0.0003 mg Fe^{3+} ions may be dissolved in the filtrate. The separation factor would then be $3 \cdot 10^6$, thus clearly exceeding the minimum value of 10^6 given above, although only 90% of the zinc was separated from the iron and the separation would therefore be insufficient for analytical purposes.

The separation factor is therefore of limited applicability, and one must specify the yields of both components of the mixture or their impurities for definite statements.

If the analytical sample contains more than two substances (which is usually the case), several separation factors must be specified according to the number of components, which can make the labeling of the separation effect quite complicated.

Therefore, when evaluating a separation process, the *selectivity* must also be taken into account, i.e. the number of substances from which the component sought is separated at the same time. The greater this number, the more efficient the process obviously is. Ideally, even all the impurities in question are sufficiently removed in a single separation operation; then there is a *specific* separation. Some examples of separations that have historically been considered specific are given in Tab. 1.1 (some of these separations require additional masking reactions).

Nowadays, the term *specific* appears to be of little value in the light of the high detection sensitivity of modern analytical methods. Even supposedly highly pure reagents or solvents still contain detectable trace substances.

Tab. 1.1: Known *specific* separations (examples).

Separated element	Compound formed	Separation process
F	$(C_2H)_3SiF$	Shake out with $CHCl_3$ a. o.
Ge	$GeCl_4$	Shake out with $CHCl_3$ a. o.
H	H_2	Diffusion through palladium
Hg	2-Methylthiophene-5-mercury acetate	Shake out with $CHCl_3$ a. o.
Pd	Dimethylglyoxime compound	Shake out with $CHCl_3$ a. o.
Tl	$Tl(C_5H_5)$	Shake out with CH_2Cl_2 etc.

It is much used in immunology; there it tries to describe the (almost) exclusive interaction of antigens with antibodies. But even there, the concept of so-called *cross-reactivity* is better used to describe the achievable selectivity or specificity.

General literature

R.E. Langman, The specificity of immunological reactions, Molecular Immunology *37*, 555–561 (2000).

B.-A. Persson & J. Vessman, The use of selectivity in analytical chemistry – some considerations, TrAC Trends in Analytical Chemistry *20*, 526–553 (2001).

D. Thevenot, K. Tóth, R. Durst & G. Wilson, Electrochemical biosensors: recommended definitions and classification, Pure and Applied Chemistry *71*, 2333–2348 (1999).

2 Classification of separation processes

2.1 Separations due to different distribution between two immiscible phases

In a large group of separation methods, the substance mixture is distributed between two immiscible phases (e.g. the analysis sample is dissolved in water and one of the components is selectively shaken out with an organic liquid). In the ideal case (which, however, can never be achieved), one substance is completely in phase 1 and the other completely in phase 2 after equilibration (cf. Fig 2.1).

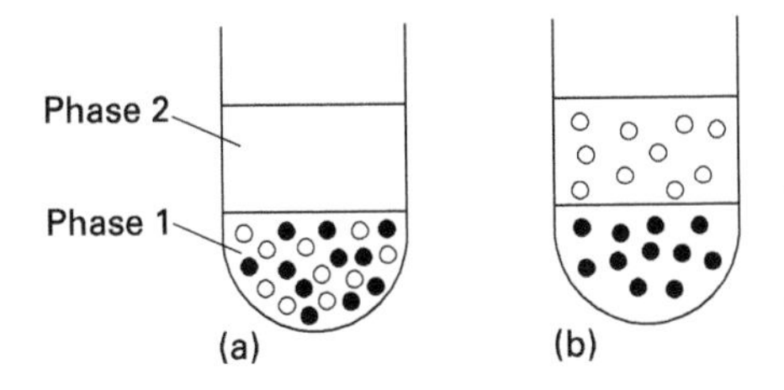

Fig. 2.1: Separation of two substances by distribution between two immiscible phases (principle).
a) Initial mixture before distribution;
b) partition after distribution.

2.2 Separations due to different migration rates in one phase

In the second group of separation methods, the substances to be separated are allowed to migrate in one direction in a homogeneous phase. If the migration velocities of the analytes to be separated are different, the mixture will be separated after passing a certain distance (cf. Fig 2.2). So to speak, a *spatial juxtaposition* (in the mixture) is transformed into a *serial arrangement by time.*

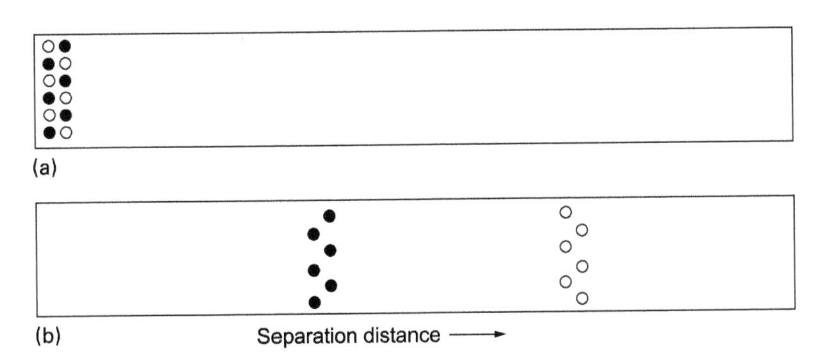

Fig. 2.2: Separation of two analytes by different migration rates in one phase (principle).
a) Initial mixture before separation;
b) conditions after passing a sufficient separation distance.

https://doi.org/10.1515/9783111181417-002

General literature

M. Gupta, Methods for Affinity-Based Separations of Enzymes and Proteins. Birkhäuser Verlag, Basel (2002).

Kirk-Othmer Separation Technology, 2 Volume Set, John Wiley & Sons, Hoboken (2008).

J. Kutter & Y. Fintschenko, Separation Methods in Microanalytical Systems, CRC Press LLC, Boca Raton (2006).

F. Pena-Pereira & M. Tobiszewski, The Application of Green Solvents in Separation Processes. Elsevier, Amsterdam (2017).

M. Purkait, R. Singh, P. Mondal & D. Haldar, Thermal Induced Membrane Separation Processes. Elsevier, Amsterdam (2020).

S. Ramaswamy, H.-J. Huang & B. Ramarao, Separation and Purification Technologies in Biorefineries. John Wiley & Sons, Chichester (2013).

G. Smejkal & A. Lazarev, Separation Methods in Proteomics, CRC Press LLC, Boca Raton (2006).

T. Uragami, Science and Technology of Separation Membranes. John Wiley & Sons, New York (2017).

3 Chemical reactions during separations

3.1 Reactions of the substance to be separated

In many cases, separations are improved or even made possible in the first place by transferring the substance to be separated into a suitable compound form (*change of state*). In this process, the compounds identified according to the scheme

$$AB + CD \rightarrow AC + BD$$

double conversions or also reactions according to the scheme

$$A + B \rightarrow AB$$

are applied; furthermore, oxidation and reduction reactions play a major role.

Two different working methods can be distinguished: Either one performs the reaction before the separation, or one runs it during the separation.

The second method is used mainly for separations by partitioning between two phases; the analyte is in one phase, the reagent in the other, and the reaction proceeds simultaneously with the separation across the phase interface. If such reactions are double conversions, they are referred to as *exchange reactions*. A special case are exchange reactions of complex ligands, which occur according to the scheme

$$AB + C \rightarrow AC + B$$

play out. This reaction is called *ligand exchange*.

3.2 Masking of disturbances

An important aid in separations is the *masking* of interferences by complexing agents. The dissociation of the newly formed complex must be so low that the interfering reaction of the originally present substance no longer occurs. Tab 3.1 lists masking reagents for inorganic ions.

3.3 Separation by transport reaction in the gas phase

If only one partner in the reaction mixture, preferably the analyte to be separated, is changed in its physical properties by a chemical reaction, there are possibilities for separation in a gas phase transport reaction. In this process, the analyte is converted into a component with a higher vapor pressure. This can be used both in static systems and in the flow tube for separation by convection flow and/or diffusion.

https://doi.org/10.1515/9783111181417-003

Tab. 3.1: Masking reagents for inorganic ions (examples)[1].

Ion	Complexing agent
Ag^+	NH_3; CN^-; $S_2O_3^{2-}$; SCN^-; thiourea; diethylenetriamine.
Al^{3+}	F^-; citrate; tartrate; gluconate; oxalate; sulfosalicylate; triethanolamine; Tiron[2]; EDTA[3].
As^{3+}	2,3-dimercaptopropanol; pyrocatechin; citrate; tartrate.
Au^{3+}	Cl^-; Br^-; I^-; CN^-; $S_2O_3^{2-}$; thiourea.
BO_3^{3-}	F^-; pyrocatechin; mannitol and other polyalcohols; hydroxy acids.
Ba^{2+}	Polyphosphates; EDTA.
Be^{2+}	F^-; citrate; tartrate.
Bi^{3+}	I^-; citrate; tartrate; oxalate; EDTA; 2,3-dimercaptopropanol; thiourea; triethanolamine; Tiron.
Ca^{2+}	Polyphosphates; EDTA.
Cd^{2+}	NH_3; CN^-; I^-; $S_2O_3^{2-}$; citrate; sulfosalicylate; tartrate; EDTA; 2,3-dimercaptopropanol; o-phenanthroline.
Co^{2+}	NH_3; CN^-; NO_2^-; $S_2O_3^{2-}$; citrate; sulfosalicylate; tartrate; EDTA; ethylenediamine; 2,3-dimercaptopropanol; o-phenanthroline; Tiron.
Cr^{3+}	Acetate; citrate; tartrate; sulfosalicylate; EDTA; triethanolamine.
Cu^{2+}	NH_3; $S_2O_3^{2-}$; citrate; tartrate; sulfosalicylate; EDTA; Tiron; 2,3-mercapto-propanol; thiosemicarbazide; thiourea; o-phenanthroline.
F^-	Be^{+2}; Al^{+3}; Zr^{+4}; borate.
Fe^{2+}	o-Phenanthroline; *α, α'*-dipyridyl.
Fe^{3+}	F^-; CN^-; citrate; tartrate; sulfosalicylate; oxalate; EDTA; Tiron; thioglycolic acid; triethanolamine; 2,3-dimercaptopropanol.
Ga^{3+}	Tartrate; oxalate; citrate; EDTA.
Ge^{4+}	F^-; oxalate; pyrocatechin.
Hg^{2+}	Cl^-; Br^-; I^-; CN^-; SCN^-; SO_3^{2-}; citrate; tartrate; EDTA; triethanolamine; thiourea; thiosemicarbazide; 2,3-mercaptopropanol.
In^{3+}	Br^-; sulfosalicylate; EDTA; thiourea.
Ir^{4+}	SCN^-; citrate; tartrate; thiourea.
Mg^{2+}	Polyphosphate; EDTA.
Mn^{2+}	NH_3; tartrate; citrate; oxalate; Tiron; sulfosalicylate; EDTA; triethanolamine; o-phenanthroline.
Mn^{3+}	F^-; PO_4^{3-}; 2,3-dimercaptopropanol.
Mo^{6+}	F^-; PO_4^{3-}; H_2O_2; citrate; tartrate; oxalate; EDTA; Tiron; pyrocatechol.
NH_4^+	Formaldehyde.
Nb^{5+}	F^-; H_2O_2; citrate; tartrate; oxalate; Tiron.
Ni^{2+}	NH_3; CN^-; citrate; tartrate; sulfosalicylate; EDTA; o-phenanthroline; 2,3-dimercaptopropanol.
Pb^{2+}	Acetate; citrate; tartrate; EDTA; diethylenetriamine; 2,3-dimercaptopropanol.
Pd^{2+}	NH_3; I^-; CN^-; $S_2O_3^{2-}$; NO_2^-; citrate; tartrate; EDTA; triethanolamine; thiourea.
Pt^{2+}	CN^-; SCN^-; NH_3; citrate; tartrate; EDTA; thiourea.
Pt^{4+}	Cl^-; EDTA.
Rh^{3+}	Citrate; tartrate; thiourea.
Ru^{4+}	CN^-; thiourea.
S^0	S^{2-}; CN^-; SO_3^{2-}.
SO_3^{2-}	Hg^{2+}; formaldehyde.
Sb^{3+}	F^-; I^-; citrate; tartrate; 2,3-dimercaptopropanol.
Sb^{5+}	Citrate; tartrate.
Sc^{3+}	Citrate.
Se^0	S^{2-}; CN^-; SO_3^{2-}.
Se^{4+}	Diaminobenzidine.

Tab. 3.1 (continued)

Ion	Complexing agent
Si^{4+}	F^-; MoO_4^{2-}; WO_4^{2-}; citrate; pyrocatechin.
Sn^{4+}	F^-; Cl^-; Br^-; I^-; citrate; tartrate; oxalate; 2,3-dimercaptopropanol.
Sr^{2+}	Polyphosphate; EDTA.
Ta^{5+}	F^-; H_2O_2; citrate; tartrate; oxalate.
Te^{4+}	I^-.
Th^{4+}	Citrate; tartrate; EDTA; triethanolamine.
Ti^{4+}	F^-; H_2O_2; tartrate; sulfosalicylate; EDTA; triethanolamine; Tiron.
Tl^{3+}	Cl^-; Br^-; CN^-; EDTA.
U^{6+}	Carbonate; H_2O_2; citrate; tartrate; oxalate; sulfosalicylate.
V^{5+}	PO_4^{3-}; H_2O_2.
WO_4^{2-}	F^-; PO_4^{3-}; H_2O_2; citrate; tartrate; oxalate; Tiron; pyrocatechin.
Zn^{2+}	NH_3; CN^- citrate; tartrate; EDTA; o-phenanthroline; 2,3-dimercapto-propanol.
Zr^{4+}	F^-; H_2O_2; citrate; tartrate; oxalate; EDTA; triethanolamine; Tiron.

1) In some of these masking reactions, oxidation state changes of the central atom occur, e.g., Co^{2+} is oxidized to Co^{3+} upon complexation with NO_2^-. Such oxidation state changes are not indicated.
2) 3,5-Pyrocatecholdisulfonic acid disodium salt monohydrate.
3) Ethylenediaminetetraacetic acid, Na salt; a number of chemically similar complexing agents are not listed.

Depending on the position of the thermodynamic equilibrium, the transported (volatile) component is reformed back into the reactant at a point spatially separated from the starting point (sink). The best-known technical example is the Mond-Langer process for producing high-purity nickel by forming transportable nickel tetracarbonyl at 50 °C (T_1) and decomposing it into elemental nickel at a hot surface (T_2 = 200 °C). Tab. 3.2 shows some examples.

Tab. 3.2: Transport reactions with temperature gradient (examples).

Transport reaction	Temperature gradient in the reaction tube
$Ni^0 + 4\ CO \rightarrow Ni(CO)_4\ (g)$	$T_1 > T_2$
$W^0 + 3\ Cl_2 \rightarrow WCl_6\ (g)$	$T_1 > T_2$
$Pt^0 + O_2 \rightarrow PtO_2\ (g)$	$T_2 > T_1$
$Si^0 + SiCl_4 \rightarrow 2\ SiCl_2\ (g)$	$T_2 < T_1$
$IrO_2 + 0.5\ O_2 \rightarrow IrO_3\ (g)$	$T_2 < T_1$

3.4 Destruction of contaminants

One technology frequently used in process technology for the purification of media should be mentioned here: The oxidative degradation of contaminants, for example by photochemical or electrochemical oxidation of organic trace substances, e.g. in water technology. Since water often has to be reused, such processes for the removal

Tab. 3.3: Destructive methods for cleaning liquid or gaseous media.

Decomposition method	Contaminant to be removed
UV (combined with ozone and/or H_2O_2)	Pesticides, VOCs, polycyclic aromatic hydrocarbons, pharmaceuticals, in water
Supercavitation	Cellulose content in paper industry wastewater
Fenton's reagent	Org. traces in water
Photocatalysis (visible light & TiO_2)	Org. traces in gas or water
Electrooxidation	Org. traces in water
Plasma (microwave; dielectric barrier discharge)	Org. traces in gas

of undesirable contaminants are gaining in importance (see Tab. 3.3). Of course, this presupposes that the substance to be purified resists the actual removal process. Similarly, inert process gases can be removed from interfering gas traces. On the one hand, irradiation can split accompanying molecules into their constituents. On the other hand, combination processes in water technology (e.g. UV/ozone or UV/H_2O_2) generate intermediate highly reactive hydroxyl and peroxyl radicals with a high oxidation potential. The degradation kinetics that can be achieved in this process determines the technical implementation (one-pot process (*batch*) or flow tube). In this process, the complex interfering substances are degraded to simple basic building blocks such as H_2O, CO_2 and mineral acids.

Processes that use catalytic degradation principles are of particular interest. For example, TiO_2 nanoparticles (anatase) can be used for photocatalytic degradation of organic substances in the liquid or gas phase under the influence of sunlight.

These techniques are summarized under the term *advanced oxidation process.*

An application in analytical chemistry is currently not known, except unpublished experiments with monochromatic UV laser light for analytical reagent purification. However, it should be possible in principle to use selectively acting chemical oxidation or destruction reactions to remove accompanying contaminants, e.g. in the production of high-purity substances.

General literature

M. Binnewies, R. Glaum, M. Schmidt & P. Schmidt, Chemical vapor transport reactions – A historical review, Zeitschrift füer Anorganische und Allgemeine Chemie *639*, 219–229 (2013).

D. Monticelli, A. Castelletti, D. Civati, S. Recchia & C. Dossi, How to efficiently produce ultrapure acids, International Journal of Analytical Chemistry, 5180610 (2019).

D. Perrin, Masking and Unmasking in Analytical Chemistry. Treatise Anal. Chem., Part 1, (Vol. 2), 599–643. Wiley, New York (1979).

T. Oppenländer, Photochemical Purification of Water and Air: Advanced Oxidation Processes (AOPs): Principles, Reaction Mechanisms, Reactor Concepts, Wiley-VCH, Weinheim (2003).

N. Serpone & A. Emeline, Suggested terms and definitions in photocatalysis and radiocatalysis, International Journal of Photoenergy 4, 91–141 (2002).

L. Sirés, E. Brillas, M. Oturan, M. Rodrigo & M. Panizza, Electrochemical advanced oxidation processes: Today and tomorrow. A review, Environmental Science and Pollution Research 21, 8336–8367 (2014).

4 Analytical application of incomplete separations

4.1 General

As indicated above, quantitative separations, as normally required in analytical chemistry, require a yield of at least 99.9% for each of the separated substances. Under certain conditions, however, such an extensive yield can be dispensed with.

A substance may have been only incompletely separated from a mixture, but the relevant portion may have been completely pure. If, in addition to the separated quantity, the yield during separation is also determined, the total quantity originally present in the initial mixture can be calculated (e.g., for 50% yield by multiplying by the factor 2). Several methods are available for determining the yields.

4.2 Empirical yield determination

In empirical yield determination, the yield obtained for the substance to be determined by the selected separation method is determined using synthetic mixtures as similar as possible to the analysis sample; the value obtained is used to correct the results of subsequent analyses.

The method is occasionally used for the determination of trace components, e.g. pesticide residues in food and other biological material. Since the yields are usually poorly reproducible, the method usually gives only approximate values; this applies especially to separations with very low yields.

4.3 Determination of yield with the aid of partition coefficients

In the processes involving equilibrium between two immiscible phases, there are generally certain laws according to which the concentrations are set. If the *distribution coefficient* (see below) and the volume ratio of the two phases are known, the yield can be calculated from the quantity of substance separated; however, any concentration dependence of the distribution coefficient must be taken into account.

The procedure is used in several variants, which will be discussed below in the corresponding sections.

https://doi.org/10.1515/9783111181417-004

4.4 Isotope dilution method

4.1.1 Definitions

The *isotope dilution method*, which goes back to Hahn (1923) and v. Hevesy (1932), uses the addition of isotopes of the element to be determined or of labeled compounds to the starting substance. Preferably, mixtures of radioactive and inactive elements (or compounds) are used, since the activities can be measured particularly easily and sensitively. However, it is also possible to start from mixtures of stable isotopes, but then the isotope ratios must be determined after the separations using the more complex mass spectrometric method.

The isotope dilution method is performed in several variants; the following symbols will be used when discussing them:

$$m_x = \text{sought analyte within sample}$$

$$m_o = \text{added analyte mass (isotope)}$$

$$m_1 = \text{mass of separated pure part}$$

$$a_x = \text{activity of sought analyte (eg. imp./min)}$$

$$a_o = \text{activity of added analyte mass}$$

$$a_1 = \text{activity of separated pure part}$$

$$S_x = \frac{a_x}{m_x} = \text{specific activity}^1 \text{ of sought analyte}$$

$$S_o = \frac{a_o}{m_o} = \text{specific activity of added analyte}$$

$$S_1 = \frac{a_1}{m_1} = \text{specific activity of separated pure analyte.}$$

4.4.2 Simple isotope dilution with labeled substance (so called *direct* isotope dilution)

In *direct isotope dilution*, a radioactive isotope of the element to be determined (or the sought compound in radioactively labeled form) is added to a weighed quantity of analysis sample. The amount added must be negligibly small compared to the amount to be determined, but the activity must be so high that it can be conveniently measured. Then separate any part of the sought substance together with the added isotope in pure form and determine the activity of this part. Since the isotopes of an element

1 Specific activity is the activity per unit weight A.

behave the same way in chemical separations[2] (the same is true for inactive and labeled compounds), the loss of sought substance is as great as the loss of active substance, and it holds:

$$m_X = m_1 \cdot \frac{a_0}{a_1}. \tag{1}$$

If the added amount of active isotope is not negligibly small compared to the amount of substance sought, the dilution of the sample by the addition must be taken into account; in addition to the activity a_o, the amount m_o of added isotope (e.g., in grams) must then be known. The calculation is based on the following consideration:

The added radioactive substance quantity m_o with the specific activity S_o contains the total activity $a_o = S_o \cdot m_o$. After mixing with the inactive analytical sample containing the amount m_X of sought substance (= dilution), the total activity remains the same, but the amount m_o of substance has increased by the amount m_X. Consequently, the specific activity of the added isotope has decreased. This specific activity S_1, which is now present, is determined by separating a portion of the element sought with the active addition in pure form. It holds:

$$S_0 \cdot m_0 = S_1 \cdot (m_0 + m_X);$$

$$m_X = \frac{S_0 - S_1}{S_1} \cdot m_0 = \left(\frac{S_0}{S_1} - 1\right) \cdot m_0. \tag{2}$$

For very small m_o, S_o becomes very large versus S_1; the latter can be neglected, and eq. (2) merges into eq. (1).

The error becomes smaller for direct isotope dilution the larger the amount of substance m_X present compared to the amount m_o of active isotope added.

The method is therefore particularly suitable for determining larger substance quantities m_X.

4.4.3 Simple isotope dilution with unlabeled substance

The method of simple isotope dilution with unlabeled substance is also called *inverse isotope dilution.* This method is used when radioactive substances are to be determined which are present in such small quantities that they cannot be isolated, but their activities are sufficient for measurement.

An inactive isotope of the element to be determined is added to the present mixture in a larger, known amount m_o and then a part of it is separated in pure form. The separated part m_1 is determined and its activity a_1 is measured.

2 Only in the case of hydrogen isotopes can noticeable deviations occur.

The percentage loss of addition is equal to the percentage loss of activity (and quantity) of the element being searched for, and the following holds true

$$a_x = a_1 \cdot \frac{m_o}{m_1}. \tag{3}$$

In contrast to the previous method, only the activity a_x of the sought element present in the analytical sample can be determined here, not its quantity m_x.

In the main application of this method, neutron activation analysis, m_x is determined by a separate determination of the specific activity $S_x = \frac{a_x}{m_x}$ of the sought element is determined: Together with the analysis substance, one irradiates a pure sample of the sought element, on which the specific activity S_x achieved under the irradiation conditions applied is then measured.

The method is more accurate the larger the added quantity m_o is compared to the sought quantity m_x; it is therefore particularly suitable for determining very small quantities of radioactive substances.

4.4.4 Double isotope dilution with unlabeled substance

The method described in the previous section allows directly only the determination of the activity a_x of the sought substance; to determine its amount, a second determinant (S_x) must be used.

The quantity sought can still be obtained by a second method: Two aliquots of *equal size* are taken from the analysis sample. To aliquot I, add the amount m_o, to aliquot II, add the amount of m_o' of unlabeled substance. From each of these two aliquots, a subset of the sought substance (m_1 or m_1') is isolated in pure form and its specific activity S_1 resp. S_1' is determined.

Since in both aliquots the sought substance quantity m_x[3] is equal, the following holds:

$$S_x \cdot m_x = S_1 \cdot (m_x + m_o) = S_1' \cdot (m_x + m_o') \tag{4}$$

It follows:

$$m_x = \frac{S_1' \cdot m_o' - S_1 \cdot m_o}{S_1 - S_1'}. \tag{5}$$

The method is mainly used in the study of metabolic processes with the help of radioactive labels. The same compound in labeled form is added to the biological material

3 Note that in this example m_x is not the total amount of the searched substance in the analysis sample, but only in the aliquot part. The actual value searched for must be calculated from this.

in which the path of a specific substance is to be followed. The specific activity of the total amount now present is therefore not known, but this amount can be determined by the described double admixture of unlabeled compound.

The errors of the method become small if the added substance amount m_o is large versus m_x and if furthermore m_o' is large versus m_o.

4.5 Separations with substoichiometric reagent addition

4.5.1 Principle

In the methods with *substoichiometric reagent addition*,[4] the substance to be determined is mixed with an insufficient amount of reagent for complete conversion. Either a partial amount is thereby converted into a compound suitable for separation, or a partial amount is masked so that only the unmasked portion can be separated. A prerequisite for the applicability of the method is that the amount of substance to be determined is known in terms of magnitude.

4.5.2 Saturation analysis

The simplest embodiment of this method is the so-called *saturation analysis*. To label the analysis sample, the substance to be determined is added in radioactive form, then a suitable reagent is added in insufficient quantity and the reacted part (or the non-masked part) is separated.

If the amount of reagent and the stoichiometry of the reaction are known, the separated amount m_1 is given, and furthermore the amount m_o and the activity a_o of the additive used for labeling are known. By measuring the activity a_1 of the separated part, the sought amount m_x of the component to be determined can be calculated according to eq. (1) or eq. (2).

The method can thus be regarded as a variant of direct isotope dilution. The advantage over the original method is that even very small amounts m_x of unknown substance can be determined, provided that chemical reactions are available which are practically quantitative even at extremely low concentrations.

4 The expression originates from Růžička & Starý (1961), the method was in principle first used by Zimakov (1958) & Rozhavskii.

4.5.3 Double isotope dilution with labeled substance

Occasionally, it is possible to separate a subset of the substance to be determined using substoichiometric reagent addition, but not to accurately determine the absolute amount of this fraction. This means that the amount m_1 is not known. The second determination equation then required can be obtained as follows:

An aliquot of the analytical sample is mixed with the amount m_0 of the sought substance in radioactive form (specific activity S_o). The specific activity of the mixture is then

$$*S = \frac{S_0 \cdot m_0}{m_0 + m_x} \tag{6}$$

To a second, *equally large* aliquot portion of the sample, add the amount of m_o' of the same radioactive isotope (specific activity S_o). The specific activity of this mixture is

$$*S' = \frac{S_0 \cdot m_0'}{m_0' + m_x} \tag{7}$$

Now isolate a portion of the sought substance from each of the two mixtures using the substoichiometric reagent addition; proceed in such a way that the isolated subsets m_1 and m_1' are equal in size ($m_1 = m_1'$). Finally, the activities a_1 and a_1' of the subsets are measured. It holds:

$$a_1 = *S \cdot m_1 = \frac{S_0 \cdot m_0}{m_0 + m_x} \cdot m_1 \tag{8}$$

$$a_1' = *S' \cdot m_1 = \frac{S_0 \cdot m_0'}{m_0' + m_x} \cdot m_1 \tag{9}$$

This gives two equations with two unknowns (m_x and m_1), from which m_x can be calculated by eliminating m_1:

$$m_x = \frac{m_0 \cdot m_0'(a_1' - a_1)}{a_1 \cdot m_0' - a_1' \cdot m_0} = \frac{m_0'(a_1' - a_1)}{\frac{m_0'}{m_0} \cdot a_1 - a_1'} \tag{10}$$

The error of this method becomes small when $m_o = m_x$ and $m_o' \gg m_x$.

4.5.4 Method according to Růžička & Starý

Another variant of separation with substoichiometric reagent addition was given by Růžička & Starý (1961). The sample, which contains the element to be determined only in a very small amount m_x, is mixed with an approximately equal amount m_o of radioactive isotope of the specific activity S_o. Then, with the help of a substoichiometric reaction, a part m_1 is separated and its activity a_1 is measured; eq. (2) is valid:

$$m_x = \left(\frac{S_o}{S_1} - 1\right) \cdot m_o \tag{11}$$

Now, the same amount of substance m_1 is separated from the radioisotope that was added to the sample using the same substoichiometric reagent amount. This has the activity a_1' and the specific activity S_o.

Since $S_o = a_1'/m_1$ and $S_1 = a_1/m_1$, we get:

$$m_x = \left(\frac{a_1'}{a_1} - 1\right) \cdot m_o \tag{12}$$

This method again saves the determination of the separated amount m_1, which can be difficult to determine in the presence of very small amounts of substance to be determined.

4.5.5 Scope of application and effectiveness of the method

Sub-stoichiometric methods may involve precipitation with insufficient amount of reagent, electrolysis with insufficient amount of current, or complexation reactions with insufficient amount of complexing agent in combination with various separation methods, e.g., shake-out or ion exchange methods, and others. The execution can be automated.

The advantage of the substoichiometric reagent addition is primarily that the isotope dilution methods can be extended to very small amounts of substance. In addition, in many cases the separated subsets are obtained much more pure, i.e. the separations proceed much more selectively.

An example is the separation of mercury from copper-containing solutions by shaking out the dithizone compound. In the usual method of working with excess reagent, both elements go into the organic phase. If, on the other hand, such a small amount of dithizone is used that the mercury is only partially reacted, it is obtained practically free of copper, since the complexation constant of the latter with dithizone is several orders of magnitude smaller than that of mercury.

A disadvantage of the substoichiometric methods is that an approximate knowledge of the present amount of substance to be determined is required. In addition, it is often difficult to find suitable reactions that are quantitative even at extremely low concentrations.

4.6 Standard addition method

In a method for the determination of radioactive substances given by Alian (1968), not an inactive isotope of the element in question is mixed in, but the active isotope itself; the procedure is as follows:

The element to be determined is isolated from an aliquot part of the analysis solution in unknown yield x, but pure form. The activity a of the separated portion is then determined.

A known activity a_s of the element to be determined is added to a second aliquot part of the analytical solution of the same size and, in the same way as for the first part, a partial quantity is isolated in pure form. Finally, the activity a_m of this portion is also measured.

The yield x in the separation is given by

$$x(\%) = \frac{a_m - a}{a_s} \cdot 100 \tag{13}$$

The prerequisite is that the yield x is the same for both separations. The method can be transferred to inactive substances by substituting the substance quantities (e.g. in grams) for the activities in eq. (13).

The addition of stable isotope-labeled analytes with known content to an aliquot of an unknown sample enables the absolute determination of the analyte. The concentration determination before and after addition of the stable isotope labeling substance is performed by mass spectrometry. This technique is now widely used for residue determination in food. A disadvantage is the time-consuming synthesis of appropriately labeled analytes. The sample must also be measured twice (with and without standard addition).

Literature to the text

C. Cervino, S. Asam, D. Knopp, M. Rychlik, & R. Niessner, Use of isotope-labeled aflatoxins for LC – MS/MS stable isotope dilution analysis of foods, J. Agric. Food Chem. *56*, 1873–1879 (2008).

E. Ciccimaro & I.A. Blair, Stable-isotope dilution LC-MS for quantitative biomarker analysis, Bioanalysis *2*, 311–341 (2010).

R.E. Hamon, D.R. Parker, & E. Lombi, Advances in isotopic dilution techniques in trace element research: A review of methodologies, benefits, and limitations, Advances in Agronomy *99*, 289–343 (2008).

S. Holgersson, Hands-on model of the principle of isotope dilution analysis for use in an interactive teaching and learning classroom exercise, Journal of Chemical Education *98*, 1208–1220 (2021).

K. Kabytaev & A. Stoyanov, Quantitative proteomics with isotope dilution analysis: principles and applications, Current Proteomics *13*, 61–67 (2016).

K. Kristjansdottir & S.J. Kron, Stable-isotope labeling for protein quantitation by mass spectrometry, Current Proteomics *7*, 144–155 (2010).

M. Rychlik & S. Asam, Stable isotope dilution assays in mycotoxin analysis, Analytical and Bioanalytical Chemistry *390*, 617–628 (2008).

N. Tretyakova, M. GogginD. Sangaraju & G. Janis, Quantitation of DNA adducts by stable isotope dilution mass spectrometry, Chemical Research in Toxicology *25*, 2007–203 (2012).

W. Xia et al., Advances of stable isotope technology in food safety analysis and nutrient metabolism research, Food Chemistry 408, 135191 (2023).

5 Concentration data

A confusing variety of designations and units of measurement are used in the literature to indicate the concentrations of components of a mixture of substances; because of the fundamental importance of the concept of concentration for all areas of chemistry, an overview of the principal possibilities and the most important expressions is given. The considerations are mainly carried out for solutions, but they can easily be transferred to other mixtures, e.g. solid substances, and to multicomponent systems.

First, one must distinguish whether the amount of solute is related to a specific amount of solvent or to solution, i.e., whether one wants to express the quantity ratio m_1/m_2 or $\frac{m_1}{m_1+m_2}$ (the latter is more usual).

Then the quantities can be specified in units of weight or volume. The resulting possibilities are shown in Tab. 5.1.

Tab. 5.1: Solubility data.

A – 1	B – 1
$\frac{\text{Weight of solute}}{\text{Weight of solvent}}$	$\frac{\text{Weight of solute}}{\text{Weight of solution}}$
A – 2	**B – 2**
$\frac{\text{Weight of solute}}{\text{Volume of solvent}}$	$\frac{\text{Weight of solute}}{\text{Volume of solution}}$
A – 3	**B – 3**
$\frac{\text{Volume of solute}}{\text{Volume of solvent}}$	$\frac{\text{Volume of solute}}{\text{Volume of solution}}$

From these 6 basic types, depending on the choice of quantity, numerous concentration designations are derived, of which only some of the most important are listed:

A – 1. Often one finds the indication *g solute/100 g solvent*, more rarely *g solute/1000 g solvent*. The designation *mole dissolved/1000 g* H_2O = *molality* should also be mentioned here.

B – 1. The concentration indication *g dissolved/100 g solution*, defined as *weight percent*, is widely used, for which often only the designation % is used.[1]

For very low concentrations one finds the indications *ppm* = parts per million = parts in 10^6 parts (= g/ton, mg/kg, µg/g or 10^{-4}%), *ppb* = parts per billion = parts in 10^9 parts

1 In contrast to the German-language literature, the designation % is also used in the Anglo-Saxon language area for the concentration indication *g dissolved/100 ml solution*. To avoid ambiguity, the dimension is then usually quoted (*g/g* or *g/v*; for the indication in Tab. 5.1 under B - 3 correspondingly *.v/v*.

https://doi.org/10.1515/9783111181417-005

(= mg/ton, µg/kg, ng/g or 10^{-7}%), *ppt* = parts per trillion = parts in 10^{12} parts (= µg/ton, ng/kg, pg/g or 10^{-10}%) and *ppq* = parts per quadrillion = parts in 10^{15} parts (= ng/ton, pg/kg, fg/g or 10^{-13}%).

Furthermore, the so-called *mole fraction* = mole dissolved/mole mixture and the indication *mole percent* (or *atom percent*) = mole dissolved/100 mole mixture should be mentioned here.

A – 2. The weight of solute is sometimes referred to 100 ml or 1 l of solvent, *g solute/100 ml solvent* or *g solute/l solvent.*

B – 2. Very frequently used designations are *g solute/100 ml solution* and *g solute/1 solution*, from which the *molar* and *normal* solutions are also derived (= *mole solute/1 solution* and *equivalent solute/1 solution*, respectively). These designations are summarized under the term *volume concentration.*

A – 3. The designation method given under A – 3 is used for mixtures of gases, for solutions of gases in liquids and for mixtures of liquids, e.g. *ml dissolved/100 ml solvent.* For mixtures of liquids, instead of referring to 100 ml, the ratio is often given as follows: Liquid 1/Liquid 2 = a: b or a + b, e.g. acetone/water 7: 3 or 7 + 3.

B – 3. Also for mixtures of gases and of liquids, the designation *volume dissolved/volume of solution* is used, e.g. *volume dissolved/100 volumes solution* = volume percent; for gas mixtures: *volume of a gas/100 volumes gas mixture.*

Concentration values that contain volume units are temperature-dependent. Furthermore, in contrast to masses, volumes do not behave additively during mixing, so that the volume of a mixture cannot be calculated from the volumes of the individual components. In the case of gas mixtures, however, deviations from additivity are usually negligible.

The frequently asked task to convert concentrations of the dimension *g/g* into concentrations of the dimension *g/v* (and vice versa) requires knowledge of the density of the solution.

Literature to the text

IUPAC, Compendium of Chemical Terminology (the *Gold Book*), ISBN 0-86542-684-8) (2012).

Part II: **Separations due to different distribution between two immiscible phases**

6 Introduction

6.1 Features of the single step – auxiliary phases – partition coefficient – partition isotherm

Decisively important characteristics of separation processes by distribution between two immiscible phases are the nature and the aggregate state of the phases concerned as well as the regularities to which the distribution is subject. These characteristics are referred to as the *properties of the single step or the single stage.* Tab. 6.1 shows the phase pairs used in this group of separation methods and the fundamentals of the associated separations.

Tab. 6.1: Phase pairs and fundamentals of separations by distribution between two immiscible phases.

Phase pair	Separation principle	Separation process (examples)
Liquid – Liquid	Different partition coefficients	Solvent extraction; Distribution chromatography
Gas – Liquid	Different solubility of gases in liquids	Absorption; Gas Chromatography
Gas – Solid	Different adsorption of gases on solids	Gas chromatography
Liquid – Solid	Different adsorption of solutes on solids	Adsorption; Adsorption Chromatography
Liquid – Solid	Different ion exchange coefficients	Ion exchange; Ion exchange chromatography
Liquid – Solid	Different solubility in liquids and melts	Precipitation; Co-precipitation; Extraction
Liquid – Solid	Different temperature dependence of the solubility	Crystallization; Zone melting
Gas – Liquid	Different vapor pressures	Distillation
Gas – Solid	Different sublimation pressures	Sublimation

Auxiliary phases. The phase pairs required for separations of this group can be formed from the substances to be separated themselves. However, they can also consist of added foreign substances, so-called auxiliary phases, by which the separation is facilitated or even made possible in the first place. The number of auxiliary phases can be one or two (rarely three); processes with 2 auxiliary phases have become particularly important.

https://doi.org/10.1515/9783111181417-006

Examples:
- Procedure without auxiliary phase: separation of CCl_4 and $CHCl_3$ by distillation.
- Method with an auxiliary phase: separation of $KClO_4$ and $NaClO_4$ by extraction of the solid mixture with ethanol. Auxiliary phase: Ethanol
- Procedure with two auxiliary phases: Separation of Fe^{3+} and Al^{3+} by shaking out the iron from aqueous 6 N HCl solution with ether. Auxiliary phases: Aqueous HCl solution and ether.

Partition coefficient. If a substance A is distributed between phases 1 and 2, a distribution coefficient α_A can be specified in each case as the equilibrium constant of the distribution equilibrium:

$$\alpha_A = \frac{\text{Concentration of A in phase 1}}{\text{Concentration of A in phase 2}} = \frac{c_1}{c_2}. \tag{1}$$

The partition coefficient depends not only on the nature of the two phases and the distributed substance, but also on several other variables: On temperature and pressure, and often also on the concentration of substance A and on the nature and concentration of other substances in the system in question.

The distribution coefficient α is often preferred to more descriptive percentage values, but these depend on the volume ratio V_1/V_2 of the two phases.

One obtains for the percentage of substance that goes to phase 1 in the distribution the expression

$$\%\text{ in phase 1} = \frac{100 \cdot \alpha \cdot V_1/V_2}{1 + \alpha \cdot V_1/V_2} \tag{2}$$

If the volumes of both phases are equal after distribution, eq. (2) simplifies, and the so-called *percentage distribution P* is obtained:

$$P = \frac{100 \cdot \alpha}{1 + \alpha}, \tag{3}$$

which is often used as particularly descriptive, especially in the distribution between two liquids.

Distribution isotherm. Of greatest importance for separations by distribution is the behavior of the distribution coefficient α when the concentration of the distributed substance A changes. This function is called the *distribution isotherm* because it is determined at a certain temperature. Ideally, α is independent of the concentration of A, in which case there is a so-called *Nernst distribution* (an analogous behavior for solutions of gases in liquids had been found earlier by Henry (1803)). In the majority of distribution methods, however, other regularities occur.

The Nernst distribution (1891) is the most favorable form of the distribution isotherm for separations; for other isotherms, one works as far as possible in concentration ranges where the Nernst distribution is approximated.

The distribution isotherm is often represented graphically, and various types of representation can be used. Usually, one plots the concentration c_1 in one phase against the concentration c_2 in the other phase after equilibration; the Nernst distribution then gives a straight line starting from the coordinate starting point (cf. Fig 6.1b). Furthermore, the change of α or of P with increasing initial concentration c_o in one of the two phases can be given (cf. Fig 6.1a).

If a distribution isotherm of the form

$$\alpha = \frac{c_1^n}{c_2} \tag{4}$$

then a plot of log c_1 against log c_2 is favorable; a straight line is then obtained from whose slope n can be determined.

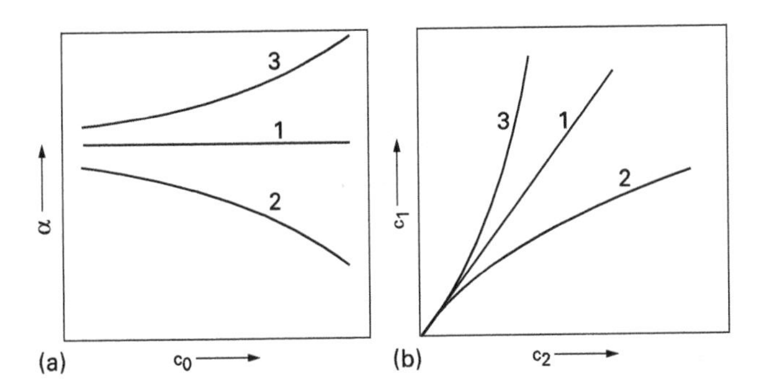

Fig. 6.1: Graphical representation of distribution isotherms.
a) Dependence of the distribution coefficient α on the initial concentration c_o of the distributed substance in a phase; 1 = Nernst distribution;
b) dependence of concentration c_1 in phase 1 on c_2 in phase 2 (concentrations after equilibration); 1 = Nernst distribution.

6.2 Effectiveness of separations by distribution – separation factor – graphical representation of effectiveness

If substances are to be separated by distribution between two immiscible phases, their distribution coefficients must be different, otherwise there is no separation effect. By analogy with Fig. 1.1 in Part 1, the result of such a separation can be shown where the two parts into which the mixture is split are formed by the phases (Fig. 6.2).

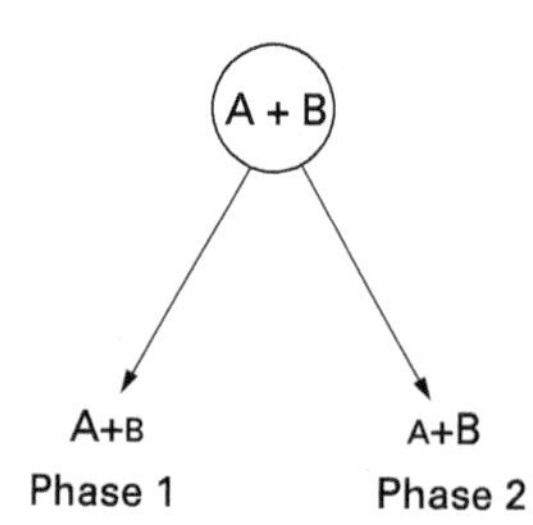

Fig. 6.2: Separation of two substances by distribution between two phases.
○ = Separation operation.

Accordingly, the separation factor β is defined:

$$\beta = \frac{(\text{Conc. of A/Conc. of B}) \text{ in phase 1}}{(\text{Conc. of A/Conc. of B}) \text{ in phase 2}}. \tag{5}$$

This expression is identical to the ratio of the partition coefficients α_A and α_B of substances A and B:

$$\beta \equiv \frac{\alpha_A}{\alpha_B}. \tag{6}$$

Again, the objections made above to the overly broad application of the separation factor must be considered, and one must specify the impurities in each phase or the yields for definite statements about the effectiveness of a separation.

Another formulation results from the already mentioned requirement that 99.9% of substance A should be in phase 1 and 99.9% of substance B in phase 2. The distribution coefficients are then (with equal volume of both phases)

$$\begin{aligned} &\alpha_A = 999 \text{ and } \alpha_B = \frac{1}{999}, \text{ are symmetrical to 1, eg..} \\ &\alpha_A \cdot \alpha_B = 1. \end{aligned} \tag{7}$$

When specifying separation factors, the at least approximate validity of this relationship is usually implicitly assumed.

The behavior of the separation factor with changes in concentration of the substances to be separated results from the concentration dependencies of the individual distribution coefficients, i.e. from the distribution isotherms. Ideally, a Nernst distribution is present for all components of a mixture; then the separation factors are independent of both the concentration ratio and the absolute concentrations of the substances concerned.

Furthermore, the distribution coefficients should not influence each other as far as possible and should not be changed by the presence of further components, as otherwise the ratios in a system would become very confusing. However, these requirements are often not met.

Graphical representation of the efficiency. The ratios for the separation of two-substance mixtures without auxiliary phases can be graphically represented; the concentration of substance A in phase 1 is plotted as abscissa against the concentration of the same substance A in phase 2 as ordinate in a square diagram (see Fig. 6.3). The concentrations are given in percent by weight, mole percent or mole fractions.

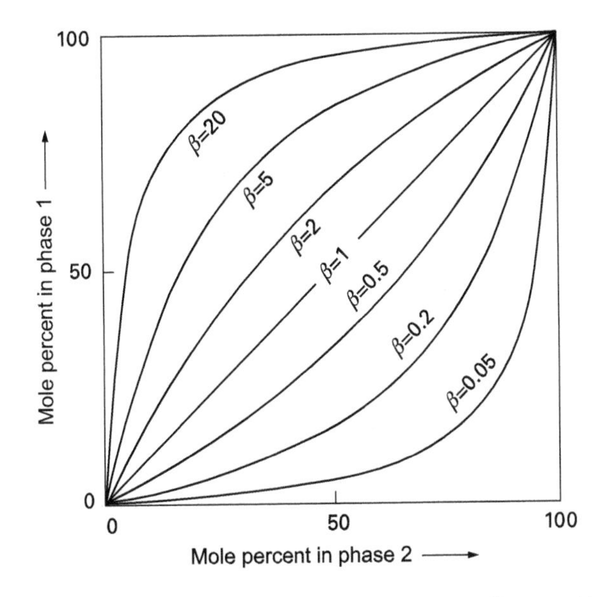

Fig. 6.3: Representation of separations in the square diagram with different separation factors for two substances each. Nernst distribution.

In systems with auxiliary phases, the added auxiliary substances are not taken into account when specifying the concentrations and the sum of the concentrations of the two substances to be separated in each phase is set equal to 100% (or equal to 1 when specifying mole fractions). However, this procedure leaves one variable, the total concentration, out of consideration.

In this way, one obtains diagrams in which for $\beta = 1$ all values lie on the diagonal, while for $\beta \neq 1$ in the ideal case (Nernst distribution) hyperbolas result which are further away from the diagonal the more the separation factor β differs from the value 1, i.e. the better the separation.

If there are no Nernst distributions for the substances to be separated, more or less asymmetric curves occur which may even intersect the diagonal (cf. Fig 6.4).

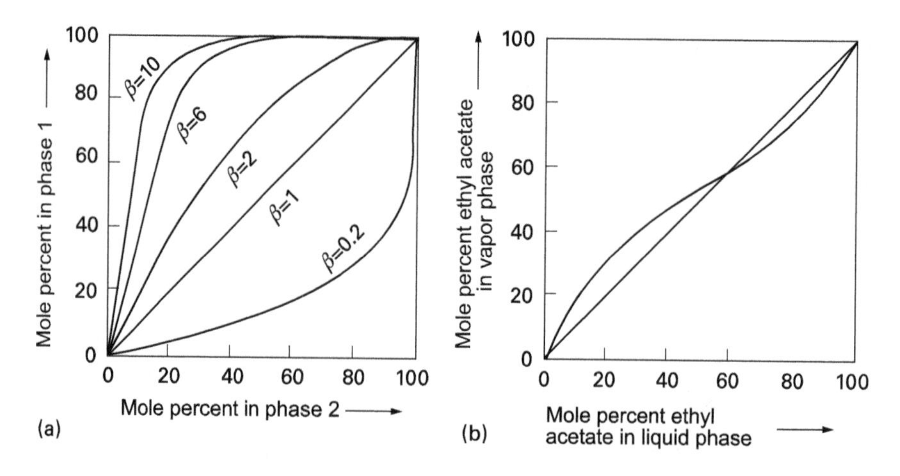

Fig. 6.4: Representation of separations in the square diagram for two cases of non-ideal distribution. a) So-called *logarithmic* distribution;
b) formation of an azeotrope (acetic acid ethyl ester-ethanol system at 760 Torr).

6.3 Practically achievable separation factors

Most of the known separations do not meet the analytical requirements with respect to the size of the separation factor and the level of the yields (cf. Tab. 6.2); therefore, one has to try to improve insufficient separation effects.

Tab. 6.2: Separation factors for some separations.

Separation	Separation factor
Isotope separations by diffusion	approx. 1.001–1.005
Separation of rare earth elements	1,1–1,5
Laser isotope separation (SILEX)	2–20
Zr-Hf separation	3–15
Crystallization of NaCl and KCl (technical)	approx. 1000
Mass spectrometry (preparative)	10–10 000
Quantitative analysis (99.9%)	approx. 1 000 000
Cu-Zn (electrolytic)	> 80 000 000
Absolute separation	∞

Several options are available to improve the effectiveness of separations:
- Increase the separation factor by favorable choice of conditions during separation;
- Insertion of auxiliary substances;

- Exploiting different rates of chemical reactions;
- Repeat the separation operation.

Of these processes, the latter is by far the most important.

6.4 Improving separations through optimized choice of conditions

Often the separation factors can be significantly improved by changing the external conditions during separation (pressure, temperature, concentration). Particularly effective is the conversion of the component to be separated into a favorable compound (e.g. into a volatile or poorly soluble substance) and the masking of individual components of the mixture (cf. Chap. 3 in Part 1).

The most favorable conditions must be determined empirically; because of the almost infinite number of possible systems, general rules can hardly be established.

6.5 Improving separations by inserting auxiliary substances

As already observed by v. Scheele (1893), lanthanum and praseodymium can be separated particularly well by fractional crystallization of the ammonium double nitrates if the mixture still contains cerium, which is pushed between the two elements in this separation method. Later, Urbain (1909) was able to improve the europium-samarium separation by crystallization of the magnesium double nitrates by sandwiching magnesium bismuth nitrate between the rare earths.

Such intercalation of auxiliary substances has also occasionally been recommended for other separation methods (e.g. adsorption, ion exchange, electrophoresis and distillation processes). However, it is limited in applicability and cumbersome, since the added substance must be subsequently separated again using a different separation method.

6.6 Improve separations by exploiting different rates in setting the distribution equilibria or different reaction rates

If there are major differences in the rates of adjustment of the distribution equilibria for different components of a mixture, these can be exploited by working quickly to improve the separation effect. The same applies if chemical reactions of different rates occur before or during separation; it may be possible to separate the faster reacting substances before the slower ones have reacted to any appreciable extent.

6.7 Improving separations by repeating the single step

6.7.1 General – discontinuous and continuous mode of operation – closed-loop process

The most important and general method of increasing the separation effect is the repetition of the single step. This leads to the so-called *multistep (multiplicative) separation methods*. Various schemes are used for these, the most common of which in analytical chemistry will be derived and discussed below.

In terms of process technology, multiplicative separations can be carried out discontinuously in individual separation steps or continuously.

A special variant is represented by the so-called *recirculation processes*; in these, an auxiliary substance is recirculated after leaving the apparatus and used again; the advantage is that the required amount of auxiliary phase is substantially reduced, and furthermore substance losses can be prevented by working in a closed system.

6.7.2 One-sided repetition

A frequently used type of repetition of separations shall be called *one-sided repetition*. For example, consider a mixture of 100 parts of substance A and 100 parts of substance B (Fig. 6.5). After the first separation, there may be 99.9% of A with 10% of B in one phase and 0.1% of A with 90% of B in the other phase. If the separation is repeated only with the respective left part and the separated parts are combined (right in Fig. 6.5), the final result after two repetitions is 99.7% of A + 0.1% of B in one phase and 0.3% of A + 99.9% of B in the other phase. Thus, on the left side of the scheme, the purity of A has been increased at the expense of the yield, and on the right side, the yield of B has been increased at the expense of the purity.

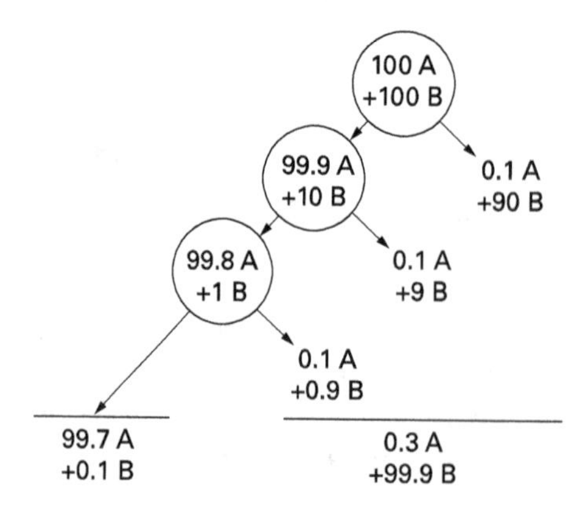

Fig. 6.5: One-sided repetition of a separation. ○ = Separation operation.

A prerequisite for the application of the method in analytical chemistry is that after each separation the yield on one side of the scheme is very high (if possible 99.9%), while the substance may still be highly contaminated. (This inevitably results in high purity with poor yield on the other side).

As will be shown below, unilateral repetition plays a significant role in distributional procedures, but it is often carried out in a somewhat modified manner.

In Fig. 6.6, the circles are intended to represent vessels in which the separation operations are carried out. After separation, one substance remains partially in the reaction vessel (low yield), while the other is transferred almost completely to the next vessel (high yield). After passing through a number of vessels, the first substance is completely separated from the second substance not retained in the vessels.

Instead of the discontinuous mode of operation in individual vessels, the process can also be made continuous, although at least one auxiliary phase must be used. This is placed in a column-shaped arrangement (example: a tube filled with a solid adsorbent) and the mixture of substances to be separated is allowed to pass through the column, possibly with a second auxiliary phase. Separation is achieved by fixing portions of one substance in different sections of the column (Fig. 6.7), while the other substance is not noticeably retained.

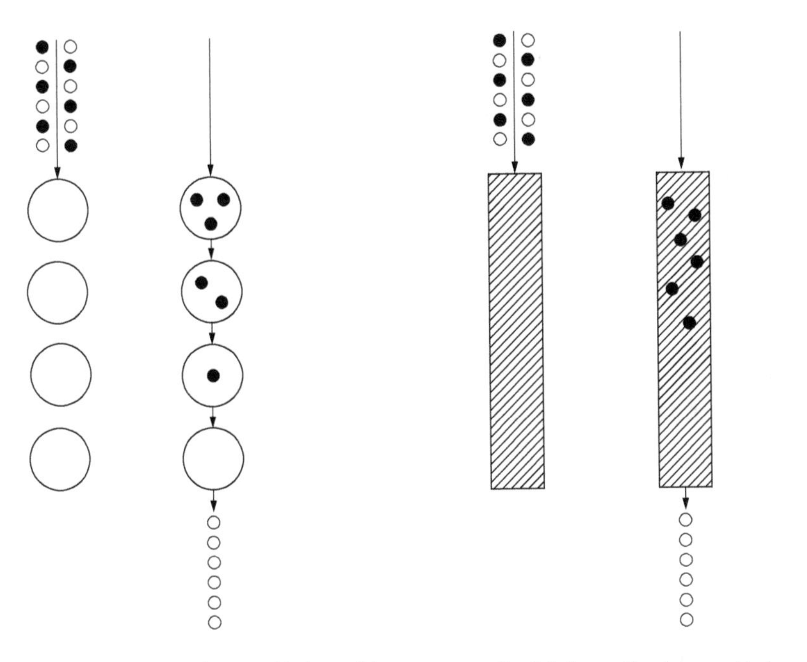

Fig. 6.6: Separation by one-sided repetition (vessel row).
○ = vessel or separation operation.

Fig. 6.7: Separation by one-sided repetition (continuous operation in one column).

6.7.3 Systematic repetition (cascade)

If a separation is repeated not only on one side as in Fig. 6.5, but also with the other part, a so-called *cascade* is obtained. For example, a mixture of 100 parts A and 100 parts B is split during the separation in such a way that 90 parts A with 10 parts B go into phase 1 and 10 parts A with 90 parts B go into phase 2 (Fig. 6.8, I. step). The separation factor is then $\frac{90/10}{10/90} = 81$. Now the separation is repeated with each of the two parts, and the separation factor should again be 81. (Fig. 6.8, II. stage). The two outer fractions then contain 81 parts A + 1 part B and 1 part A + 81 parts B, respectively. The separation factor between these two fractions is

$$\beta = \frac{81/1}{1/81} = 81^2 = 6561,$$

however, two middle fractions were still formed, where the separation factor is 1 and which lower the yields on the two final fractions.

In general, ideally, in such a scheme, the total separation factor in the final fractions is

$$\beta_{\text{gesamt}} = \beta^n, \tag{8}$$

when n *is* the number of separation stages.

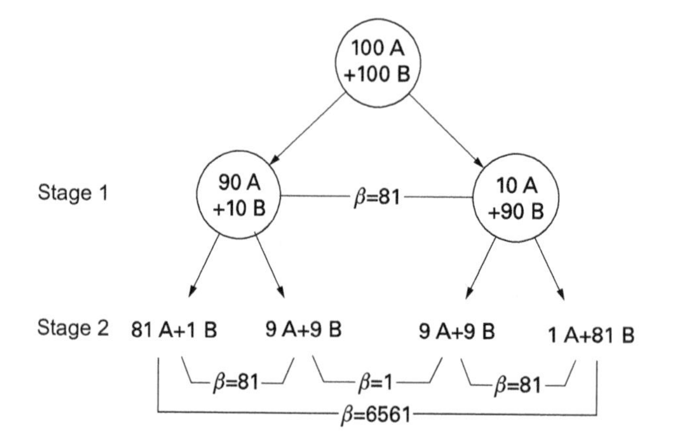

Fig. 6.8: Systematic repetition of a separation: cascade.
○ = Separation operation.

In order to assess the number n of separation steps required for analytical separations at a given separation factor β, it is recommended to follow the increase of the total separation factor β_{total} as n increases. In Fig. 6.9, this function is plotted for different values of β. As can be seen, it is very important that the single stage has a separation factor that is not too small, otherwise the number of separation stages must be extraordinarily large to bring β_{total} to the required value of about 10^6 However, there are procedures in which the repetitions can be carried out automatically, whereby thousands of separation stages (currently a maximum of several hundred thousand, see below) can be achieved without great difficulty.

According to the cascade scheme, separations can be repeated as often as desired; however, the disadvantage is that the number of fractions increases very sharply, with yields of the purest final fractions dropping. The process is therefore of no significance for analytical chemistry; it is described here only for reasons of systematics.

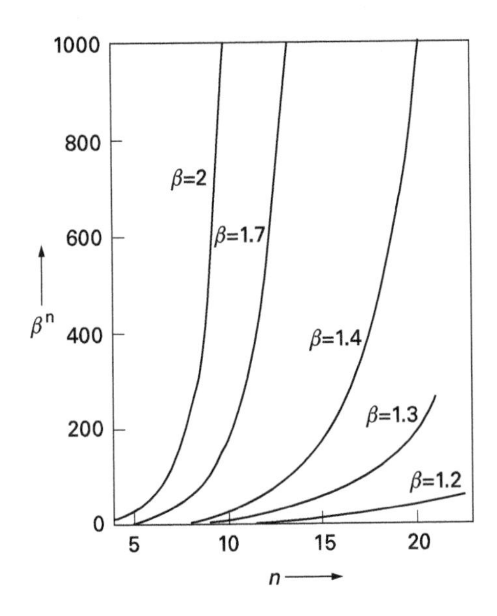

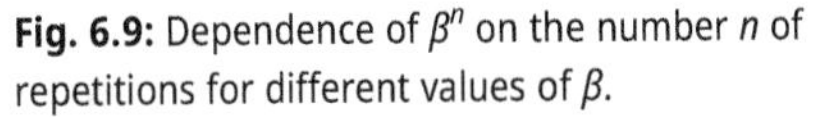

Fig. 6.9: Dependence of β^n on the number n of repetitions for different values of β.

6.7.4 Systematic repetition (triangular scheme – separation series)

The *triangular scheme* introduced by Pattinson (1833) for the separation of rare earths by fractional crystallization is much more favorable. The method is similar to the cascade method, but the middle fractions are combined (Fig. 6.10).

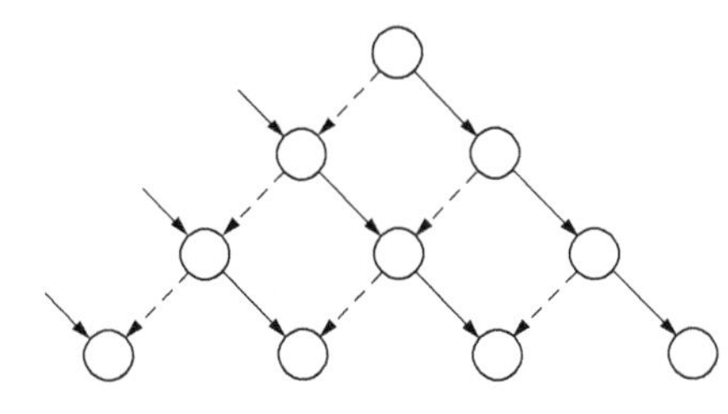

Fig. 6.10: Systematic repetition of separations in the triangular scheme.
○ = separation operation; ↘ phase 1 path; ↙ phase 2 path.

In this method, too, the individual separation operations can be repeated as often as desired; the number of fractions also increases, but not nearly as much as in the cascade scheme.

The number of fractions can now be limited by removing the final fraction at the right side of the scheme after a sufficient number of rows for separation, i.e. stopping the separation here (Fig. 6.11).

The same method is shown in a slightly different representation (rotation of the triangular scheme by 45°) in Fig. 6.12. If one now interprets the circles shown not as individual separation operations but as vessels in which the separations are carried out, there is obviously a *series of vessels* or *separation series* through which the one phase is carried through in portions.

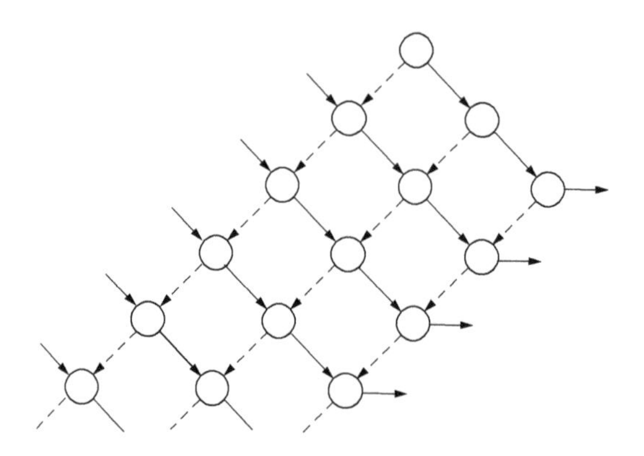

Fig. 6.11: Operation in the triangular scheme with termination in the 3rd row.
○ = separation operation; → phase 1 path; ↙ phase 2 path.

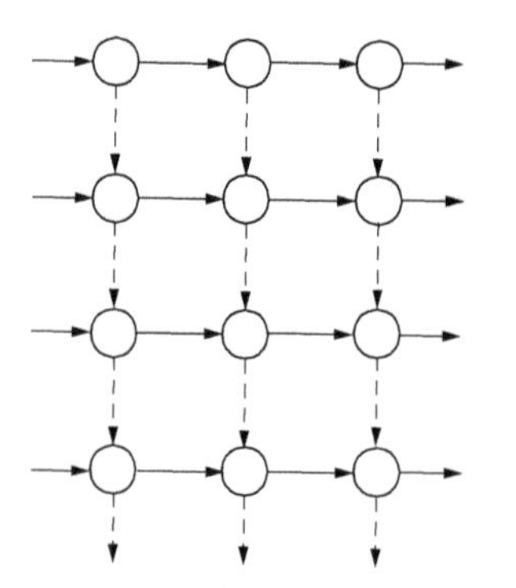

Fig. 6.12: Working method in the triangular scheme as in Fig. 6.11; interpretation as vessel row.
○ = vessel for the separation operation; → path of phase 1; ↓ path of phase 2.

In the case of fractional crystallization, the solid phase (the formed crystals) would be in the vessels through which the solution flows discontinuously at the end of each crystallization process. Of course, the solid phase need not be transferred to new vessels each time, but can be left in the initial vessels, i.e. the row of vessels can be moved symbolically downwards in the diagram in Fig. 6.12.

In the selected example of fractional crystallization, only one auxiliary phase, the solvent, is used. The disadvantage of this is a constant reduction in the amount of solid phase and a continuous change in the ratio of the two phases in the course of the separation.

This can be avoided by using two auxiliary phases, but otherwise proceeding as in the scheme of Fig. 6.12.

Let a number n of individual vessels be arranged in a row, each vessel containing a certain amount of the one auxiliary phase (Fig. 6.13). This auxiliary phase remains in the vessels throughout the separation process and is therefore referred to as the *stationary phase.*

The mixture to be separated and a portion of the second auxiliary phase are then added to the first vessel and the distribution equilibrium is established in it. The second auxiliary phase in vessel no. 1 is then transferred to vessel no. 2 and equilibrated with the stationary phase already present. Furthermore, a fresh portion of the second auxiliary phase is brought into vessel no. 1 and the distribution equilibrium is also set here (1st repetition in Fig. 6.13).

In contrast to the stationary phase, the second auxiliary phase continues to migrate during separation; it is therefore referred to as the *mobile phase* or *moving phase.* The stationary phase can be liquid or solid, the moving phase gaseous or liquid.

The separation can be repeated n times by moving all portions of the moving phase one vessel to the right each time and adding a fresh portion of the same phase to the first vessel in the separation series. After $n - 1$ repetitions, both stationary and moving phases are present in each vessel, and the series is established.

If the separation is continued as described, the moving phase at the end of the vessel series moves out of the system (*n-th*; ($n + 1$)-th . . . repeated in Fig. 6.13).

The result of such a repeated distribution with a series of 100 vessels is shown in Fig. 6.14. It was assumed that two substances A and B with distribution coefficients $\alpha_A = 2$ and $\alpha_B = 0.5$ are to be separated, that they are present in equal amounts, and further that Nernst's law is valid. The distribution of A and B among the different vessels in the series is plotted after 10, 30, and 100 repetitions, with the ordinate reflecting the concentration of the components in the vessels and the abscissa the numbers of the vessels. The repetitions of the separation operations were not performed so often that the substances migrated out of the series at the end.

After 10 separation steps, A and B are still barely separated, after 30 separation layers they are already quite well separated, and after 100 separation layers they are practically completely separated. In the process, the two substances are distributed over an progressively increasing number of vessels, and the concentrations decrease more and more.

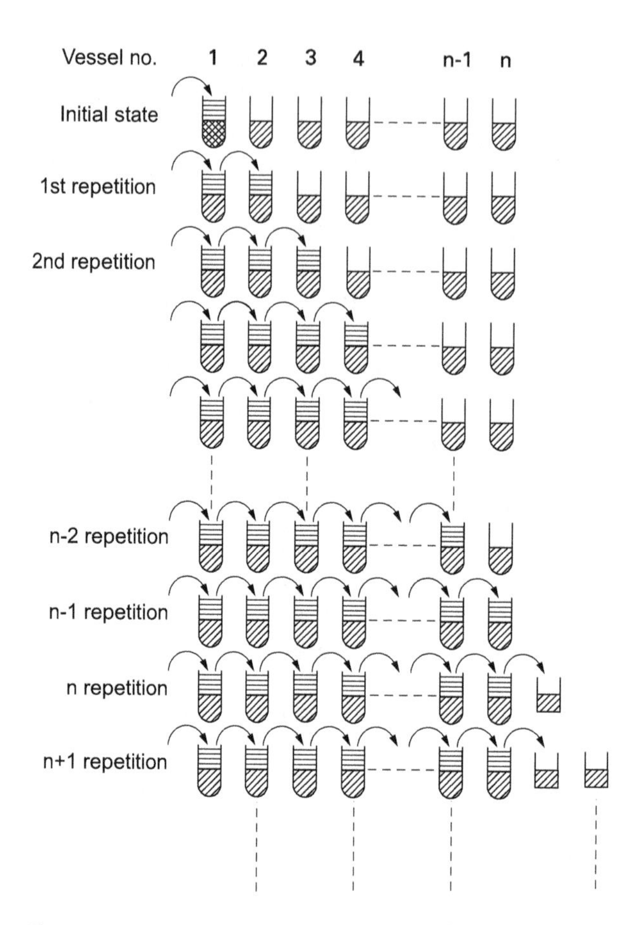

Fig. 6.13: Distribution series as in Fig. 6.12; working method with two auxiliary phases.

The main characteristics of this scheme are:

- Introduction of the substances to be separated into the initial vessel of the separation series;
- discontinuous arrangement of the separation stages in a series of vessels;
- intermittent (discontinuous) conveyance of the moving phase.

With the same arrangement of the stationary phase in individual vessels, the moving phase can also flow continuously through the system. However, because of the relatively large experimental effort involved, this variant is hardly ever used in analytical chemistry.

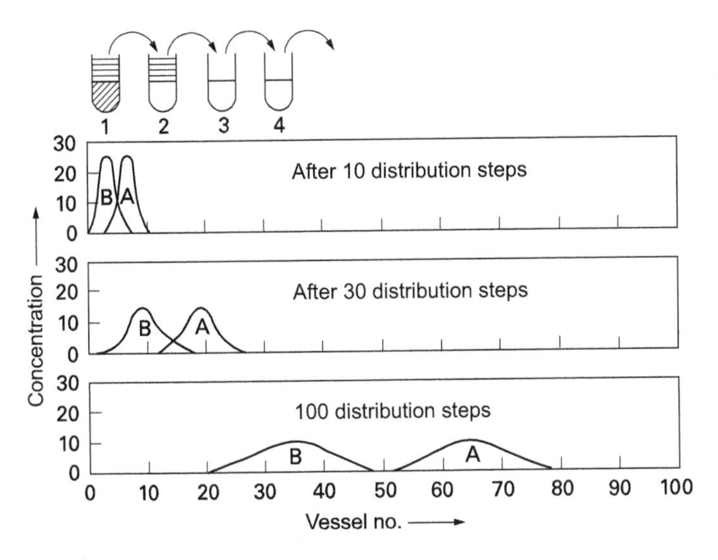

Fig. 6.14: Distribution of substances A and B after 10, 30 and 100 separation steps. $\alpha_A = 2$; $\alpha_B = 0.5$.

6.7.5 Systematic repetition (separation columns)

Of great importance is a further development in which the stationary phase is no longer arranged in individual vessels but in the form of an elongated, continuous separation bed in a column. One can derive this method from a perpendicular separation series (Fig. 6.15).

The substance mixture is fed onto the head of the column (corresponding to the input into the first vessel of a separation series) and the agitated phase is allowed to flow continuously through the column. In the process, the substances to be separated are entrained to varying extents depending on their distribution coefficients and are separated to a greater or lesser extent as they pass through the stationary phase in accordance with the separation factors. The original substance zone splits into individual bands in the process.

Separations in such arrangements are called *chromatography* after the first investigations of colored compounds carried out with them. The name was retained even after the process was extended to the separation of colorless substances.

In the case of separations with such columns, the following working methods must be distinguished:

- the development process;
- the elution process;
- the frontal analysis and
- the displacement technique.

Of these, the development and elution processes are by far the most important.

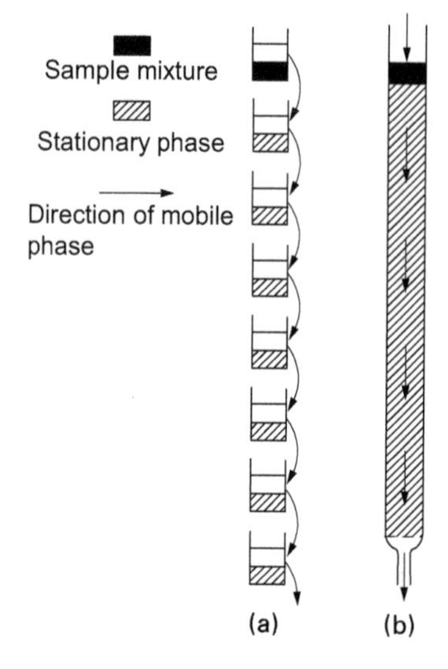

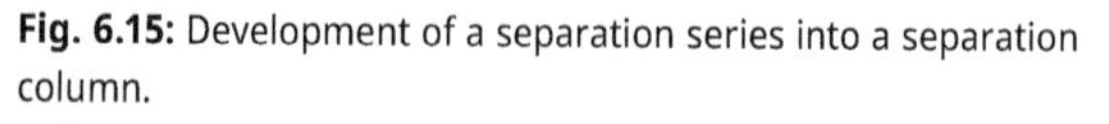

Fig. 6.15: Development of a separation series into a separation column.
a) Separation series (discontinuous single stages, discontinuous onward transport of the moving phase);
b) separation column (continuous separation bed, continuous flow of the moving phase).

Separations by developing the chromatogram. The splitting of the original substance zone during chromatographic separation is called *developing* the chromatogram; it corresponds to the process shown in Fig. 6.14 in a row of vessels. If the separation is stopped at this point, the separated substances are present as individual zones on the stationary phase. To isolate them, the stationary phase must be removed from the column or the column must be cut (one can also use columns made from sections that can be disassembled). Since this method of operation is cumbersome, it plays only a minor role in chromatographic separations in columns, but it is used almost exclusively in the thin-layer methods to be discussed below.

Separations by elution. The second method of performing chromatographic separations is to allow moving phase to flow through the column until the individual components of the mixture to be separated are flushed out of the end of the column one by one. The process is called *elution* and the moving phase leaving the column is called *eluate*.

During elution, the separated substances can each be recovered separately by collecting the eluate fractionally – most expediently with an automatic fraction collector. Often, however, individual components are not isolated and only their quantities or concentrations in the eluate are continuously determined using suitable measuring and recording devices. This provides the so-called *elution curve* of the mixture used.

Elution curves. If the respective total amount of eluted substances is measured during elution and plotted against the eluate volume, the *integral* elution curve is obtained. It is more common to continuously follow the concentrations in the eluate instead of the total amount; the *differential* elution curve is then obtained from the former by differentiation (Fig. 6.16).

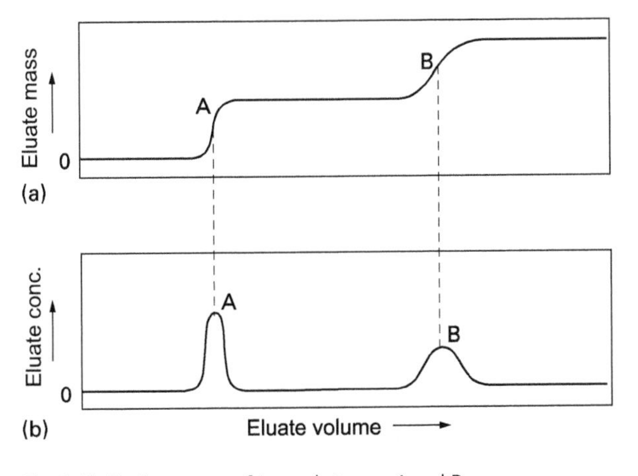

Fig. 6.16: Elution curves of two substances A and B.
a) Integral elution curve;
b) differential elution curve.

In the following, only the differential elution curves are discussed, since the usual instruments record almost exclusively these.

Shape of the elution curves. Ideally, the elution curves have the form of a *Poisson distribution* if the separation column has only a few separation stages (Fig. 6.17). When the number of separation steps is increased, i.e. when the efficiency of the column is improved, it changes to the form of a normal distribution (*Gaussian distribution*).

If the distribution isotherm does not correspond to a Nernst distribution, deviations from the ideal shape of the elution curves occur, which worsen the separations (cf. Fig 6.18).

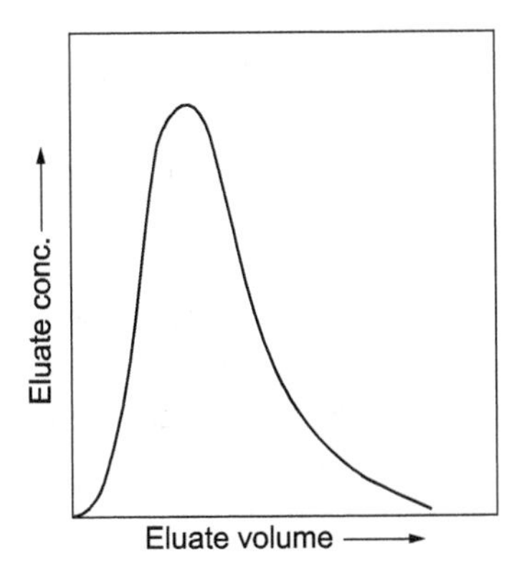

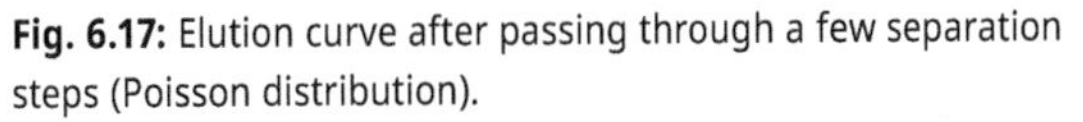
Fig. 6.17: Elution curve after passing through a few separation steps (Poisson distribution).

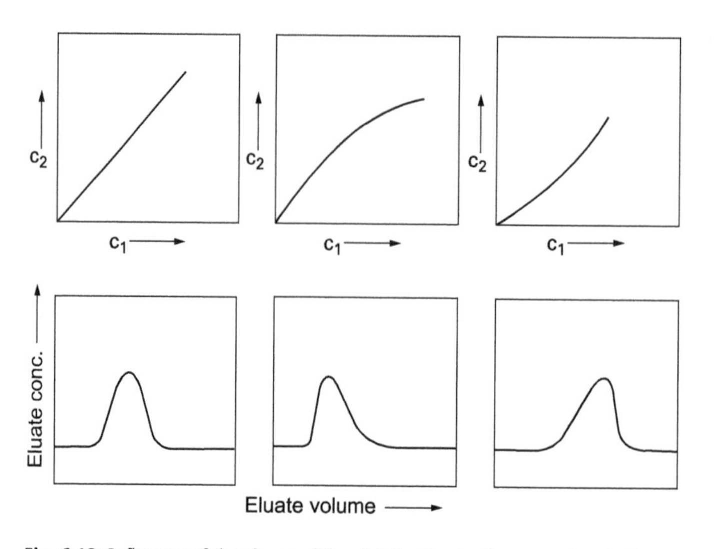

Fig. 6.18: Influence of the shape of the distribution isotherm on the elution curves.

Stepwise elution. If substances with very high and very low partition coefficients are present in a mixture of substances at the same time, one will leave the column quickly, while the other will only leave after a large amount of eluent has passed through; the elution will therefore be tedious. In such cases, the components that migrate rapidly through the column can be eluted first with a suitable liquid, and then a second, possibly third, etc., eluent can be applied. In such cases, the components moving rapidly through the column can be eluted first with a suitable liquid and then a second or third eluent, etc., can be used to remove the remaining substances from the column in a shorter time. In the graphical representation of such elution curves, the change of eluent is usually indicated (cf. Fig 6.19).

Gradient elution. If the eluent is not improved stepwise but continuously, the so-called *gradient elution* is present, which was introduced by Mitchell and Desreux (1949). As with stepwise elution, the distribution coefficients of the slowly eluting substances are changed in favor of the moving phase.

In *gradient elution*, a second better eluting liquid is continuously added to a liquid eluent with relatively poor eluting properties. Apparatively, this is easily accomplished by mixing two computer-controlled motorized burettes to produce the preselected gradient by varying propulsion speeds in a T-piece. Depending on whether the change in composition is linear or nonlinear, a distinction is made between different types of gradient (Fig. 6.20), and almost any gradient can be obtained by using several mixing chambers (up to 9) simultaneously.

The shape of the gradients applied in a separation is usually plotted as an increasing line on the elution diagram.

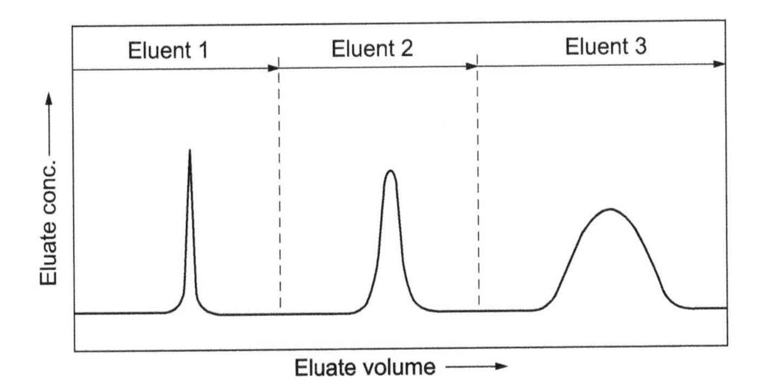

Fig. 6.19: Stepwise elution with three eluents.

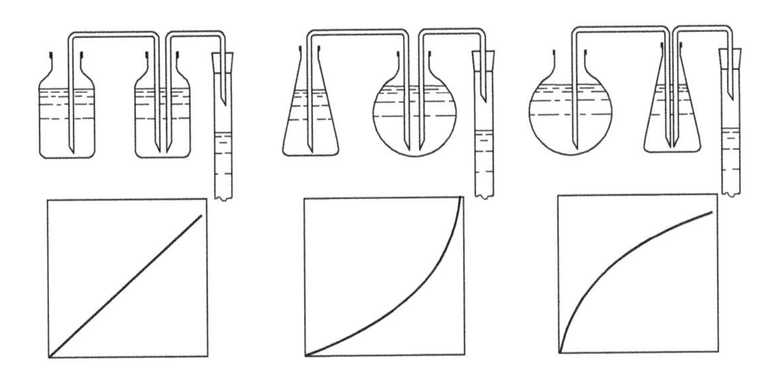

Fig. 6.20: Generation of concentration gradients of different shapes.

In addition to the composition of the eluent, the temperature, the pH (by mixing two buffer solutions) or the flow rate of the moving phase (by changing the pressure or the column cross-section) can be varied. In addition to concentration gradients, pH and temperature gradients in particular have become more important.

Gradient elution combines several favorable effects: First, the duration of separation and the amount of eluent are reduced (just as with stepwise elution). Since the elution bands become broader and flatter with increasing elution time (see below), the acceleration of the elution causes a sharpening of the bands with an increase in the concentration maxima (cf. Fig 6.21), which is particularly advantageous for the components eluted last and increases the detection limit of minor components of the analysis sample.

In addition, there is an improvement in the shape of elution bands which exhibit so-called *tailing* as a result of non-straight distribution isotherms. The too slowly eluted parts of the band are pushed forward towards the band maximum by the improvement of the eluent (Fig. 6.22).

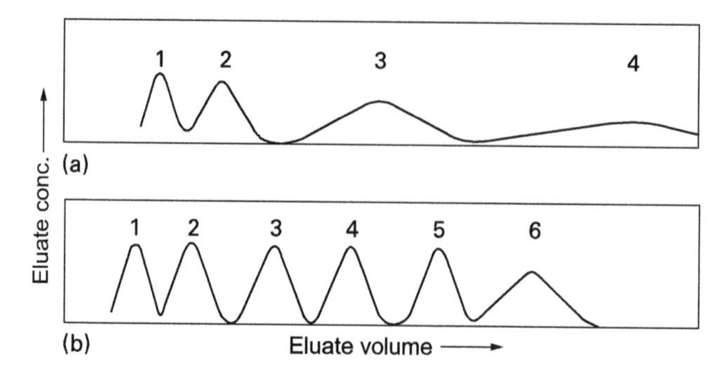

Fig. 6.21: Shortening of elution diagrams by gradient elution.
a) Elution without gradient;
b) gradient elution.

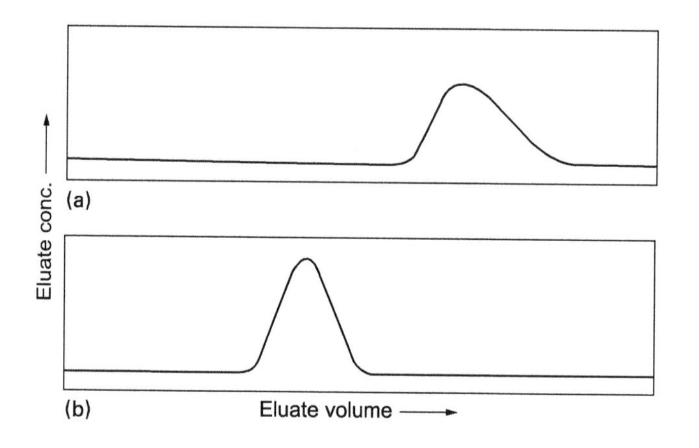

Fig. 6.22: Improvement of distorted elution bands by gradient elution.
a) Elution with *tailing*;
b) gradient elution of the same substance.

Retention. Due to the different distribution between the two phases of the separation column, the components to be separated are retained to a greater or lesser extent by the stationary phase. The extent of this retardation on the way through the column (*retention* is, to a certain extent, a characteristic parameter for each of the substances in a mixture, which can be used to identify unknown substances. These parameters are determined using two different methods and are referred to as the R_f *value* and the *retention volume.*

R_f value. The R_f value introduced by LeRosen (1942) is defined as the ratio of the distance a traveled by a substance i during separation to the distance c traveled by the moving phase (cf. Figure 6.23):

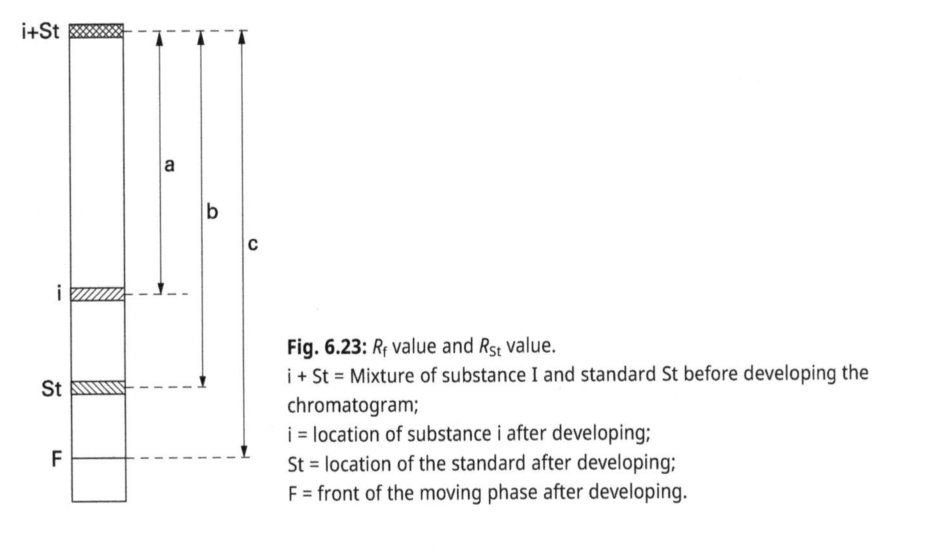

Fig. 6.23: R_f value and R_{St} value.
i + St = Mixture of substance I and standard St before developing the chromatogram;
i = location of substance i after developing;
St = location of the standard after developing;
F = front of the moving phase after developing.

$$R_f = \frac{a}{c}. \tag{9}$$

The R_f value can only be determined after developing a chromatogram, not after eluting the components. Since the separated compounds can migrate at most as far as the front of the moving phase, the R_f value reaches at most the value 1.

R_f values are often difficult to reproduce; the reproducibility can be improved by referring to a reference substance St which is added to the sample and allowed to run during the separation. The ratio of the R_f values of both substances i and St is called the R_{St} value (cf. Figure 6.23):

$$\begin{aligned} R_{f(i)} &= \frac{a}{c} \quad R_{f(St)} = \frac{b}{c}; \\ R_{St} &= \frac{R_{f(i)}}{R_{f(St)}} = \frac{a}{b}. \end{aligned} \tag{10}$$

Retention volume. If the chromatogram is not developed but the components of the mixture to be analyzed are eluted, the retention volume V_i is used as a measure of the delay in passing through the column. The retention volume is the volume of eluate required to carry substance i through the separation column until the maximum of the elution band just leaves the column (cf. Fig 6.24).

If the test duration, e.g. in minutes, is plotted on the abscissa instead of the eluate volume, the *retention time* t_i of component i is obtained analogously.

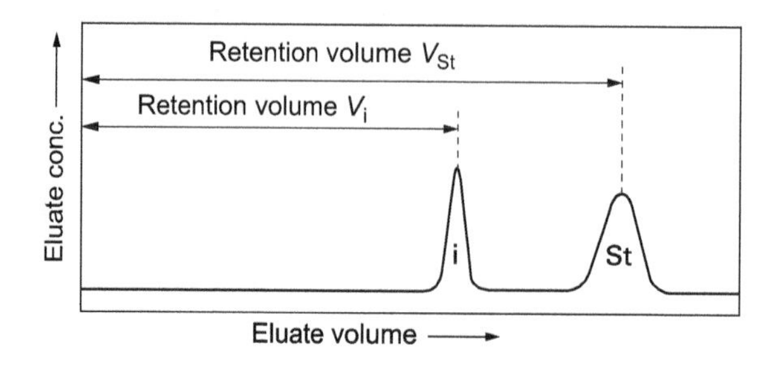

Fig. 6.24: Definition of the retention volume.

Here, too, reproducibility is improved by referring to a standard substance; by dividing the retention volumes of substance i and standard St, the so-called *relative retention volume* $V_{i,St}$

$$V_{i,St} = \frac{V_i}{V_{St}} \tag{11}$$

and correspondingly the *relative retention time* $t_{i,St}$

$$t_{i,St} = \frac{t_i}{t_{St}}. \tag{12}$$

Corrected retention volume. The determination of the R_f value and the retention volume differ in one important detail: To determine the R_f value, the substance is placed on the dry separation bed and the mobile phase is allowed to flow through. Its front is used as a reference line, i.e. as a measure of the amount of mobile phase.

The retention volume is obtained by placing the substance on the top of a column which is not only filled with the stationary phase but also already contains the mobile phase in the interstices. The quantity of the mobile phase is called the *dead volume* of the column. When elution is started after substance addition, mobile phase immediately flows out of the end of the column. The retention volume obtained for a given substance is thus increased by the dead volume compared to the value that would have resulted if the column had been dry at the beginning of the experiment (i.e. had not yet contained any mobile phase).

If the dead volume V_t is subtracted from the found retention volume V_i of substance i, the *corrected retention volume* is obtained V_i^o

$$V_i^o = V_i - V_t. \tag{13}$$

Accordingly, the relative corrected retention volume $V_{i,St}$ of substance i, related to standard St, is obtained:

$$V_{i,St}^o = \frac{V_i - V_t}{V_{St} - V_t}. \tag{14}$$

Not the experimentally directly determined, but the corrected retention volume plays the decisive role in theoretical considerations, in the identification of unknown components and in the comparison of experimental results of different observers.

Breakthrough curves. If a substance dissolved in the eluent is continuously applied to a separation column, pure eluent will flow off until the absorption capacity of the stationary phase for the substance is exhausted and the substance leaves the column with it. If the concentration of the added substance in the eluate is followed in such an experiment, the so-called *breakthrough curve* is obtained (Fig. 6.25).

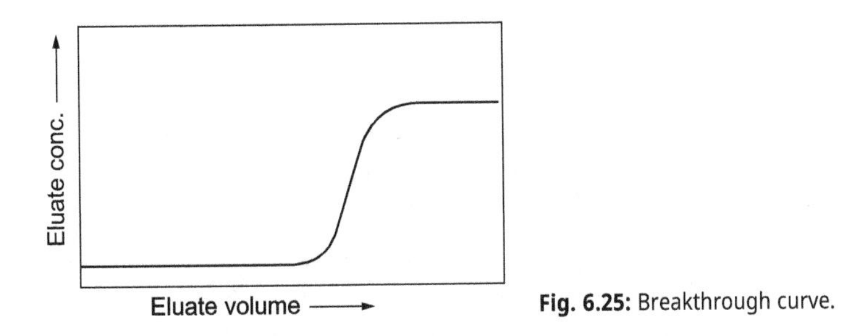

Fig. 6.25: Breakthrough curve.

Such breakthrough curves are used, among other things, to assess the capacity of separation columns. The capacity is the maximum amount of substance that can be taken up by the stationary phase of the column; although this quantity can usually only be determined approximately, it is an important parameter because it gives an indication of the amount of substance that can be applied to the column in an elution experiment without overloading it.

Frontal analysis. If a mixture of substances dissolved in the eluent is continuously applied to the column rather than a single substance, as is the case when determining the breakthrough curve, several breakthrough curves follow one another. The component least retained by the stationary phase is eluted first – and initially in pure form – but is then overlaid by the following ones as soon as these also leave the column. Therefore, by the technique of *frontal analysis*, in principle only a part of the most rapidly migrating compound can be obtained purely; subsequently, mixtures are always eluted, which become progressively more complicated (Fig. 6.26).

The successive steps correspond to the elution bands in ordinary elution and thus give an indication of the composition of the mixture; furthermore, the height of the steps corresponds to the concentrations in the starting material. The method is to be counted among the incomplete separations, in which the concentrations in the starting mixture can be determined with the aid of the distribution coefficients (here, however, only indirectly) (cf. Part 1, Section 4.3). However, this method has no advantages over normal elution and, as a precursor of chromatographic separations in effect, has almost only historical significance (Goppelsröder; 1861).

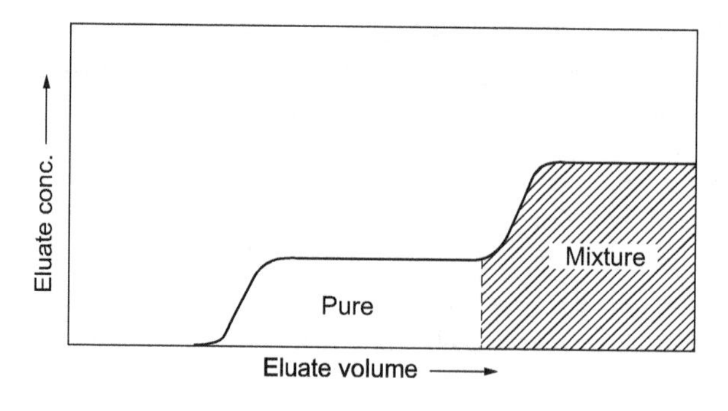

Fig. 6.26: Frontal analysis (principle).

Displacement technique. The displacement technique given by Tiselius (1942) can be used when substances are held very firmly by the stationary phase of a separation column and thus require extremely large amounts of moving phase for elution. In such cases, one elutes with a liquid containing a compound held even more firmly by the stationary phase. The *displacer* pushes the components of the analysis sample ahead of it on the column; these separate in the process into zones which, under favorable conditions, contain only one substance, but are immediately adjacent to one another, so that clean separation after leaving the column is difficult. The process is therefore also of no great significance.

Efficiency of separation columns. As already mentioned, a separation column can be thought of as being divided into individual sections which correspond to the vessels of a separation series and in each of which an adjustment of the distribution equilibrium takes place. The number n of such sections, called *separation plates* is a measure of the column's efficiency.

Two further terms are derived from the number of stages n, the separation efficiency n' = number of plates per meter of column length, and the height H of a plate,[1] which is usually given in cm, occasionally also in mm. For a given column length, the lower the height of an individual separation plate, the greater the number of separation plates and thus the greater the efficiency. To characterize separation columns, the height of the separation stage is used above all.

The efficiency of a column depends on the following variables in addition to the separation factors given by the nature of the phases and the substances to be separated:
- Length of the column;
- diameter of the column;
- flow rate of the moving phase;

1 In the Anglo-Saxon literature, the stage height is designated as *HETP (= height equivalent to a theoretical plate).*

- grain size of the stationary phase;
- loading of the column;
- temperature.

When the column length is increased, the number of separation plates and thus the efficiency of the column increases. On the other hand, the flow resistance increases, which usually reduces the flow velocity of the mobile phase; since the separation efficiency deteriorates if the flow is too slow (see below), the column length cannot be increased arbitrarily.

In a first approximation, the *column diameter* has no influence on the number of separation plates. However, it has been shown that as the diameter increases, it becomes increasingly difficult to distribute the stationary phase uniformly over the entire column. Therefore, as the mobile phase flows through, the disturbing effect of *channeling* occurs increasingly, causing partial streams to travel longer distances in the column without adjusting the distribution equilibria. This effect sets limits to the increase of the column diameter.

This disadvantage is countered by the fact that the capacity of the column increases with increasing diameter, so that larger substance quantities can be separated.

The greater the *flow rate* of the mobile phase, the more incomplete the distribution equilibria will be, and the lower the column efficiency will be.

In the case of extremely slow flow of the mobile phase, however, the diffusion of the substance components located in the moving phase becomes disturbingly noticeable. This diffusion runs randomly in all directions; the components of the diffusion running in the direction of the flow and those running against the flow lead to a broadening of the substance zones in the column and thus to a deterioration of the separations, while the diffusion perpendicular to the flow direction has no influence.

Accordingly, a certain flow velocity of the mobile phase will result in an optimum separation effect, and both a reduction and an increase in this velocity will have an unfavorable influence on the separation because of the then more pronounced influence of one of the two disturbances in each case.

The *particle size of the stationary phase* is of decisive importance for the separation efficiency of the column, as can be seen from the following consideration: At least one horizontal position of the particles is required to set the distribution equilibrium once; the height of a separation plate thus becomes equal to the particle diameter at best. The smaller the particle diameter, the greater the number of separation plates for a given column length.

In practice, however, the separation plate height is always considerably greater than the particle diameter, since a complete equilibrium setting per particle layer cannot be achieved. Qualitatively, however, the above consideration remains valid.

However, an arbitrary reduction of the particle diameter is not possible due to the increasing flow resistance.

The shape of the particles also plays a role; the best results are usually obtained with spherical particles of uniform diameter that are packed as uniformly as possible.

The *loading of the column*, i.e. the amount of substance mixture added, must have a noticeable influence on its efficiency if a considerable part of the stationary phase is occupied from the beginning, since this part of the column is lost for the separation. Therefore, only a relatively small amount of the substance mixture should be added. Usually, no more than about 10%, at most 20% of the column is occupied.

Finally, the separation effect of a column also depends on the *temperature*; an increase has a favorable effect in that the distribution equilibria are established more quickly, but this effect can be overcompensated by a decrease in the separation factors. The influence of temperature on a given separation problem can therefore generally not be predicted, but must be determined empirically.

Separation factors, partition coefficients and efficiency of separation columns. For the separation of a given mixture of substances, not only the number of separation plates of the column used is decisive, but also the separation efficiency of the individual plate; only when the separation factors achieved with a given combination of stationary and mobile phases are sufficiently large can the mixture be completely separated.

However, not only the separation factors, but also the individual distribution coefficients are important, as the following consideration shows: Assume that the task is to separate two compounds whose separation factor is relatively large, but which both pass quite predominantly into the mobile phase (e.g., let the percent distribution be 99.90 and 99.99%, respectively, in favor of the mobile phase; the separation factor would then be about 10). Despite the sufficient separation factor, both substances will pass through the column almost unhindered and will elute practically unseparated.

On the other hand, the compounds to be separated may be held very tightly by the stationary phase; then separation is possible, but elution, as mentioned above, requires very large amounts of mobile phase, a considerable amount of time, and the elution bands become very broad.

One will therefore choose the conditions for column separations as far as possible so that no extremely high or extremely low partition coefficients occur.

Determination of the number of separation steps. As shown in Fig. 6.14, the width of a band increases with the number of separation steps passed through and thus also with the distance traveled in a column. However, this broadening is not proportional to the path length, but is smaller; if the number of separation steps through which the substance passes is doubled, the band only widens by a factor of $\sqrt{2}$. This behavior is the basic prerequisite for separation to take place.

An illustration of the dependence of the band width on the number of separation steps is given in Fig. 6.27. There are three separation series with 10, 100 and 1000 vessels, respectively, and the maximum of a substance band is supposed to migrate to

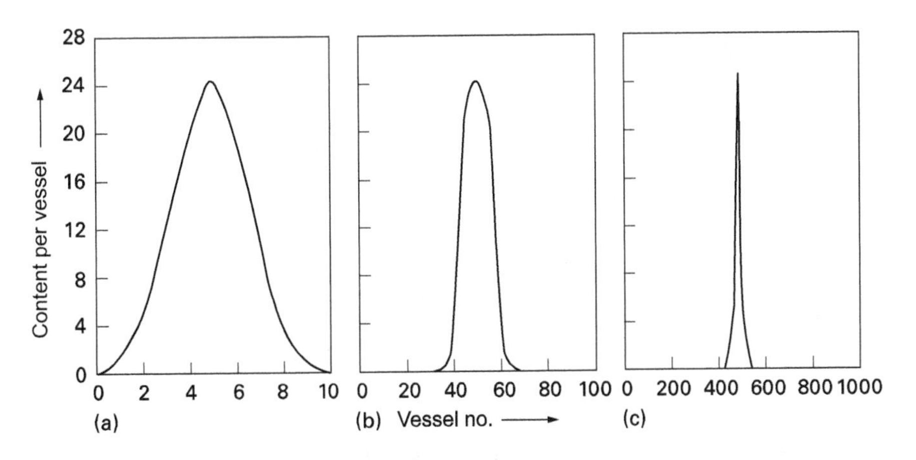

Fig. 6.27: Relative sharpening of zones with increasing number of separation steps.
a) Distribution of a substance among the vessels of a separation series after 10 separation steps;
b) distribution after 100 separation steps;
c) distribution after 1000 separation steps.

the center of the respective vessel series. The abscissa scales are chosen in such a way that the same lengths result for the numbers of vessels which differ by a factor of 10.

As can be seen, noticeable amounts of substance are still present in about 8 out of 10 vessels in the first series, in about 32 out of 100 in the second, and in only about 100 out of 1000 in the third. The relative number of vessels used drops from about 80% to about 10%.

A corresponding behavior is observed when a substance migrates through a column (one can imagine the vessel series of Fig. 6.27 as three columns of the same length, but with 10, 100, and 1000 separation steps, respectively). However, an important difference compared to the observation of the vessel series is that the retention volume (and not the number of vessels containing the substance) is measured during elution.

The retention volume is determined by the distribution of the substance under investigation between the stationary and moving phases. The more it is retained by the stationary phase, the greater the volume of moving phase required for elution. As a result, the concentration of a substance with a large retention volume in the eluate is reduced and thus the band is broadened, even if the number of separation stages passed is the same. This effect has already been shown in Fig. 6.21.

There is thus a relationship between the width of an elution band, the number of separation steps n of the column and the retention volume V_i of substance i, and one can use this regularity to determine n (Martin & Synge (1941); James & Martin (1952).

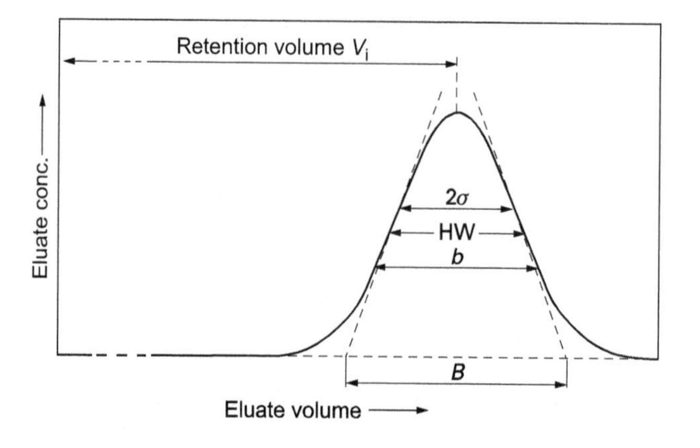

Fig. 6.28: Definition of the width of an elution curve (Gaussian distribution curve).

The number of separation plates can also be calculated from the slope of a breakthrough curve according to Glückauf (1955).[2]

One arrives at somewhat different formulas for the number of separation plates. depending on the definition of the width of the elution band. Since this ideally represents a Gaussian distribution curve, it is obvious to take the distance 2 σ of the two inflection points of this curve as the width (i.e. the width in 60.7% of the total height, cf. Fig 6.28). For n this results in

$$n = 4\left(\frac{V_\mathrm{i}}{2\sigma}\right)^2. \tag{15}$$

With the *half width HW* (= width in 50% of the total height) you get

$$n = 8 \cdot \ln 2\left(\frac{V_\mathrm{i}}{\mathrm{HW}}\right)^2 = 5,54\left(\frac{V_\mathrm{i}}{\mathrm{HW}}\right)^2; \tag{16}$$

with the width b in $\frac{1}{\mathrm{e}} \cdot h = 36,8\ \%$ the total height h results in

$$n = 8\left(\frac{V_\mathrm{i}}{b}\right)^2 \tag{17}$$

and finally with the base width B (= distance of the intersection points of the turning tangents on the base line = 4 σ)

2 In addition to the approach used here, a kinetic theory has been developed that is more powerful but less descriptive and more mathematically complex (Giddings, 1965).

$$n = 16\left(\frac{V_i}{B}\right)^2.$$

Since the number of separation plates is obtained by dividing two measured variables of the same dimensions, the retention time t_i or the corresponding band width measured in time units can also be used instead of the retention volume V_i and the band width measured in volume units.

The number of separation plates of a column cannot be determined very accurately in the way described, since the prerequisite of bell-shaped, completely symmetrical elution bands is usually not fulfilled. Therefore, different values for n are often obtained for a column, depending on whether the number of separation plates is determined for a substance with a large or a small retention volume. In general, substances with a large retention volume should be used, since the width of the band can then be measured more accurately.

Resolution. Retention volume V and band width 2 σ can still be used to define the *resolving power R*. For two adjacent bands of substances i and k with retention volumes V_i and V_k the following applies

$$R = \frac{(V_i - V_k)}{2(\sigma_i + \sigma_k)} = \frac{2(V_i - V_k)}{B_i + B_k}. \tag{18}$$

$(B = \text{base width})$.

The definition is given in Fig. 6.29. Two bands are considered resolved when R = 1, but higher values, i.e., better resolutions, must be available for quantitative analyses.

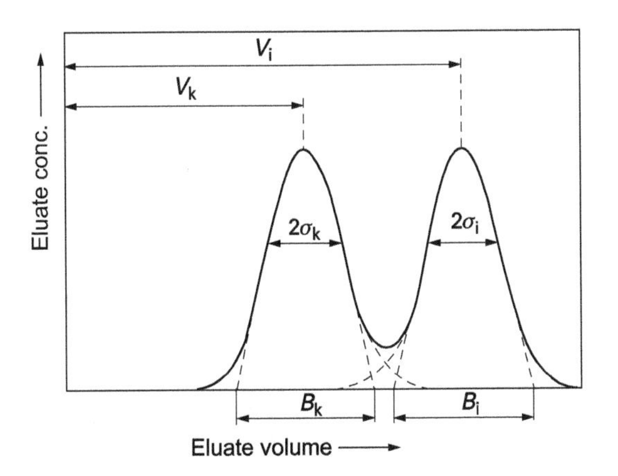

Fig. 6.29: Definition of the resolution.

6.7.6 Systematic repetition (thin film technology)

In so-called *thin-layer chromatography* the stationary phase is not arranged in a column but in the form of a more or less thin plate. The method, which can be thought of as being derived by cutting a section out of a column (cf. Fig 6.30), was first worked out by Ismailov and Schraiber (1938) for separations by adsorption. The adsorbent is fixed on a solid support, e.g., a glass plate, and the mixture of substances to be separated is applied in dissolved form to the end of the plate. The method was later improved, most notably by Stahl (1956). A special version uses strips of paper as a thin layer (*paper chromatography* according to Liesegang (1943) and Consden et al. (1944)).

The chromatograms are developed either descending from top to bottom or ascending. In the latter case, the layer is placed in a trough filled with the mobile phase; the capillary forces cause it to be sucked up into the porous layer (cf. Fig 6.31). To prevent evaporation losses, the whole arrangement is placed in a closed chamber.

Substances can also be eluted in thin-layer chromatography, but this method of operation has not gained any importance compared to the development technique.

The method is also used in the form of so-called *circular chromatography*. The substance mixture is applied to the center of a larger square or round thin-film plate. By dripping on the mobile phase, the chromatogram is developed radially outwards, and the separated components are obtained as concentric rings (Fig. 6.32). Evaporative loss of solvent can be prevented by covering with a second plate; the mobile phase is then added through a hole in its center.

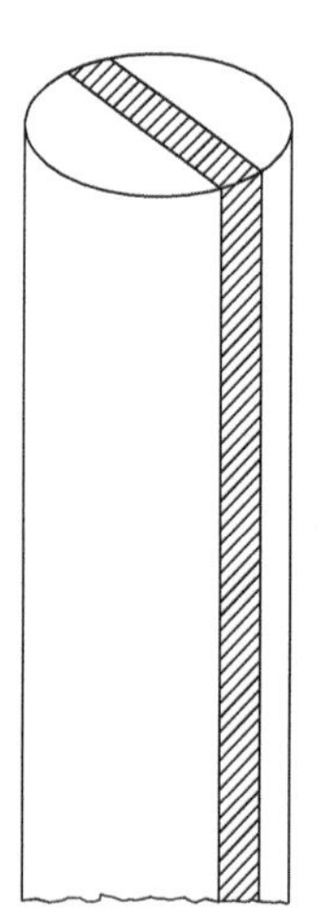

Fig. 6.30: Derivation of thin-layer chromatography from column chromatography.

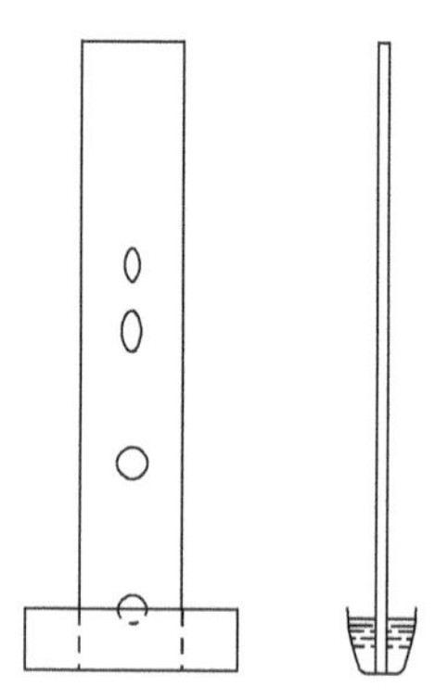

Fig. 6.31: Thin-layer chromatography with ascending development of the chromatograms.

Two-dimensional thin-layer chromatography. If a mixture is only incompletely separated by thin-layer chromatography, so that the individual substance spots still contain

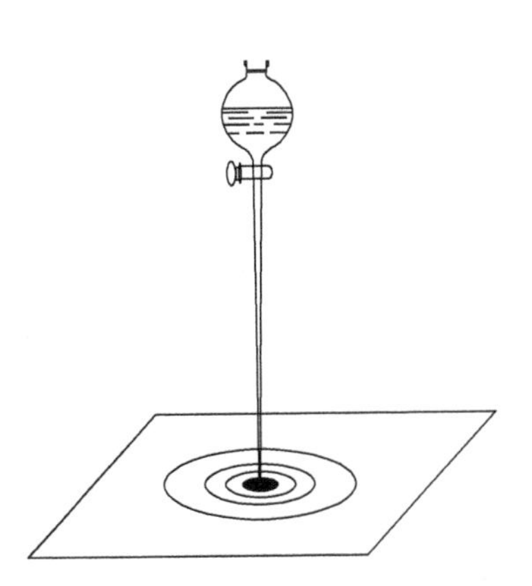

Fig. 6.32: Circular chromatography.

several components, further separation can be achieved by *two-dimensional* operation. The sample is placed in one corner of the square thin-film plate and initially developed in one direction. Then the plate is rotated 90° and developed a second time with a different mobile phase (cf. Fig 6.33). The method corresponds to a repeated separation in a separation column while changing the mobile phase; it is not to be confused with the two-dimensional methods to be discussed later, in which the substance mixture can be fed continuously.

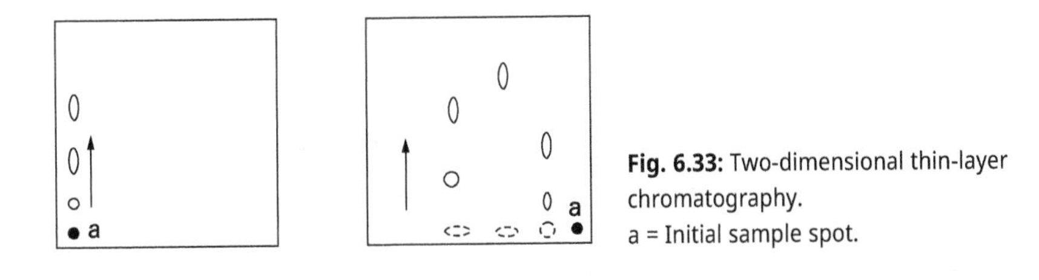

Fig. 6.33: Two-dimensional thin-layer chromatography.
a = Initial sample spot.

6.7.7 Countercurrent method

A further development of the methods described are the *countercurrent methods,* in which two immiscible phases flow against each other in a vessel system or in a separation tube (cf. Fig 6.34). The mixture of substances to be separated can be fed several times in individual portions or continuously in the middle of the system; depending on the distribution coefficients of the individual components, part of the mixture is discharged upwards and part downwards. In contrast to the previous processes, this achieves a separation of the starting material into only two parts, so that mixtures with more than two components are only incompletely separated.

In the countercurrent methods, the number of separation plates is increased for a given length of apparatus compared with the separation column. However, due to experimental difficulties, these methods do not play a major role in analytical chemistry – with the exception of distillation.

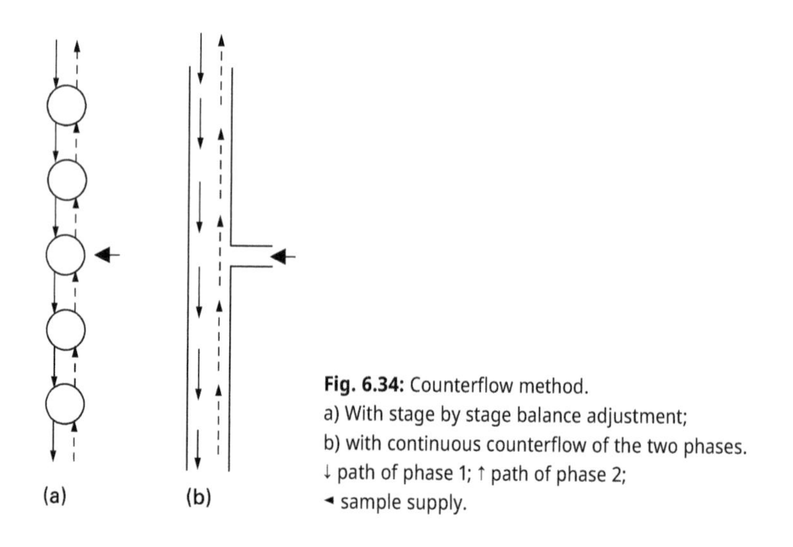

Fig. 6.34: Counterflow method.
a) With stage by stage balance adjustment;
b) with continuous counterflow of the two phases.
↓ path of phase 1; ↑ path of phase 2;
◂ sample supply.

6.7.8 Cross-flow method

If the two phases are not allowed to flow against each other, but as broad currents at an angle of 90° to each other, the so-called *cross-flow processes* result. With these, the substance mixture can be fed continuously; the greater the distribution coefficient of a component in favor of the phase flowing horizontally to the right, the more it is deflected from the perpendicular (Martin, 1949).

Despite the advantage of continuous substance feed and the possibility, in principle, of separating numerous substances simultaneously, the process has so far been of little significance; it is experimentally difficult to make two immiscible phases flow against each other without causing non-uniformities in the flow or vortex formation, which negates the separation effects.

6.8 Overview

The separations due to different distribution between two immiscible phases can be ordered according to the following aspects:

Characteristics of the single step (type of phases, auxiliary phases, partition coefficient, partition isotherm, separation factor) and *way of performing the separation:*

a) Separation by one-time equilibrium adjustment;
b) separations with repetition of the single step (one-sided repetition, separation series, separation column or thin film method, countercurrent method, cross-current method).

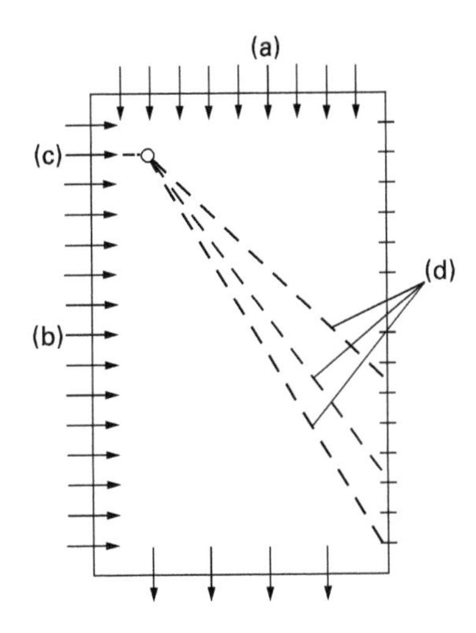

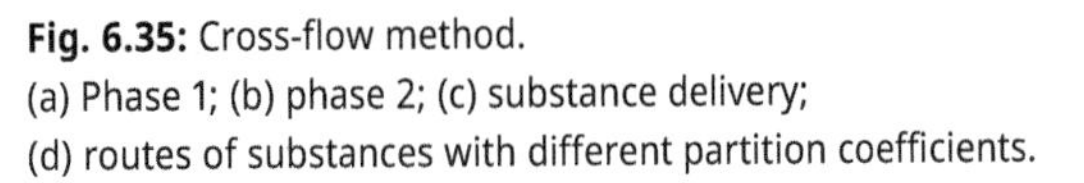
Fig. 6.35: Cross-flow method.
(a) Phase 1; (b) phase 2; (c) substance delivery;
(d) routes of substances with different partition coefficients.

The extraordinary progress in this field has been achieved mainly by the development and extensive elaboration of the separation column and thin film methods with two auxiliary phases. In addition, the older methods with one-time equilibration and one-sided repetition still play an important role. The other methods have not yet acquired a more general significance for analytical chemistry.

Literature to the text

Thin film technology

B. Fried, Practical Thin-layer Chromatography, CRC Press, Boca Raton (2017).
C. Poole, Instrumental Thin-layer Chromatography, Elsevier, Amsterdam (2014).
J. Sherma & B. Fried, Handbook of Thin-Layer Chromatography, Marcel Dekker, New York (2003).
P. Wall, Thin-Layer Chromatography: A Modern Practical Approach. Royal Society of Chemistry, Cambridge (2005).

Countercurrent method

A. Berthod & S. Alex, Industrial applications of CCC, in: J. Cazes, Encyclopedia of Chromatography, Marcel Dekker, New York *2*, 192–1197 (2010).
P. Fedotov, Untraditional applications of countercurrent chromatography, Journal of Liquid Chromatography & Related Technologies *25*, 2065–2078 (2002).
J. Friesen, J. McAlpine, S.-N. Chen & G. Pauli, Countercurrent separation of natural products: An update, Journal of Natural Products *78*, 1765–1796 (2015).
Y. Ito, Golden rules and pitfalls in selecting optimum conditions for high-speed counter-current chromatography Journal of Chromatography A *1065*, 145–168 (2005).

Gradient elution

K. Broeckhoven & G. Desmet, Theory of separation performance and peak width in gradient elution liquid chromatography: A tutorial, Analytica Chimica Acta *1218*, 339962 (2022).

L. Snyder, J. Kirkland & J. Dolan, Introduction to Modern Liquid Chromatography, John Wiley & Sons, New York (2010).

Cross-flow method

M. Marioli & W. Kok, Continuous asymmetrical flow field-flow fractionation for the purification of proteins and nanoparticles, Separation and Purification Technology *242*, 116744 (2020).

T. Shimokusu, V. Maybruck, J. Ault & S. Shin, Colloid separation by CO_2-induced diffusiophoresis, Langmuir *36*, 7032–7038 (2020).

Separation plate number of columns (HETP)

V. Berezkin, I. Malyukova, & D. Avoce, Use of equations for the description of experimental dependence of the height equivalent to a theoretical plate on carrier gas velocity in capillary gas-liquid chromatography, Journal of Chromatography A *872*, 111–118 (2000).

F. Gritti & G. Guiochon, Mass transfer kinetics, band broadening and column efficiency, Journal of Chromatography A, *1122*, 2–40 (2012).

G. Guiochon, The limits of the separation power of unidimensional column liquid chromatography, Journal of Chromatography A, *1126*, 6–49 (2006).

J. Roberts & G. Carta, Relationship between HETP measurements and breakthrough curves in short chromatography columns, Biotechnology Progress *37*, e3065 (2021).

N. Snow, Is Golay's famous equation for HETP still relevant in Capillary Gas Chromatography? Part 1: A common view of HETP, LC-GC North America *40*, 22–26 and 31 (2022).

7 Distribution between two liquids

7.1 General

7.1.1 Historical development

The first analytical application of the distribution between two liquids[1] was given by Rothe (1892) (etherification of Fe^{3+} from strongly hydrochloric acid solution). A significant extension of the possible applications of this method resulted primarily from the study of diphenylthiocarbazone compounds (*dithizone compounds*), which demonstrated the usefulness of the method for separations in the trace range. Later, Seaborg used radioactive isotopes to prove that separations could be made at extremely low concentrations.

The multiplicative distribution in a separation series was worked out by Frenc (1925) as well as by Jantzen (1932), and later also by Craig (1949); the mode of operation in columns is from Martin & Synge (1941).

7.1.2 Auxiliary phases – distribution isotherm – rate of equilibration–separation factors

In the case of distribution between two liquids, two auxiliary phases, the two solvents, are usually present. If the analytical sample is itself a liquid (e.g. a mixture of hydrocarbons), it can be shaken out directly with a solvent that is immiscible with it; in such cases, only one auxiliary phase is thus required. This method of operation plays an important role in technical procedures, but is hardly ever used in analytical chemistry.

Ideally, the distribution isotherm runs in a straight line (*Nernst's distribution law*):

$$\frac{c_1}{c_2} = \alpha = \text{const.} \tag{1}$$

The form of Nernst's distribution law given in eq. (1) applies only when the molecular size of the distributed substance is the same in both phases; if association occurs in one of the two phases, the more general eq. (2) applies

1 For separations by distribution between two liquids (*solvent extraction*), the term *extraction* or *liquid-liquid extraction* is also often found; however, the term *extraction* will be used in this book exclusively for the separation of individual components from mixtures of solid substances.

https://doi.org/10.1515/9783111181417-007

$$\frac{c_1^n}{c_2} = a = \text{const.} \qquad (2)$$

where n *is* the degree of association. This is determined by plotting log c_1 against log c_2 from the slope of the straight lines obtained (cf. Fig 7.1).

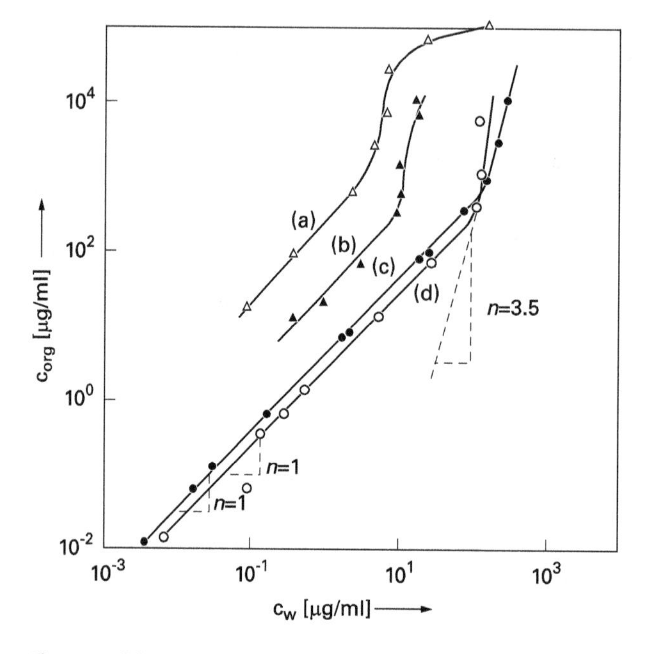

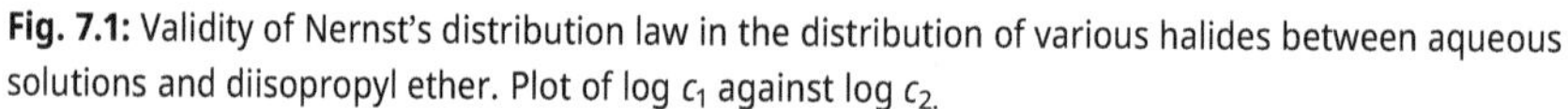

Fig. 7.1: Validity of Nernst's distribution law in the distribution of various halides between aqueous solutions and diisopropyl ether. Plot of log c_1 against log c_2.
(a) Gallium chloride, 7 mol/l HCl; (b) Ferric chloride, 7 mol/l HCl; (c) Indium bromide, 4.5 mol/l HBr; (d) Thallium(III) chloride, 3 mol/l HCl.

As already mentioned, Nernst's distribution law often applies over a very large concentration range, so that the elaboration of effective separation processes is possible especially in the range of extremely low concentrations. However, deviations are occasionally observed at very high and at very low concentrations (cf. Figs. 7.1 and 7.2).

The partition coefficient can also be used to identify unknown substances.

The *rate of equilibration* is generally quite high; if the two phases are shaken vigorously, equilibrium is usually reached after about 20–30 s, but to be on the safe side it is customary to shake for 2–3 min. With careful rocking back and forth of a vessel with the two liquids, 40–50 rocking movements are usually sufficient.

The distribution can be much slower if a sluggish complexation reaction determines the rate (see below, section 2.1.8).

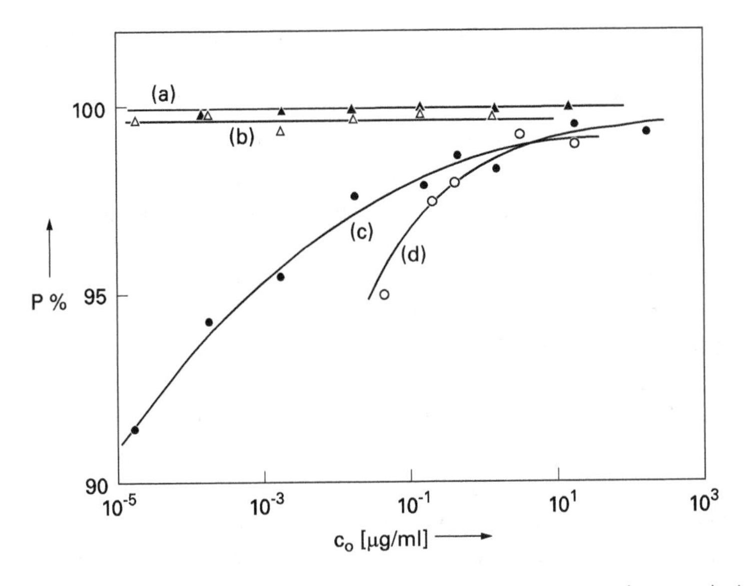

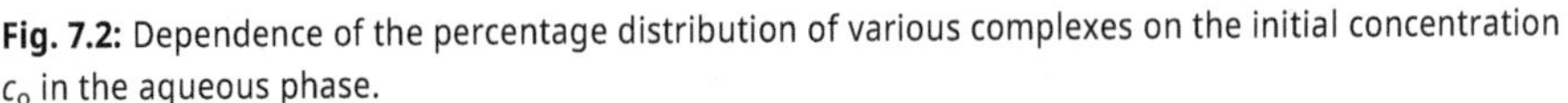

Fig. 7.2: Dependence of the percentage distribution of various complexes on the initial concentration c_o in the aqueous phase.
(a) Cobalt diethyldithiocarbamate compound; org. phase CCl_4;
(b) Cobalt diphenylthiocarbazone compound; org. phase CCl_4;
(c) Cobalt-8-hydroxyquinoline compound; org. phase $CHCl_3$;
(d) Molybdenum-toluene-3,4-dithiol compound; org. phase isoamyl acetate.

Separation factors. Very high separation factors (e.g. 10^6–10^7) are often observed in the distribution between two solvents, but effective separations can be achieved by repeating even in the presence of much less favorable conditions (see below).

If Nernst's law applies to all substances to be separated, the separation factors are independent of the absolute concentrations and the concentration ratios (provided there is also no mutual influence).

7.1.3 Solvents – solvent mixtures – synergistic effect – extracted compounds – chemical reactions during distribution

The partition coefficient of any substance is essentially determined by the choice of solvent pair. Analytical chemistry is dominated by methods in which compounds are shaken out of aqueous solution with an organic solvent; immiscible organic solvent pairs have played only a minor role to date.

When shaking out purely inorganic compounds (e.g. chloro-, bromo- or iodo-complexes, nitrates, thiocyanates etc.) from aqueous solutions, a rule stated by W. Fischer (1939) is useful: The more *water-like* the organic solvent used, the more inorganic compounds can be extracted with it, but the less selective the separations become and the worse the separation effects; the more *water-unlike* the organic solvent, the fewer inorganic compounds can be extracted with it, but the more effective the separations become.

Water-like solvents are mainly lower alcohols, ketones and esters, while *water-unlike solvents* are, for example, benzene, CCl_4, aliphatic hydrocarbons and others. Lower ethers, higher molecular esters, alcohols, etc. occupy an intermediate position.

However, exceptions to this rule often occur; it does not apply when complexes of inorganic ions are extracted with an organic complexing agent. Here, solvents such as CH_2Cl_2, $CHCl_3$, benzene and others are usually most favorable.

Difficulties arise in the separation of high molecular weight natural products, especially proteins, due to partitioning between aqueous solutions and organic solvents. It is generally not possible to shake them out with the organic liquids commonly used, and furthermore denaturation often occurs at the phase interface. Attempts have been made to achieve separations using water-soluble liquids, such as glycol, but the formation of two phases must be forced by adding salts to the water layer.

A method indicated by Albertsson (1958) has proved to be more effective, in which both phases consist of aqueous solutions containing different polymers (*polymer phase separation*). Even at polymer concentrations of a few percent, segregation can occur in such systems (Tab. 7.1), so that each phase contains about 90–95% water at equilibrium.

Mixed solvents. If a compound is shaken out of aqueous solution with a mixture of organic solvents consisting of a good and a bad component, the distribution will decrease with increasing proportion of bad component. The decrease needs not be linear, however, but can occur in a variety of ways: If the drop in the distribution coefficient occurs for one of the substances to be separated already after the addition of a little *bad* solvent, and for the other only after the addition of a lot of *bad* solvent, the separation factor reaches a maximum at the average ratio of the two organic liquids (cf. Fig 7.3).

Tab. 7.1: Phase pairs in polymer phase separation (examples).

Phase 1	Phase 2
Dextran + H_2O	Polyethylene glycol + H_2O
Dextran + H_2O	Polyvinyl alcohol + H_2O
Dextran + H_2O	Methylcellulose + H_2O
Dextran + H_2O	Polyvinylpyrrolidone + H_2O

As an example of this effect, the separation of cobalt and nickel by extracting the thiocyanates is mentioned; with diethyl ether-amyl alcohol mixtures (25:1), the separation is considerably improved compared to pure amyl alcohol (Rosenheim).

Micellar extraction. The extraction of organic trace substances from aqueous media without the aid of organic solvents is achieved by so-called *micellar extraction*. Watanabe (1982) was the first who, after introducing water-soluble surfactants into the

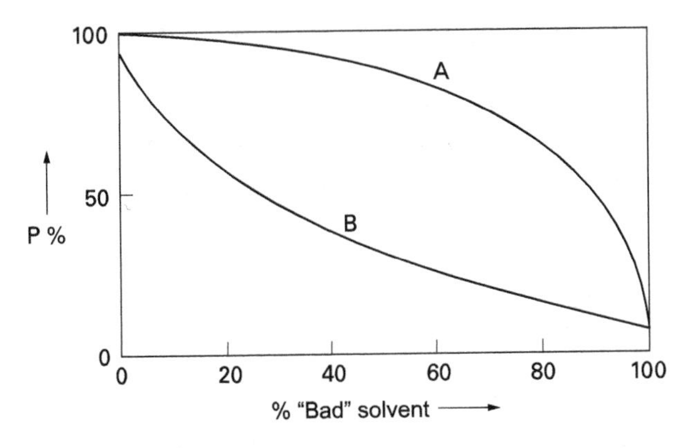

Fig. 7.3: Distribution of two compounds A and B as a function of the mixing ratio of a good and a poorly effective organic solvent.

water phase, made it possible to separate trace substances by enrichment in a newly forming, pure micelle phase. The background for this is the formation of a few nm-sized micelles (cf. Fig 7.4) in the water phase.

A micelle dispersed in water is formed by aggregation of the hydrophobic regions of the surfactant molecules involved. The hydrophilic regions of the surfactant molecules thereby protrude into the water volume. This creates stable nano-compartments, which are available for liquid-liquid distribution. The inner regions of a micelle act as a hydrophobic phase in which, for example, organic trace substances can accumulate.

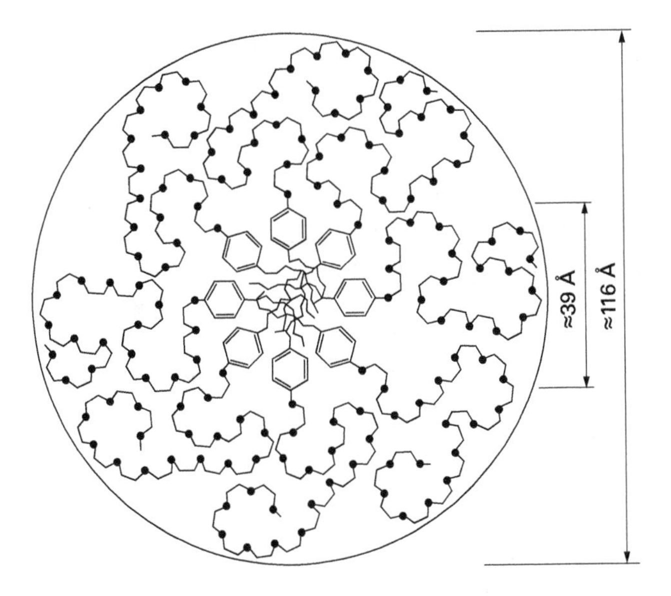

Fig. 7.4: Idealized micelle formed by addition of the surfactant polyoxyethylene – nonylphenyl ether; with the associated dimensions. Both the inner hydrophobic core and the outer hydrophilic shell are shown.

The simplest way to separate the micelles from water is to trigger the formation of the micelle phase after the critical micelle concentration has been exceeded. (cf. Fig 7.5). Micelle phase formation can be triggered, for example, by a moderate temperature increase of a few degrees Celsius. This is indicated by the formation of an opaque phase (*cloud point*). The phases (water/surfactant micelles) can be easily separated by centrifugation.

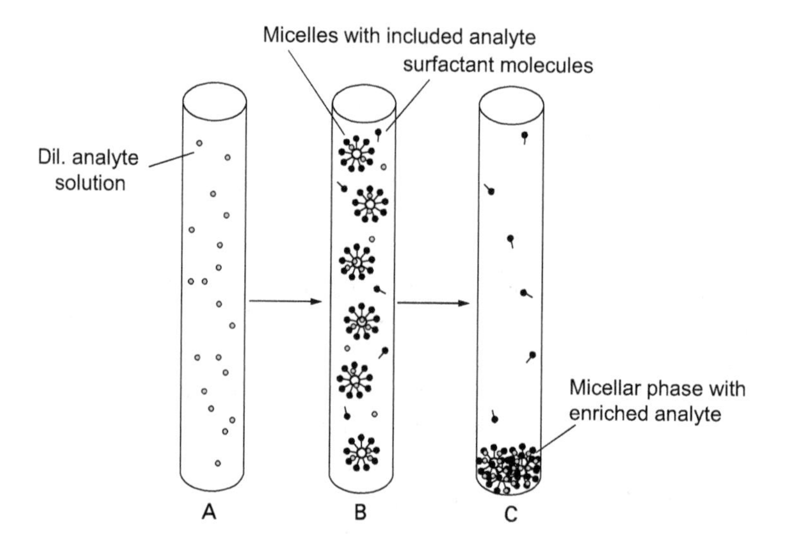

Fig. 7.5: Illustration of micellar extraction.
a) Aqueous phase with org. analyte traces;
b) after addition of surfactant in higher concentration, micelles with hydrophobic interior are formed;
c) after forcing the micelle phase formation, e.g. by heating, the org. traces are enriched in the newly formed micelle phase.

A variety of differently structured surfactants enables the formation of graded hydrophilic or hydrophobic nanocompartments. In addition to numerous organic analytes, heavy metal ions, for example, can also be extracted from aqueous solutions after complexation. Particularly interesting are separations of biological samples, such as enzymes, viruses, membrane proteins, vitamins and porphyrins (cf. Table 7.2).

The enrichment that can be achieved is determined by the ratio of the initial volume of the water phase containing the analyte to the volume of the micelle phase formed.

A disadvantage is the presence of high surfactant concentrations in subsequent determination procedures. Nevertheless, this separation technique is an attractive alternative for avoiding high solvent volumes.

Synergistic effect. The term *synergistic effect* (or *synergism*) refers to the phenomenon, already observed by Schikorr (1926), that the distribution of a compound can be substantially increased by the addition of a third substance.

Tab. 7.2: Micellar extractions (examples).

Ion	Ligand	Nonionic surfactant	Enrichment factor
Zn^{2+}	PAN	PONPE −7.5	40
Ni^{2+}	PAN	Triton X-100	40
Metalloporphyrins		Triton X-100	10–100
Vitamin E		C_{10} −$APSO_4$	45
PAHs		Genapol X-80	135
Napropamide		Genapol X-150	85
PCBs		Triton X-100	

For example, the distribution of the uranium(VI) compound with di-(2-ethylhexyl) phosphoric acid between aqueous solutions and kerosene can be shifted strongly in favor of the organic phase when neutral phosphine oxides, such as tri-n-butyl phosphine oxide, are added. Similar ratios were also found in numerous other systems (see Tab. 7.3). The explanation is to be sought in solvation effects and complexation reactions; the area is related to the mixed solvent area.

While usually only simple addition compounds are formed, in some systems esters (vanadium-hydroxyquinoline compound) or complexes with complicated compositions occur, as in the case of shaking out the gallium-thenoyltrifluoroacetone compound from chloride-acetate solution. Since the organic solvent also plays a role and the effect can be reversed by larger reagent additions, such systems are highly variable.

Tab. 7.3: Synergistic effect (examples). TTA = thenoyltrifluoroacetone; Ac = acetylacetone; HFA = hexafluoroacetylacetone; TOPO = tri-n-octylphosphine oxide; DBP = di-n-butylphosphoric acid; TBP = tri-n-butylphosphate; TOA = tri-n-octylamine; TPAsCl = tetraphenylarsonium chloride; Ox = 8-hydroxyquinoline.

Ion	Chelating agent	Org. solvent	Synergism through	Formed compound
Mn^{2+}	TTA	Benzene	Pyridine	$Mn(TTA)_2 \cdot 2$ Py
Ni^{2+}	TTA	Benzene	Isoquinoline	$Ni(TTA)_2 \cdot 2$ Chin
Zn^{2+}	TTA	Hexane	TOPO	$Zn(TTA)_2 \cdot$ TOPO
Co^{2+}	Ac	Benzene	γ-Picolin	$Co(Ac)_2 \cdot 2$ Pic
Cu^{2+}	HFA	Benzene	TOPO	$Cu(HFA)_2 \cdot$ TOPO
UO_2^{2+}	DBP	CCl_4	TBP	$UO_2(DBP)_2 \cdot$ DBP · TBP
UO_2^{2+}	TTA	Benzene	TOPO	$UO_2(TTA)_2 \cdot$ TOPO
PuO_2^{2+}	TTA	Cyclohexane	TBP	$PuO_2(TTA)_2 \cdot$ TBP
Pu^{3+}	TTA	Cyclohexane	TBP	$Pu(TTA)_3 \cdot 2$ TBP

Tab. 7.3 (continued)

Ion	Chelating agent	Org. solvent	Synergism through	Formed compound
Am^{3+}	TTA	Benzene	TOA	$Am(TTA)_3 \cdot TTA \cdot TOA$
Ga^{3+}	TTA	o-Dichlorobenzene	TPAsCl	$[Ga(TTA)_2 (CH_3COO)Cl] \cdot (C_6H_5)_4As$
VO^{3+}	Ox	Benzene	n-Butanol	$VO(Ox)_2 \cdot OC_4H_9$

Extracted compounds. As a rule, the less polar the structure of the molecules, the better inorganic and organic compounds are shaken out from aqueous solutions with organic solvents. Thus, compounds such as Cl_2, Br_2, I_2, ClO_2, OsO_4, $HgCl_2$, $AsCl_3$, $GeCl_4$, SnI_4 and the like even go into nonpolar or less polar solvents such as CCl_4, $CHCl_3$ or benzene.

On the other hand, less polar compounds can be converted into more polar ones and then shaken out of organic solution with water; for example, weak organic acids or phenols are separated from organic solvents with aqueous alkali metal hydroxide solutions to form the alkali metal salts.

To extract more polar compounds such as $HFeCl_4$, $HAuCl_4$, $HTlCl_4$, $HTlBr_4$, H_2TaF_7, $H_2Ce(NO_3)_6$, $In(SCN)_3$ and others, polar solvents such as diethyl ether, methyl iso-butyl ketone, tri-n-butyl phosphate and others must be used.

Often, reactions of inorganic ions with organic complexing agents containing oxygen, nitrogen or sulfur, or with organic ions of opposite charge, yield compounds that can be shaken out of aqueous solution with organic solvents (see Tab. 7.4).

Chemical reactions during distribution. If the substances to be separated are not present in a suitable form from the outset, they must be converted into compounds with favorable partition coefficients or separation factors by reagent addition, oxidation or reduction. This is often done by dissolving the reagent in the organic phase used to extract the initial aqueous solution and forming the compound of interest by an exchange reaction. Ligand exchange processes have also been described, but have not become more important. A redox reaction is used in the so-called *amalgam exchange* (see below).

7.1.4 pH distribution curves

In the above-mentioned mode of operation with double reaction during distribution, a pronounced pH dependence of the partition coefficient occurs when the organic reagent consists of a weak acid with which an inorganic cation reacts. (The same applies to inorganic anions reacted with a weak organic base).

Tab. 7.4: Organic complexing agents, acids and bases which form compounds suitable for extraction with inorganic ions.

Complexing agent	Analyte
Hexafluoroacetylacetone	Alkyl and aryl phosphoric acids
Thenoyltrifluoroacetone	Basic triphenylmethane dyes
Dimethylglyoxime	Triphenyltin hydroxide
α-Benzoinoxime	Tetraphenylphosphonium hydroxide
Salicylaldoxime	Tetraphenylarsonium hydroxide
Diphenylthiocarbazone	Tetraphenylstibonium hydroxide
Sodium diethyldithiocarbamate	Aliphatic amines (e.g. tri-n-octylamine)
8-Mercaptoquinoline	Quaternary ammonium hydroxides (e.g. tetra-n-hexyl-ammonium hydroxide)
8-Hydroxyquinoline	
α-Nitroso-β-naphthol	Di-n-butylamine
N-Nitroso-phenyl-hydroxyamine (NH_4 salt; cupron)	Diantipyrylmethane
N-Benzoyl-N-phenyl-hydroxylamine	
Benzhydroxamic acid	
Diphenylguanidine	

Under simplifying assumptions, which are, however, fulfilled with sufficient approximation for numerous practically important systems, the influence of the pH of the water layer on the distribution can be determined.

A metal ion Me^{n+} of valence n may react during distribution with n molecules of an organic acid HR, which is practically soluble only in the organic phase.

Furthermore, the compound MeR_n formed should go almost completely into the organic phase. If o and w are set as indices for the organic and aqueous phases, respectively, then:

$$(Me^{n+})_w + n \cdot (HR)_o \rightarrow (MeR_n)_o + n \cdot (H^+)_w. \tag{3}$$

The equilibrium constant K *of* this reaction is (neglecting the activity coefficients):

$$K = \frac{[MeR_n]_o \cdot [H^+]_w^n}{[Me^{n+}]_w \cdot [HR]_o^n}. \tag{4}$$

If MeR_n goes practically completely into the organic phase and the metal ion Me^{n+} remains almost completely in the water layer, this expression contains the partition coefficient α for the element in question:

$$\alpha = \frac{[MeR_n]_o}{[Me^{n+}]_w}, \tag{5}$$

and it applies:

$$K = \alpha \cdot \frac{[H^+]_w^n}{[HR]_o^n} \tag{6}$$

respectively

$$\alpha = K \cdot \frac{[HR]_o^n}{[H^+]_w^n} \tag{7}$$

Accordingly, the partition coefficient α depends, under the above conditions, on the complexation constant K valid for the compound in question, on the pH of the water layer after partitioning, and on the reagent concentration $[HR]_o$ in the organic phase. (This is not the original, but the excess concentration of HR remaining after reaction of part of the acid HR with the metal ion Me^{n+}).

By logarithmizing eq. (7), we obtain:

$$\log \alpha = \log K + n \cdot \log [HR]_o + n \cdot pH. \tag{8}$$

For simplification, it is now assumed that the volumes of both phases are equal, and for the graphical representation the clearer percentage distribution P % is chosen. Eq. (8) then results in several sets of symmetric curves with inflection points at a distribution of 50% ($\alpha = 1$); the inflection point is also denoted by $E_{1/2}$.

For a given concentration $[HR]_o$ and given n, the position of the distribution curves depends on the value of the equilibrium constant K; if K increases by one power of ten, the distribution curve shifts in parallel by one pH unit towards lower pH values (cf. Fig 7.6).

For given K and given n, the position of the distribution curve depends on the reagent concentration $[HR]_o$ in the organic phase after equilibration; increasing $[HR]_o$ by one power of ten shifts the distribution curve one pH unit toward lower pH values (Fig. 7.7).

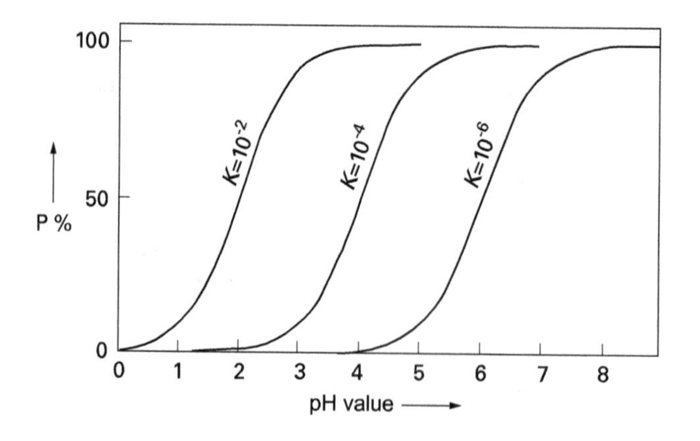

Fig. 7.6: Percent distribution of a compound consisting of a metal ion and an organic acid as a function of the pH of the water layer at different values of K ($[HR]_o$ = 1 mol/l; n = 1).

Finally, for given values of K and $[HR]_o$, the position and slope of the distribution curve depend on n; if n is increased by one unit, the curve shifts toward smaller pH values, and the slope becomes steeper (for $n = 1$, α increases by one power of ten per pH unit, for $n = 2$ by two powers of ten, and for $n = 3$ by three powers of ten; see Fig. 7.8).

Remarkably, the concentration of the extracted metal ion does not appear in eq. (7), so the partition coefficient must be independent of the concentration of the ion in question. Accordingly, a very large range of validity of Nernst's law was also observed in numerous such systems (cf. Fig 7.2).

If not an inorganic cation is extracted with an organic acid, but vice versa an inorganic anion with an organic base, the same laws apply in principle, only the course of the curves is reversed (cf. Fig 7.9).

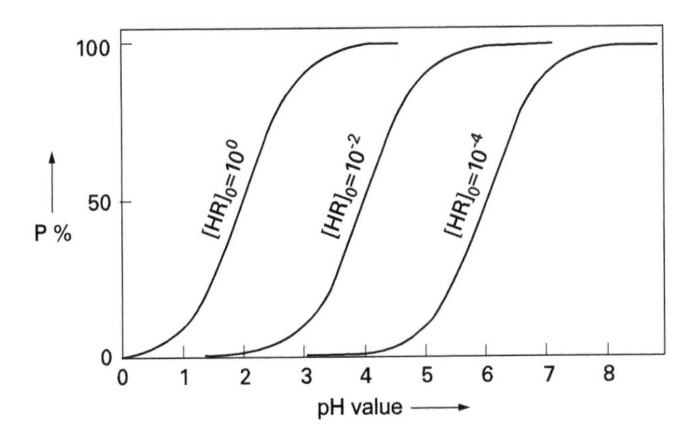

Fig. 7.7: Percentage distribution of a compound of a metal ion and an organic acid as a function of the pH of the water layer at different reagent concentrations $[HR]_o$ in the organic phase ($K = 10^{-2}$; $n = 1$).

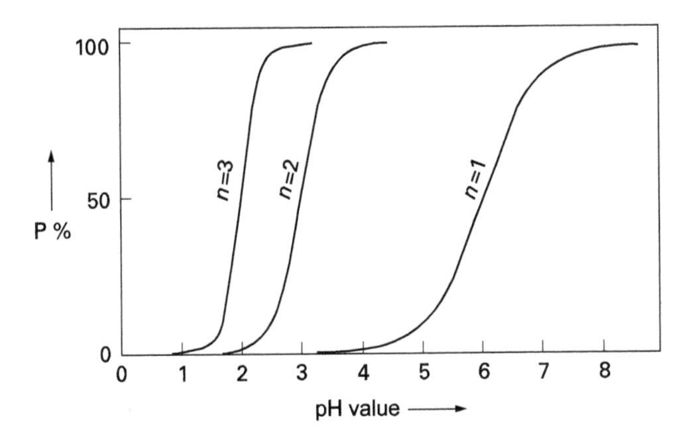

Fig. 7.8: Percentage distribution of compounds from a metal ion and an organic acid as a function of the pH of the water layer at different valences n *of* the metal ion ($K = 10^{-6}$; $[HR]_o = 1$ mol/l).

Using eq. (8), the value of n and thus the formula of the extracted compound can be obtained by two methods: Either determine the pH dependence of the partition coefficient α at constant reagent concentration $[HR]_o$ and determine n from the slope of the pH distribution curve, or change the reagent concentration $[HR]_o$ at constant pH; then plot log α against log $[HR]_o$ and a straight line with slope n is obtained.

Deviations from the ideal shape of the distribution curves. Usually, to determine a distribution curve of a compound, the reagent is dissolved in the organic phase and then a series of buffered aqueous solutions of the metal ion are shaken out with portions of the organic phase. In this process, however, the reagent concentration $[HR]_o$ does not remain constant; at low pH, only a small portion of the reagent is consumed by the ion being extracted, and the concentration $[HR]_o$ is relatively large. As the pH increases, the partition coefficient increases, and as a result of the increasing conversion of the organic reagent, $[HR]_o$ decreases. The magnitude of the change depends on the quantity ratio of metal ion and organic acid. A somewhat distorted distribution curve is obtained due to this effect (Fig. 7.10).

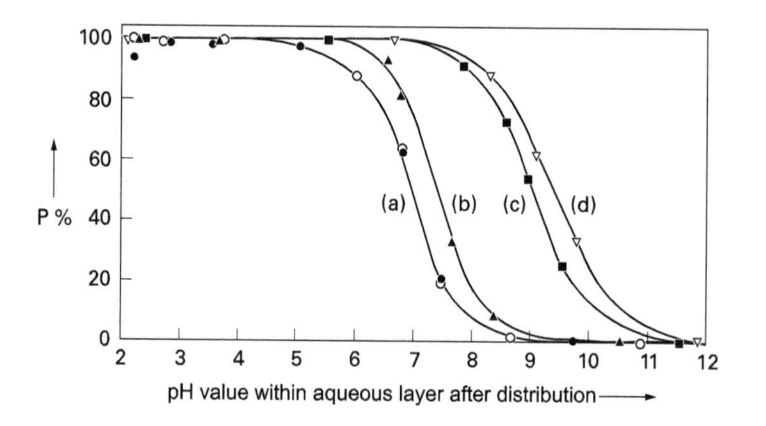

Fig. 7.9: Distribution of tetraphenylstibonium compounds of inorganic anions between aqueous solutions and chloroform as a function of pH.
(a) Cl^- and Br^-; (b) I^-; (c) SCN^-; (d) F^-.

Frequently, several compounds are formed between the inorganic component and the organic reagent when the pH value changes, e.g. basic or acidic complexes. Under certain circumstances, very irregular distribution curves may occur, which may even show maxima and minima (Fig. 7.11).

Deviations still occur if the distribution equilibrium is not reached due to insufficient shaking time, which is to be feared above all with small reagent concentrations. The distribution curve is then apparently shifted to higher pH values.

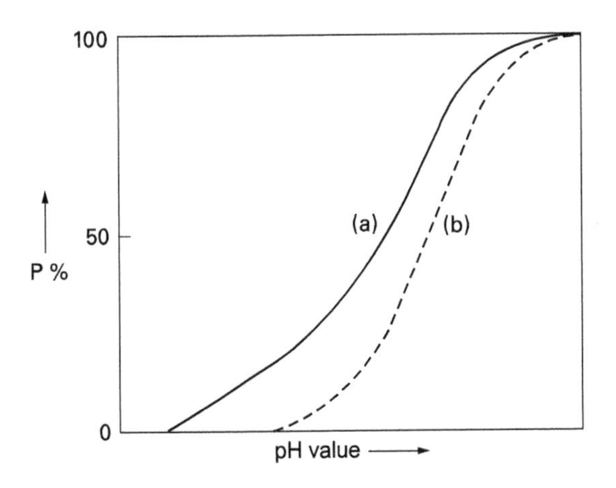

Fig. 7.10: Deviations from the ideal shape of the distribution curve due to reagent consumption.
a) Distorted curve; b) ideal curve.

7.1.5 Influence of equionic additives

Additions of equionic compounds influence the distribution of purely inorganic compounds by repressing dissociation. In addition, especially at high concentrations of the additive, water molecules are removed from the solvate shell of the compound to be shaken out, so that it is forced more strongly into the organic phase, an effect which corresponds to the salting-out effect in precipitation. As an example, Tab. 7.5 shows the effect of various additives on the precipitation of thorium nitrate.

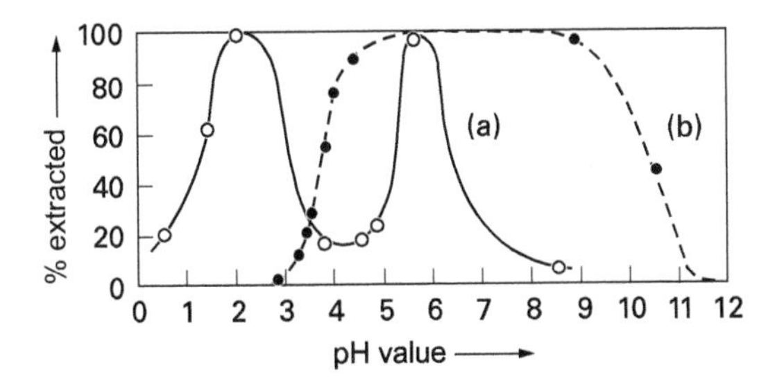

Fig. 7.11: Irregular distribution curves.
(a) Distribution of vanadium (V)-pyrrolidine-dithiocarbamate compound between aqueous solutions and chloroform;
(b) distribution of the UO_2^{2+}-8-hydroxyquinoline compound between aqueous solutions and chloroform.

Tab. 7.5: Distribution of $Th(NO_3)_4$ between aqueous 1 mol/l HNO_3 solutions and diethyl ether at saturation with different nitrates.

Added salt	% of Th^{4+} in ether
–	0,02
$LiNO_3$	56,5
$NaNO_3$	0,67
NH_4NO_3	0,36
$Mg(NO_3)_2$	43,8
$Ca(NO_3)_2$	56,8
$Sr(NO_3)_2$	0,18
$Al(NO_3)_3$	54,2

7.1.6 Influence of complexing agents

In the presence of complexing agents in the aqueous phase, a competitive reaction for the inorganic ion occurs between these and the organic reagent. The pH distribution curves are thus shifted more or less strongly to higher pH values when an inorganic cation is shaken out with an organic acid (unless the complex also goes into the organic phase) (cf. Fig 7.12). In the extreme case, a complete masking of the ion occurs, which then cannot be extracted at all.

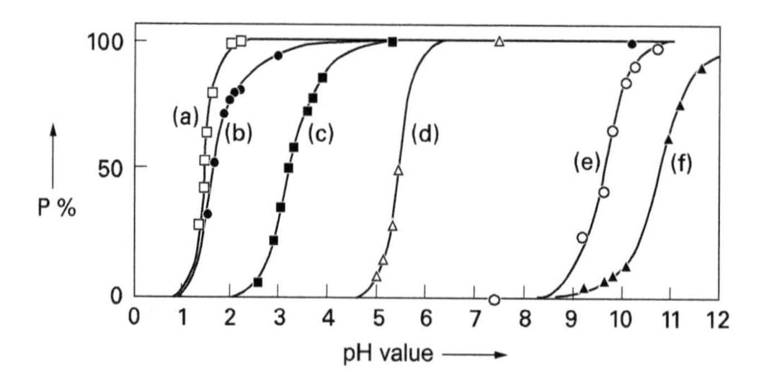

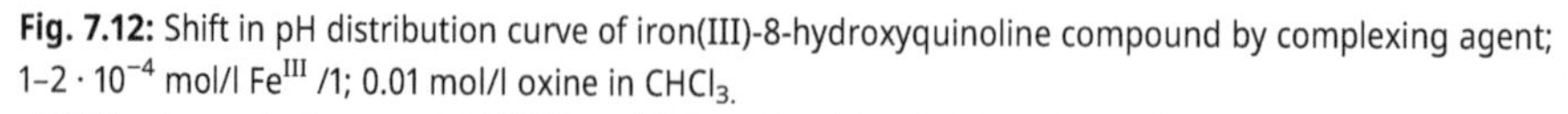

Fig. 7.12: Shift in pH distribution curve of iron(III)-8-hydroxyquinoline compound by complexing agent; 1–2 · 10^{-4} mol/l Fe^{III} /1; 0.01 mol/l oxine in $CHCl_3$.
a) Without complexing agent; b) 0.010 mol/l Tartaric acid; c) 0.010 mol/l Nitrilotriacetic acid;
d) 0.010 mol/l Oxalic acid; e) 0.010 mol/l Ethylenediaminetetraacetic acid;
f) 0.010 mol/l 1,2-Diaminocylohexanetetraacetic acid.

7.1.7 Incomplete separations by distribution between two liquids (cf. 1st part, Chap. 4)

If a substance is incompletely separated during shaking out, the yield can be determined by the direct isotope dilution method; furthermore, the substoichiometric method has been repeatedly applied.

Furthermore, quantitative determinations have been carried out by determining concentration-dependent partition coefficients.

7.1.8 Disturbances

During the distribution between two liquids, several disturbances occur which can affect the separations. One difficulty lies in completely separating the two phases after equilibration, and losses can also be caused by droplets stuck to the walls or particles sitting in the ground joints. Furthermore, colloidally distributed residues of the other liquid occasionally remain in both phases and do not settle even after prolonged standing. This disturbance can be eliminated by centrifugation; water droplets in the organic phase can also be removed by filtering through a dry or hydrophobic paper filter.

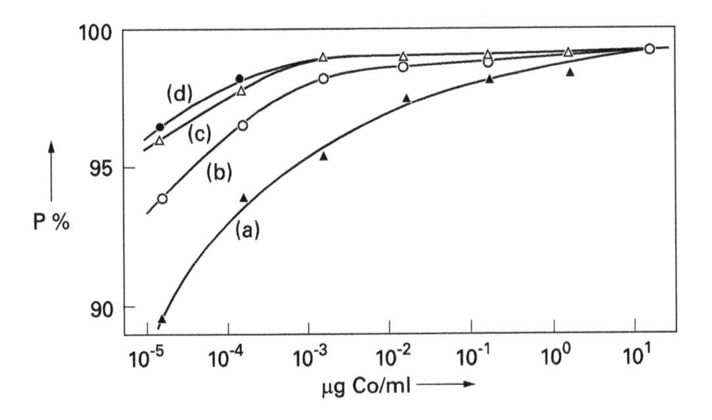

Fig. 7.13: Percent distribution of the cobalt-8-hydroxyquinoline complex as a function of cobalt concentration at different shaking times.
(a) 3 min shaking time; (b) 24 h; (c) 48 h; (d) 72 h.

A further disturbance occurs when a slow complexation reaction is the rate-determining step in the distribution. Especially at very low concentrations of the extracted substance, this has been observed and thus an apparent drop in the distribution coefficients (cf. Figs. 7.2 and 7.13).

7.2 Separations by one-time equilibrium adjustment

7.2.1 Working method and devices

Separations by one-time equilibration are usually carried out in so-called *separating funnels*, of which numerous embodiments are given (cf. Fig 7.14). The distribution is accelerated by vigorous shaking, in the case of foaming liquids by careful rocking back and forth, by stirring devices or by passing a gas stream through (cf. Fig 7.14c).

After dismounting, the two phases are separated by draining the lower one or by lifting off the upper one.

A rarely used variant consists of filtering an aqueous solution through a paper filter impregnated with an organic solvent.

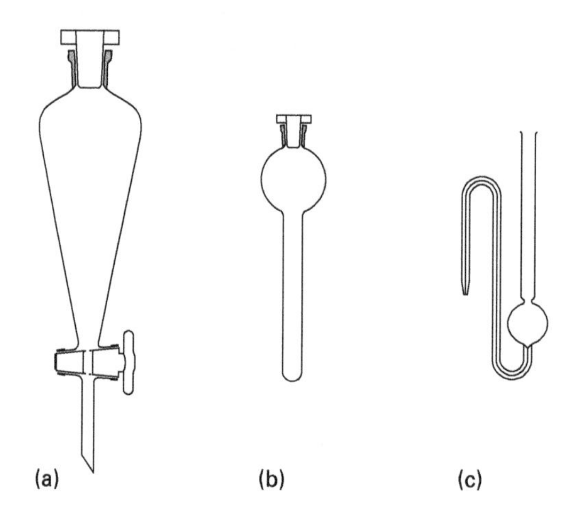

Fig. 7.14: Some types of separating funnels.
a) Ordinary separation funnel;
b) separation funnel for a smaller volume of heavy phase;
c) micro-separation funnel with movement of the phases by a gas stream.

7.2.2 Applications

Since extremely high separation factors are achieved in many systems, a single shake-out is often sufficient for analytical purposes, especially when separating trace components from solutions.

However, the above-mentioned disturbances due to incomplete phase separation and losses at the vessel wall or in sections set limits to the method.

Amalgam exchange. The so-called *amalgam exchange* occupies a special position. In this process, an aqueous salt solution is brought into contact with an amalgam; if the metal dissolved in the mercury is less noble than the metal present in ionic form in the aqueous solution, an exchange of the ions for the metal occurs in the course of a redox reaction.

The exchange reactions are essentially determined by the amalgam potentials of the two reactants, but the composition of the aqueous solution, especially the presence of complexing agents, also plays a role; for example, Ag^{+} no longer exchanges from cyanide solution, but on the other hand, the redox potential of the Hg^{2+}/Hg° system can be shifted, thus improving the conditions for the exchange of various ions.

If the aqueous solution contains ions of metals more noble than mercury, the latter can itself effect the reduction, and no other element need be dissolved in it.

Amalgam exchange has been used to separate a whole range of metal and semimetal ions from solutions (Tab. 7.6).

Tab. 7.6: Separations by amalgam exchange (examples).

Metal ion solution	Amalgam species
Cu^{2+} in 0.1 mol/l tartaric acid	Zn amalgam (1.5%)
Cu^{2+} in 0.1 mol/l tartaric acid	Cd amalgam (1.5%)
Hg^{2+}, Tl^{+}, SbO^{+} in 0.5 mol/l HNO_3	Cd amalgam (2%)
Ag^{+} in 0.8 mol/l HNO_3	Cd amalgam (2%)
Ag^{+} in 0.25 mol/l H_2SO_4	Bi amalgam (1.4%)
Ag^{+} in 0.5 mol/l $NaNO_3$ solution	Tl amalgam (3%)
Cu^{2+} –, Pb^{2+} –, Cd^{2+} –solution	Zn amalgam (10^{-3} M)
Pb^{2+}, Cf^{3+} in 7 mol/l CH_3 COONa, pH 5–6	Na amalgam (0.6%)
Au^{3+} ($HAuCl_4$ in 0.2 mol/l HNO_3)	Mercury
Ag^{+} in 1 mol/l NH_3 + 0.2 mol/l NH_4 NO_3)	Mercury
Pt^{4+} (H_2 $PtCl_6$ in 0.2 mol/l HNO_3)	Mercury
Te^{4+} in 2 mol/l HCl	Mercury
Pd^{2+} in 1 mol/l KSCN solution	Mercury

Another application of amalgam exchange is the separation of radioactive isotopes from aqueous solutions, which are not exchanged for a less noble metal, but for the same (inactive) metal in mercury solution. Thus, there is an isotope exchange. In this way, radioactive ions present in extremely low concentrations, such as Tl^{+}, Pb^{2}, Hg^{2}, Zn^{2}, Cd^{2+}, In^{3+}, Bi^{3+} and Sr^{2+} can be effectively removed from solutions and thus *decontaminated*.

The exchange equilibrium here is not determined by the potentials, but by the quantitative ratio of the metal in the mercury and the ions in the solution.

If the solution contains ions of more noble metals, these are also exchanged; the active element can be separated from these by shaking the amalgam after exchange with a strong aqueous solution of a salt of this element, thus shifting the equilibrium back to the side of the aqueous solution. In this reverse reaction, the nobler metals remain in the mercury.

Equilibria are achieved in a few minutes during amalgam exchange, provided that the two phases are intensively mixed.

7.3 Separations by one-sided repetition

The condition for the operation of one-sided repetition in separations by shaking out is that one of the substances to be separated remains practically completely in one phase, while the second goes partially into the other phase. This condition is very often present. Separations can be carried out discontinuously or continuously.

7.3.1 Discontinuous mode of operation

Unilateral repetition is usually performed by shaking out an aqueous solution in a separating funnel several times with an organic solvent.

The yield of the substance to be separated can be increased as desired in this way (unless the distribution coefficient drops sharply at low concentrations). Furthermore, the above-mentioned disturbances due to incomplete phase separation, losses on the walls and formation of colloidal droplets in the water phase are effectively eliminated.

Single droplet microextraction (SDME). Despite the simplicity and efficiency of liquid-liquid extraction, a major drawback remains in the usual design: The accumulation of larger contaminated solvent volumes. The reprocessing of the same is associated with significant costs and environmental risks.

As early as the 1990s, attempts were therefore made to achieve extraction distribution with minimal volumes (*single-drop microextraction: SDME*). Jeannot and Cantwell (1996) were the first to publish a drop of 1-octanol at the end of a PTFE rod for extracting trace organics from aqueous solutions. After the sampling period, the PTFE rod was removed from the aqueous sample solution and the solvent droplet was collected in a GC injection syringe.

Current embodiments use direct sampling: a defined amount of solvent is introduced into the water sample through a septum while hanging from the syringe needle. After equilibration, the drop is sucked back into the syringe, withdrawn through the septum and then sent for further analysis. (cf. Fig 7.15).

Various variants of this technique are now known, such as dynamic SDME, in which the absorbing solvent droplet is repeatedly sucked in and ejected during sampling for faster equilibrium adjustment. Similarly, sampling from flowing systems is well known. The combination with a third, immiscible phase for back extraction of the enriched analyte has also been widely reported.

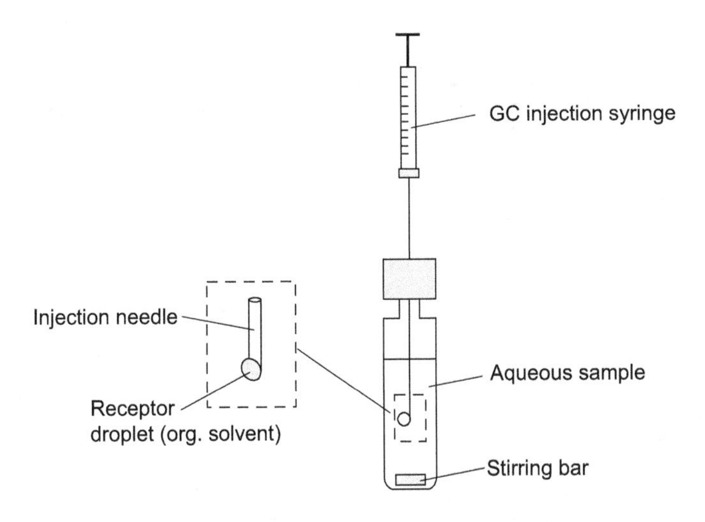

Fig. 7.15: Direct single droplet microextraction (SDME).

Combinations with complexing agents in the receiving phase also allow the separation and enrichment of heavy metals. Recently, the use of ionic liquids as a low-loss organic phase has become increasingly widespread.

By applying µl volumes of the receiving organic phase, enrichments up to a factor of 1000 are easily achieved within a few minutes. However, difficulties are caused by shearing or dissolution of the receiving droplet phase due to heterogeneous composition of the sample. The use of internal standards for loss control is therefore mandatory.

The SDME technique is particularly popular in environmental analysis. Some applications are shown as examples in Tab. 7.7.

7.3.2 Continuous mode of operation

Perforators. In the continuous mode of operation, the organic liquid is allowed to flow through the aqueous solution present, using different apparatus, so-called *perforators*, depending on the ratio of the specific weights of the two phases (Fig. 7.16).

The effectiveness of the perforators is not very great without additional stirring devices, since complete equilibration is not achieved when one liquid droplet bubbles through the other phase.

In another method of operation, short columns are used with a packing of an inert, porous support material impregnated with a suitable liquid. The components to be separated are dissolved in a second liquid immiscible with the one in the column and allowed to flow through it, quantitatively retaining substances with high partition coefficients (in favor of the phase in the column).

Tab. 7.7: Application of the SDME (examples).

Analytes	Sample	Extraction phase	Extract volume µl	Detection limit µg l^{-1}
Explosives	Water	Toluene	1	0,8–1,3
Phenols	Seawater	Decanol	2,5	0,2–1,6
Pesticides	Wine	1-Octane	2	0,003–0,05
Anesthetics	Urine	DBP	1	30–50
Pb^{2+}	Blood	Toluene	2	0,08
BTEX	Oil	Hexadecane	1	–
Organophosphorus pesticides	Juice	Toluene	1,6	< 5
Fatty acids	Blood plasma	n-Butyl phthalate	2	0,02–0,08

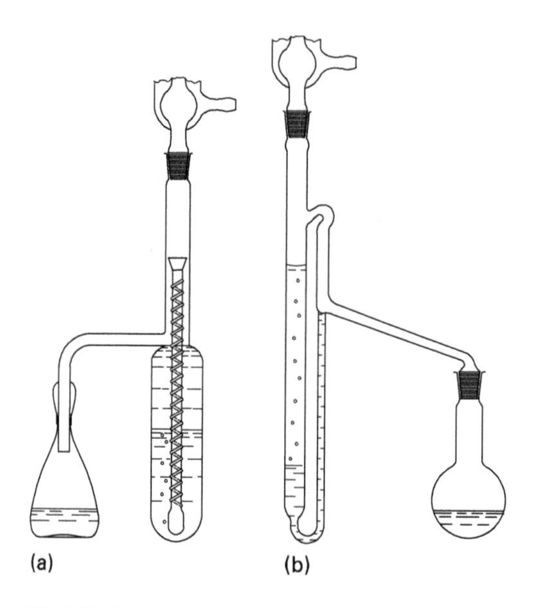

Fig. 7.16: Perforators.
a) For separation with a specifically lighter phase;
b) for separation with a specifically heavier phase.

An example of this method is the separation of Fe^{3+} and Ga^{3+} traces from strongly hydrochloric acid solutions; the aqueous phase is allowed to flow through a column with tri-n-butyl phosphate on a support. While iron and gallium remain practically completely in the column, Zn^{2+}, Cu^{2+}, Al^{3+}, In^{3+} and other ions pass through unhindered.

By using short columns (approx. 1 cm in length), very rapid separations of radioactive trace elements can be achieved; experience shows that the distribution equilibria are approximated within a few seconds.

Membrane-supported liquid/liquid extraction. *Supported liquid membranes (SLM)* enable the extraction and transport of substances from a hydrophilic phase A through a membrane covered with a hydrophobic phase into another hydrophilic phase B (see Fig. 7.17). This principle was first presented by Bloch (1970) for the removal of metal ions from aqueous solutions.

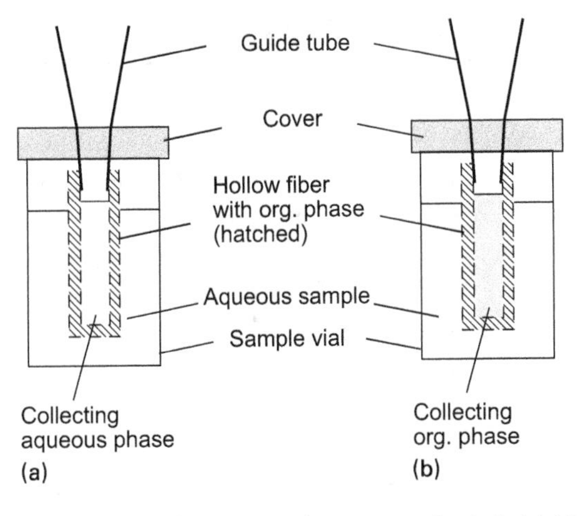

Fig. 7.17: Membrane-supported (SLM) extraction (principle) from water.
a) Aqueous receptor phase;
b) organic receptor phase.

In combination with hollow fiber bundle technology, continuous or discontinuous separation is possible. The driving force for the separation is not only the distribution in two immiscible phases (phase A/org. phase or org. phase/phase B in the membrane pore), but also the molecular diffusion of the species to be extracted through the pore space. The distribution of the species to be separated can be drastically increased by complex formation in the org. membrane phase. Advantageously, the membrane phase requires only small amounts of complexing agent and serves as a separating interface.

The separation efficiency is therefore determined by the complexation tendency, the membrane thickness, the membrane exchange area, and the diffusivity of the species to be exchanged through the membrane, as well as the removal of the exchanged species in phase B.

In the meantime, numerous applications are known in both the technical and analytical fields (cf. Tab 7.8)

Tab. 7.8: Membrane-supported SLM systems (examples).

Analyte	Inflow (Donor)	Drain (Acceptor)	Membrane phase (Transporter)
Eu^{3+}	HCl	HCl	HDEHP/Dodecane
Zn^{2+}, Cd^{2+}	HCl	CH_3COONH_4	TLA/Triethylbenzene
Co^{2+}, Ni^{2+}	CH_3COOK	HCl	Dialkylphosphoric acid
Citric acid	H_2O	Na_2CO_3	Triethylamine
Banzodiazepines	Blood	HCl	Nonalol
Chloroacetic acids	H_2O	NaOH	TOPO/Dihexylether
Phenoxyacetic acids	H_2O	NaOH	1-Octanol
Aminoalcohols	Urine	HCl	1-Octanol
Doping substances	Urine	Methanol/NH_4OH	1-Octanol

The inverse variant is also known: Water is present in the membrane pores and thus separates 2 organic phases. SLM technologies are used for the enrichment of substances prior to analytical determination. Enrichment factors of up to 10^3 are reported.

7.4 Separations by systematic repetition: Separation series

7.4.1 Discontinuous transport of the moving phase

Separation series can be set up for distribution between two liquids in a series of separating funnels, with the upper phase being moved on to the next vessel after each equilibration (cf. Fig 7.13). Since this method of operation is rather cumbersome, apparatuses are usually used for this purpose in which shaking and advancement of the agitated phase can be performed semiautomatically or fully automatically in a whole series of vessels simultaneously. As an example of various designs, a device according to Hecker (1953) is described.

The solution of the mixture to be separated is poured into the first of a series of similar distribution elements arranged one behind the other (Fig. 7.18a). In each further distribution element, enough of the stationary phase is added so that the liquid level in each case reaches the lower edge of the outlet tube A. The liquid is then poured into the outlet tube. By rotating around the axis C, the entire distribution battery is then placed horizontally and the specifically lighter moving phase is poured into the first element; the volumes of both phases should be approximately equal (Fig. 7.18b). Then the distribution equilibrium is set by prolonged rocking back and forth about the axis C (experience shows that about 40–50 rocking movements are sufficient). The compound to be separated should preferably go into the upper phase. The row of vessels is then returned to the vertical position; the upper phase flows through the outlet A into the storage vessel D (Fig. 7.18c).

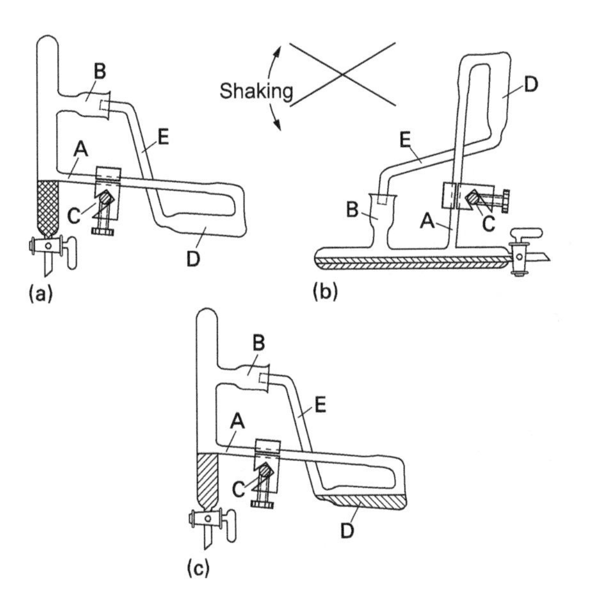

Fig. 7.18: Mode of operation of the distribution battery after Hecker.
a) Starting position;
b) filling of the moving phase and setting of the distribution equilibrium;
c) transfer the agitated phase to the storage vessel.

If the distribution vessels are finally swung back to the horizontal, the upper phase from D flows through tube E, which runs diagonally behind the drawing plane, into the filler neck of the second distribution element (not drawn in Fig. 7.18).

A fresh portion of the agitated phase is now added to the first vessel and the distribution equilibrium – this time in the first two vessels – is again established by swaying back and forth. By turning it upside down, the agitated phase then flows from vessel 1 into the first storage chamber, from vessel 2 into the second storage chamber, and when it is turned over to the horizontal, it flows from storage chamber 1 into distribution element 2, and from storage chamber 2 into element 3. In this way, all distribution elements are filled step by step with the agitated phase, and then a portion of the agitated phase is removed at the end of the series at each further distribution step (corresponding to elution in the column method).

Such devices have been developed in various embodiments with up to several hundred individual distribution elements.

If the separation factors of the substances present are large enough, the experiment can be terminated before the individual compounds leave the separation series at the end of the instrument. The pure substances present in the distribution elements can then be recovered and investigated further. As an example of this method of operation, the separation of a fatty acid mixture is reproduced in Fig. 7.19, where the improvement in separation due to the increased number of separation plates can be seen.

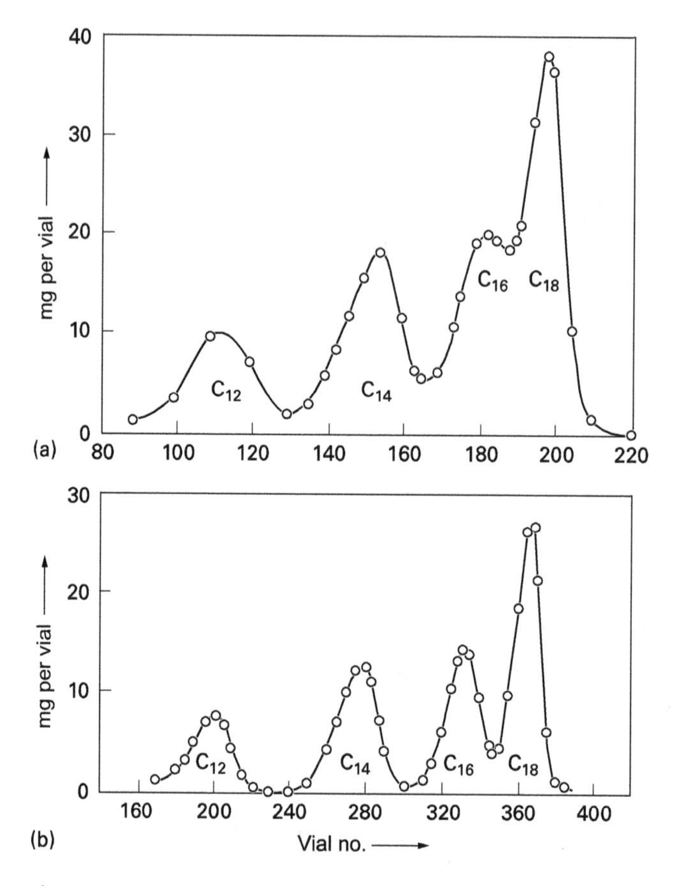

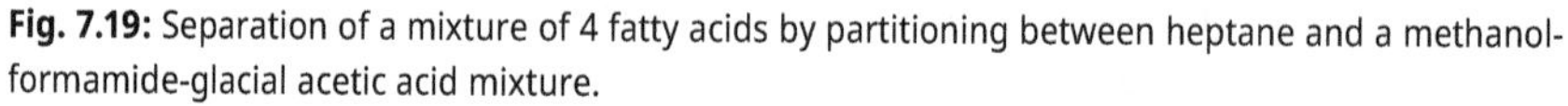

Fig. 7.19: Separation of a mixture of 4 fatty acids by partitioning between heptane and a methanol-formamide-glacial acetic acid mixture.
a) 220 separation plates;
b) 400 separation plates.

7.4.2 Continuous advancement of the moving phase

The agitated phase can be made to flow continuously through a series of distribution vessels in apparative arrangements as shown in Fig. 7.20. However, such simple devices have not proved to be very effective, since the distribution equilibria are not fully established (cf. what has been said about perforators). Better results are obtained with arrangements containing special stirrers and separate settling chambers for the two phases; however, this makes the procedure somewhat cumbersome and less suitable for analytical purposes.

7.4.3 Circulation method

It is possible without difficulty to return the moving phase from the end of a distribution battery back to the head of the arrangement, thus circulating it. The method is used mainly for the separation of substances with low partition coefficients, for which a larger quantity of moving phase is required. Several apparatuses working on this principle have been given by Fischer.

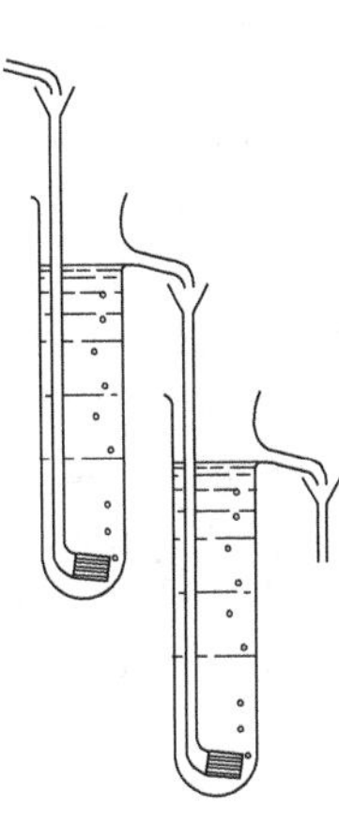

Fig. 7.20: Arrangement for continuous transport of the moving phase.

7.5 Separations by systematic repetition: Column method (distribution chromatography)

7.5.1 Principle

In liquid chromatography, the stationary phase is a liquid that is fixed by absorption in a porous solid support material. A column is filled with this material, the mixture of substances to be separated is applied to its head and the individual components are eluted by allowing a suitable second liquid to flow through the column as the moving phase.

7.5.2 Support material of the stationary phase

Chemically modified silica surfaces are often used as support materials, but other inorg. materials and various organic polymers are also in use.

A distinction is made between supports that are suitable for absorbing polar solvents such as water, lower alcohols, glycols, etc., and those that absorb nonpolar solvents such as petroleum ether, benzene, silicone oil, etc. due to their chemical nature or pretreatment; in this case, the term *reversed phase* distribution chromatography is used.

The carriers are used with particle diameters of about > 1 μm. The finer the grain size, the better the separation effect of the column, but the greater the flow resistance.

Also of importance is the loading of the support, i.e. the amount of stationary liquid on a given amount of support material. It is advisable to choose substances that allow a loading of at least 40-50% without clumping; this increases the capacity of the column and also reduces the risk of interference due to adsorption effects on the surface of the solid particles.

Often, the stationary liquid is diluted with a low-boiling solvent before being applied to the substrate, and then allowed to evaporate again. In this way, a better distribution of the liquid on the solid material can be achieved.

A monodisperse particle size distribution of the carrier, particle sizes in the μm-range and a completely uniform bedding in the separation bed are of decisive importance for the effectiveness of the columns. High-performance columns can be produced by special measures, such as dry filling with simultaneous rotation and shaking of the column or by *wet* slurrying together with the agitated phase, which is allowed to flow continuously downwards so that the solid particles are filtered off to a certain extent.

7.5.3 Column material and dimensions

The columns are usually made of glass, metal or polytetrafluoroethylene (PTFE). They contain a frit or glass wool plug at the bottom to hold the stationary phase. The surface of the separating bed can be covered with some filter paper to prevent swirling of the stationary phase during substance addition. More recent developments represent so-called monolithic columns, in which the column consists of a single porous piece.

The dimensions of the columns vary from a few millimeters to several centimeters in clear width and about 10–20 cm to several meters in length. For analytical work, diameters of 1–2 cm with lengths of 0.5–2 m are common.

7.5.4 Substance introduction

In the simplest case, the substance is added by dripping on the dissolved analysis sample; in many cases, an injection syringe is also used, the needle of which is inserted through a PTFE or plastic cap at the head of the column. This method can also be used for columns operated under pressure (see Fig. 7.21).

Sample volumes up to the ml-range are reproducibly dispensed via a dispensing loop. The substance quantities dispensed on standard analytical columns are about 50 mg, but very narrow columns (approx. 2 mm i.d.) for microgram quantities and

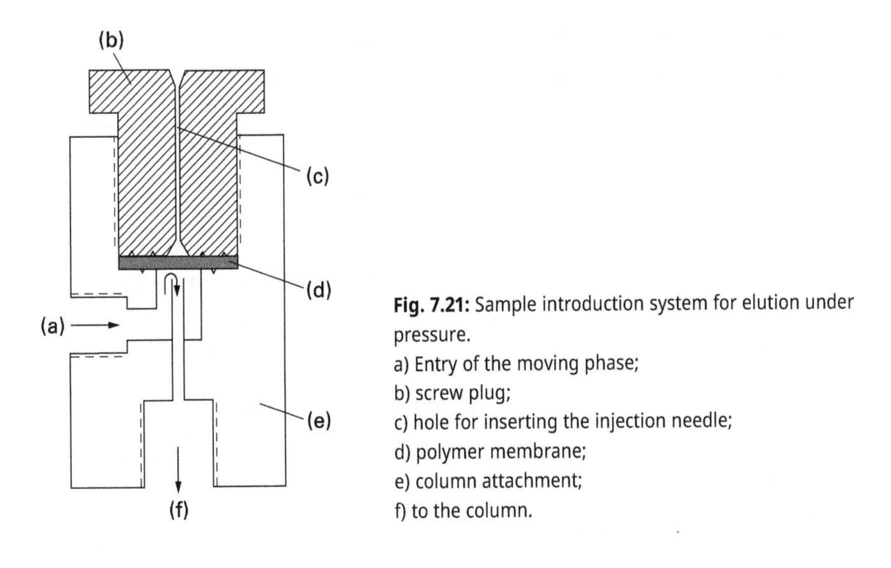

Fig. 7.21: Sample introduction system for elution under pressure.
a) Entry of the moving phase;
b) screw plug;
c) hole for inserting the injection needle;
d) polymer membrane;
e) column attachment;
f) to the column.

preparative columns for samples weighing a few grams have also been developed. Even smaller dimensions, down to the lower ng-range, are nowadays used in the microfluidic field (*lab-on-the-chip*).

7.5.5 Stationary and mobile phase

Numerous pairs of immiscible liquids can be used for partition chromatography; an aqueous phase is usually combined with an organic phase, but partition can also be carried out between two organic liquids (cf. Tab 7.9). A special case is represented by systems with ion exchangers as carrier materials and aqueous-alcoholic solutions as mobile phase. In the ion exchanger, after swelling, the ethanol concentration is lower than in the aqueous ethanol solution in contact with it, and it is possible, for example, to separate carbohydrates by partitioning between these two phases. Ionic liquids represent unusual stationary phases. These are non-molecular low-boiling liquids which can be covalently bonded to silica gel support material, among other things.

The mobile phase must be saturated with stationary phase before entering the column, otherwise it will be washed out of the support more or less quickly. It has proved useful to arrange a small pre-column as an additional safeguard before the actual separation column, in which the final saturation is achieved.

Tab. 7.9: Examples of stationary and mobile phases in partition chromatography.

Stationary phase	Mobile phase	Analytes
H_2O on silica gel	Cyclohexane	Phenols
H_2O on silica gel	$CHCl_3$ + n-butanol	Benzoic acids
H_2O on silica gel	Ethyl acetate	Digitalis glycosides
0.1 N H_2SO_4 on silica gel	$CHCl_3$ + tert. amyl alcohol	Aliphatic carboxylic acids
CH_3OH on silica gel	Petroleum ether	Carotenoids
CH_3OH + H_2O on diatomaceous earth	Hexane + benzene	Estrogens
Dimethyl sulfoxide on silica gel	Cyclohexane	Porphyrins
Dimethylformamide + Carbowax^{TM} on diatomaceous earth	i-Octane	Polycycl. aromatics
Heptane on siliconized diatomaceous earth	Acetonitrile + CH_3OH	Triglycerides
Squalane on diatomaceous earth	Acetonitrile	Hydrocarbons
Kerosene oil on siliconized diatomaceous earth	Acetone + H_2O	Fatty acids
Di-(2-ethylhexyl)phosphoric acid on siliconized diatomaceous earth	Dil. HNO_3	Rare earths
i-Octane on siliconized diatomaceous earth	CH_3OH + H_2O	Fatty acids
Di-(2-ethylhexyl)phosphoric acid on cellulose	HCl solution	Pt – Au
Cyclohexane on polymeric fluorocarbons	CH_3NO_2	Asphalt components

7.5.6 Flow velocity of the mobile phase – pressure columns

The flow velocity of the mobile phase can only be influenced imperfectly by gravity in the usual arrangements with transport, so that the time required for separation is given, among other things, by the flow resistance. This disadvantage can be eliminated by elution under pressure (*high pressure liquid chromatography*: HPLC), whereby even longer columns with high numbers of separation stages still give acceptable analysis times (cf. Fig 7.22).

Several effects contribute to the broadening of the elution bands, of which the influence of irregular packing of the carrier material and incomplete equilibration (the latter especially at high flow velocities of the mobile phase) are probably the most important. As a consequence of incomplete equilibration, an increase of the separation stage height with flow velocity is observed (cf. Fig. 7.23).

If the columns are not filled very carefully, the effect of non-uniform packing, which is independent of the flow velocity, can override the other interference effects to such an extent that the separation plate height is practically independent of the velocity of the mobile phase (cf. Fig 7.24). It is then possible to elute very rapidly without any major loss of separation efficiency.

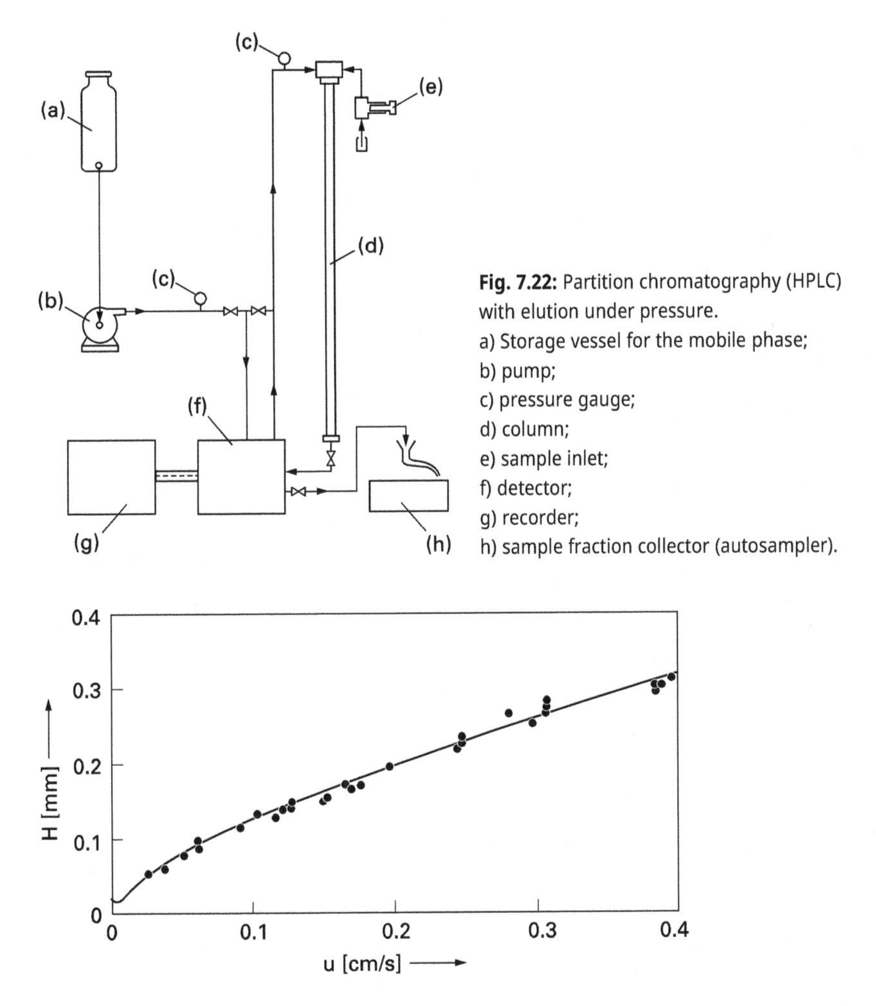

Fig. 7.22: Partition chromatography (HPLC) with elution under pressure.
a) Storage vessel for the mobile phase;
b) pump;
c) pressure gauge;
d) column;
e) sample inlet;
f) detector;
g) recorder;
h) sample fraction collector (autosampler).

Fig. 7.23: Dependence of the separation plate height *H* on the velocity *u of* the mobile phase; elution of nitrobenzene with trimethylpropane, stationary phase: triscyanoethoxypropane.

7.5.7 Influence of the temperature

Separations by partition chromatography are mainly performed at room temperature. It is important that the temperature is kept well constant; in case of fluctuations the

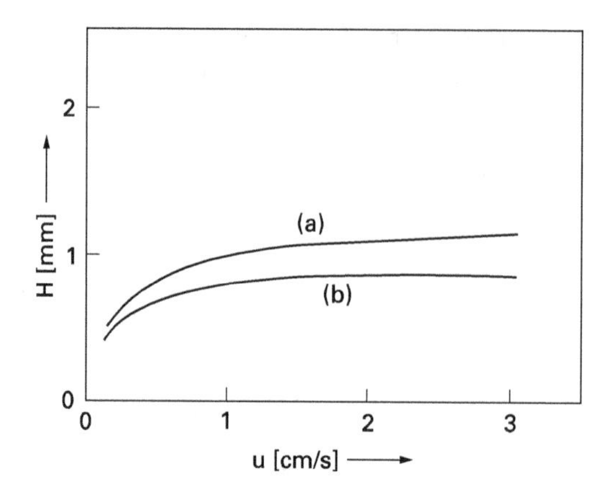

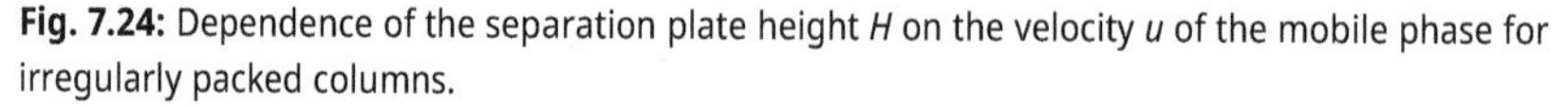

Fig. 7.24: Dependence of the separation plate height *H* on the velocity *u* of the mobile phase for irregularly packed columns.
a) Column with 3.08 mm diameter;
b) column with 1.54 mm diameter.

efficiency of the column deteriorates, furthermore the stationary phase may bleed out due to the temperature dependence of the solubility of both liquids in each other.

When the temperature is increased, the rate of mass transfer between the two phases increases. As a result, the separation plate height of the column may decrease, but according to previous experiments, this effect is small at low flow velocities of the moving phase; it only becomes noticeable at higher velocities.

7.5.8 Stepwise elution – gradient elution

If an analysis sample contains substances with very different partition coefficients, an acceleration of the separations can be achieved by stepwise elution with solvents of increasing efficiency or by gradient elution with continuous improvement of the mobile phase. For acidic and for basic substances, elution with aqueous solutions using a pH gradient can be favorable.

7.5.9 Isolation of components – elution curves – detectors

To isolate the separated components, the eluate is collected fraction by fraction and the individual substances are recovered from the fractions.

Often, in addition or on their own, elution curves are recorded by continuous measurement of refraction, light absorption, electrical conductivity, dielectric constants, polarographic diffusion current, heat of adsorption, and others. Almost all principles of chemical sensing in the liquid phase are now applied. Particularly sensitive is the coupling with mass spectrometry, which, however, must use elaborate procedures to couple the separated analyte, which is in the eluate, with the vacuum of a mass spectrometer. For this purpose, the eluent must be continuously separated from the analyte to the greatest possible extent.

The retention volume V_i of a component i is related to its partition coefficient α_i and the volumes V_{st} and V_m of stationary and mobile phases by the relation (9):

$$V_i = \alpha_i \cdot V_{st} + V_m \tag{9}$$

To identify unknown components in an elution diagram, a law stated by van Duin is occasionally useful: If one plots the logarithms of the corrected retention volumes V^o against the number of C atoms of a homologous series, straight lines are obtained (cf. Fig 7.25). If the retention volumes of some compounds of a homologous series are known for a given separation column, the number of C atoms of an unknown compound of the same series can be determined.

7.5.10 Efficacy of partition chromatography

The separation efficiency of distribution chromatographic columns is usually very good because of the straight-line distribution isotherms and the resulting symmetrical elution curves. Data on the height of a separation plate vary somewhat; the value is probably generally between 0.5 and 1.5 mm. With particularly carefully prepared

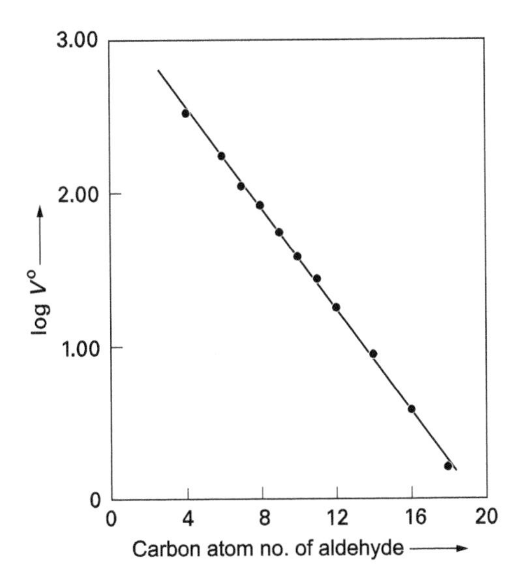

Fig. 7.25: Dependence of the corrected retention volumes of the 2,4-dinitrophenylhydrazones of various aldehydes on the number of C atoms.

columns of small diameter (in one case also with 8–11 mm diameter) even heights of 0.1–0.3 mm were achieved. Larger columns for preparative purposes have heights of 2.5–5 mm.

Distribution chromatography therefore permits separations with several thousand separation plates without great experimental difficulty; it is one of the most effective of the known separation methods.

Particular advantages are the great variability in the choice of the two phases and the gentle operation at low temperatures. This makes the method particularly suitable for the separation of high-boiling or labile volatile substances where other methods fail. Examples of applications include the separation of amino acids, fatty acids, proteins, steroids and alkaloids (see also Tab. 7.9).

7.6 Separations by systematic repetition: Thin-layer technique (thin-layer distribution chromatography)

In thin-layer distribution chromatography, a powdered solid support is impregnated with the stationary liquid phase and spread in a thin layer on a glass plate or other support; the stationary phase can also be applied to paper strips, but plates of cellulose powder or alumina have proved more convenient.

Water, aqueous solutions or hydrophilic solvents such as propylene glycol etc. are used as stationary liquids. When working with phase inversion, kerosene oil, ethyl oleate, tri-n-butyl phosphate, benzene or butanol solutions of di-(2-ethylhexyl) phosphoric acid, etc. are added to the carrier and developed with water or aqueous solutions.

The method has occasionally been used to separate alcohols, phenols, organic acids and bases, and steroids, and regularities between R_f value and constitution have also been established.

7.7 Countercurrent distribution

The countercurrent method is used in the distribution between two liquid phases mainly for separations on a preparative or technical scale; for analytical problems the method has little significance. On the one hand, it is only suitable for separating mixtures into two parts; on the other hand, the necessary apparatus is either not very effective (simple packed columns) or quite costly (columns with stirring devices; pulsating columns).

Literature to the text

General

T. Hii & H.K. Lee, Liquid-liquid extraction in environmental analysis, in: J. Pawliszyn & H.L. Lord, Handbook of Sample Preparation, 39–51, John Wiley & Sons, New York (2010).
A. Kabir, M. Locatelli & H. Ulusoy, Recent trends in microextraction techniques employed in analytical and bioanalytical sample preparation, Separations *4*, 36 (2017).
V. Kislik, Solvent Extraction, Elsevier, Amsterdam (2012).
P. Naik, N. Kumar, N. Paul & T. Banerjee, Deep Eutectic Solvents in Liquid-Liquid Extraction, CRC Press, Boca Raton (2022).
C. Poole & M. Cooke, Encyclopedia of Separation Science, Academic Press, Walthan (2000).
I. Wilson & C. Poole, Handbook of Methods and Instrumentation in Separation Science, Elsevier Science London (2009).
E. Yilmaz & M. Soylak, New Generation Green Solvents for Separation and Preconcentration of Organic and Inorganic Species, Elsevier, Amsterdam (2020).

Membrane-assisted liquid/liquid extraction

L. Chimuka, E. Cukrowska, M. Michel & B. Buszewski, B. Advances in sample preparation using membrane-based liquid-phase microextraction techniques, TrAC Trends in Analytical Chemistry *30*, 1781–1792 (2011).
J. Jonsson & L. Mathiasson, Membrane-based techniques for sample enrichment, Journal of Chromatography A 902, 205–225 (2000).
N. Kochergynsky, Q. Yang, and L. Seelam, Recent advances in supported liquid membrane technology, Separation & Purification Technology *53*, 171–177 (2007).
J. Lee, H. Lee, K. Rasmussen, and S. Pederson-Bjergaard, Environmental and bioanalytical applications of hollow fiber membrane liquid-phase microextraction. A review, Analytica Chimica Acta *624*, 253–268 (2008).
Y. Ong, K. Yee, Y.Cheng & S. Tan, A review on the use and stability of supported liquid membranes in the pervaporation process, Separation and Purification Reviews *43*, 62–88 (2014).
W. Riedl Membrane-assisted liquid-liquid extraction – Where do we stand today? Chemie-Ingenieur-Technik *91*, 1544–1553 (2019).
M. Vilt & W. Ho, Applications and advances with supported liquid membranes, in: K. Mohanty u. M.Purkait, Membrane Technologies and Applications, CRC Press, Boca Raton, 279–303 (2012).

Microextraction

A. Jain & K. Verma, Recent advances in applications of single-drop microextraction: A review, Analytica Chimica Acta *706*, 37–65 (2011).
M. Jeannot, A. Przyjazny & J. Kokosa, Single drop microextraction – Development, applications and future trends, Journal of Chromatography A *1217*, 2326–2336 (2010).
J. Pawliszyn & S. Pedersen-Bjergaard, Analytical microextraction: current status and future trends, Journal of Chromatographic Science *44*, 291–307 (2006).

E. Psillakis & N. Kalogerakis, Developments in liquid-phase microextraction, Trends in Analytical Chemistry *22*, 565–574 (2003).

M. Rezaee et al., Determination of organic compounds in water using dispersive liquid-liquid microextraction, Journal of Chromatography A, *1116*, 1–9 (2006).

A. Sarafraz-Yazdi & A. Amiri, Liquid-phase microextraction, Trends in Analytical Chemistry *29*, 1–14 (2010).

L. Xu, C. Basheer, and H. Lee, Developments in single-drop microextraction, Journal of Chromatography A *1152*, 184–192 (2007).

Micellar extraction

M. de Almeida Bezerra, M. Zezzi Arruda & S. Costa Ferreira, Cloud point extraction as a procedure of separation and pre-concentration for metal determination using spectroanalytical techniques: A review, Applied Spectroscopy Reviews *40*, 269–299 (2005).

E. Azooz, R. Ridha & H. Abdulridha, The fundamentals and recent applications of micellar system extraction for nanoparticles and bioactive molecules: A review, Nano Biomedicine and Engineering *13*, 264–278 (2021).

E. Paleologos, D. Giokas, & M. Karayannis, Micelle-mediated separation and cloud-point extraction,TrAC – Trends in Analytical Chemistry *24*, 426–436 (2005).

P. Śliwa & K. Śliwa, Nanomicellar extraction of polyphenols – methodology and applications review, International Journal of Molecular Sciences, *22*1392 (2021).

C. Stalikas, Micelle-mediated extraction as a tool for separation and preconcentration in metal analysis, TrAC, Trends in Analytical Chemistry *21*, 343–355 (2002).

A. Yazdi, Surfactant-based extraction methods, TrAC, Trends in Analytical Chemistry *30*, 918–929 (2011).

Distribution chromatography (HPLC)

R. Kaushal, N. Kaur, U. Navneet, S. Ashutosh & A. Thakkar, High performance liquid chromatography detectors – a review, International Research Journal of Pharmacy *2*, 1–7 (2011).

C. Mant & R. Hodges, High-Performance Liquid Chromatography of Peptides and Proteins – Separation, Analysis, and Conformation, CRC Press, Boca Raton (2017).

S. Moldoveanu & V. David, Selection of the HPLC Method in Chemical Analysis, Elsevier,Amsterdam (2016).

N. Morgan & P.D. Smith, HPLC detectors, Chromatographic Science Series *101*, 207–231 (2011).

V. Pino & A. Afonso, Surface-bonded ionic liquid stationary phases in high-performance liquid chromatography. A review, Analytica Chimica Acta *714*, 20–37 (2012).

L.R. Snyder, J.J. Kirkland and J.W. Dolan, Introduction to Modern Liquid Chromatography, John Wiley & Sons, New Jersey (2010).

L. Trojer, A. Greiderer, C. Bisjak, W. Wieder, N. Heigl, C. Huck, and G. Bonn, Monolithic stationary phases in HPLC, Chromatographic Science Series *101*, 3–45 (2011).

8 Solubility of gases in liquids

8.1 General

8.1.1 Historical development

For a long time, gas separations were performed almost exclusively with the aid of chemical reactions by absorption methods (Bunsen, Hempel (1913), Bunte, Orsat (et al.), since only small separation effects usually occur when gases are dissolved in liquids. Only Martin and Synge (1941) proposed to multiply minor differences of solubilities of different gases in liquids in separation columns; the first such separation was described by James and Martin (1952).

8.1.2 Auxiliary phases – distribution isotherm – equilibration rate – separation factors

In separations by dissolving gases in liquids, one works either with an auxiliary phase – a solvent which selectively dissolves out or absorbs a component – or with two auxiliary phases, in which case, in addition to the solvent, an inert gas is used to drive the gas mixture under investigation through the liquid auxiliary phase.[1]

Distribution isotherm. If no chemical reaction occurs, *Henry's Law* (1803) applies to the solubility of gases in liquids:

$$c = \alpha \cdot p \tag{1}$$

where c *is* the concentration of the gas in the liquid (usually expressed in normal cm^3 [0 °C, 760 Torr] per g or per ml of solvent), p *is* the gas pressure in Torr (in the case of gas mixtures, the partial pressure of the component concerned) and α is a constant, the so-called *Henry constant.*

Henry's law states that the distribution isotherm is straight line.

For the graphical representation, either the concentration in the solution is plotted as the ordinate against the pressure as the abscissa using ordinary coordinates (cf. Fig. 8.1), or double-logarithmic coordinates are chosen for an otherwise identical application; this results in straight lines with a slope of 1 (45°).

Since in the gas phase the pressure p *is* proportional to the concentration c (e.g. in mg/ml), Nernst's distribution theorem and Henry's law are equivalent. In both, the

1 Strictly speaking, the inert gas is not a new phase in the sense of the previous definition (2nd part, Section 1.1).

https://doi.org/10.1515/9783111181417-008

isotherms are straight lines that are ideally unaffected by the presence of other components. For gases, solubility in liquids generally decreases with temperature.

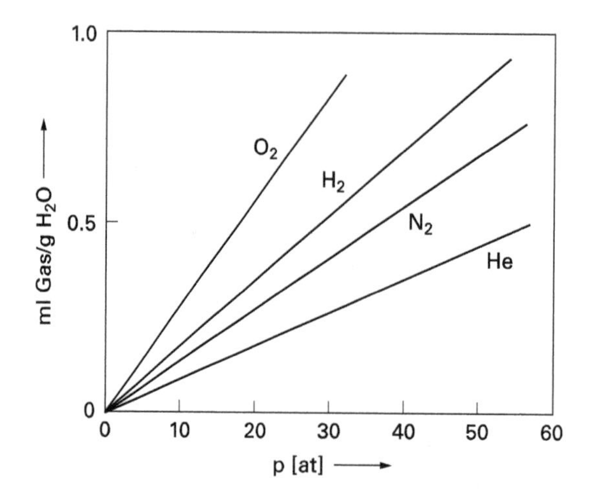

Fig. 8.1: Isotherms for the distribution of gases between water and gas phase.

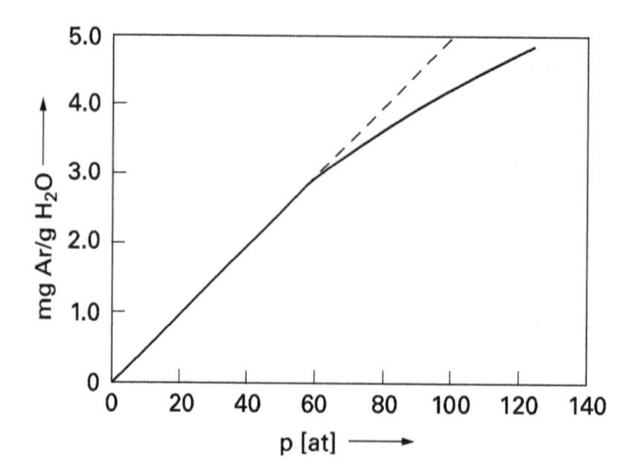

Fig. 8.2: Distribution of argon between water and the gas phase at 0.2 °C.

Henry's law is valid in the range of low pressures and small concentrations in the solution. Deviations are observed at high pressures (see Fig. 8.2).

The diffusion of the solute in the liquid phase is decisive for the rate of equilibration. A high rate of exchange between the two phases is ensured by vigorous stirring or by forming the thinnest possible liquid films.

The Henry constants of different gases generally do not differ much, so that the separation factors are usually only small. High separation factors are only achieved in such systems if a component of the mixture to be separated is absorbed by the liquid phase under chemical reaction; one then works with only one auxiliary phase (the liquid) and can achieve sufficient separations by one-time equilibration or by one-sided repetition.

8.1.3 Solvents

In the latter methods, where a component of a mixture is absorbed, aqueous solutions are generally used for analytical separations. For methods with smaller separation effects, where systematic repetition is required, a large number of different organic solvents are in use.

8.2 Separations by one-time equilibrium adjustment

8.2.1 Discontinuous mode of operation

In the so-called *classical* gas analysis, separations by absorption are usually carried out in gas pipettes or gas burettes.

In Hempel's gas pipettes (Fig. 8.3), vessel b is filled with the absorption liquid and then the gas mixture is forced from a storage burette through capillary c to b with the aid of a barrier liquid until the gas space occupies about half the volume. The displaced part of the absorption liquid enters vessel a. The absorption in b is accelerated by reshaking.

The gas mixture is poured into the Bunte gas burette (see Fig. 8.4), with a small amount of sealing liquid (e.g. water) in the lower part of the tube. This is then sucked off downwards and the absorption solution is also allowed to enter the gas chamber from below.

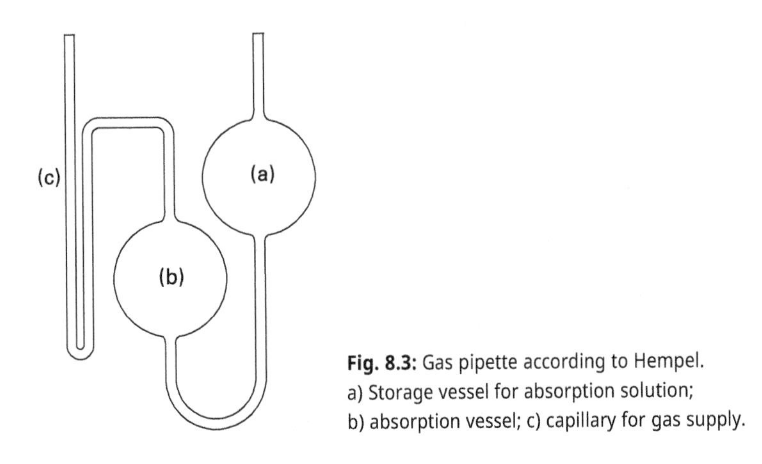

Fig. 8.3: Gas pipette according to Hempel.
a) Storage vessel for absorption solution;
b) absorption vessel; c) capillary for gas supply.

In order to be able to determine several components of a gas mixture, composite apparatuses have been developed, especially by Orsat, in which several gases are absorbed in succession in different vessels.

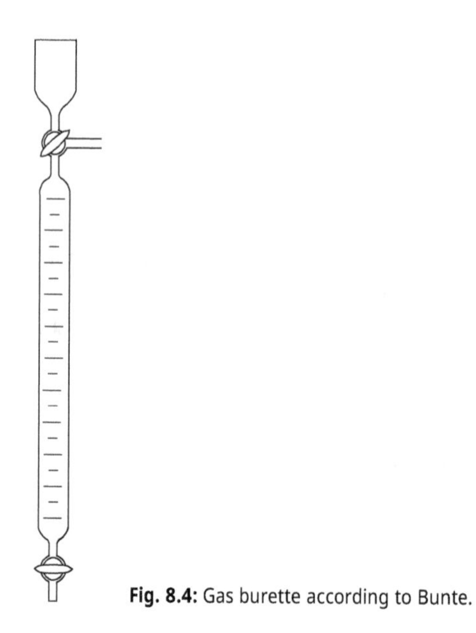

Fig. 8.4: Gas burette according to Bunte.

Since only relatively few selective solvents are available (cf. Tab. 8.1), the application of this method is limited. A fundamental flaw arises from the usually low but unavoidable solubility of all gases in the absorption liquids. The absorption methods have therefore been largely displaced by gas chromatography (see below).

Tab. 8.1: Liquid absorbents for gases (examples).

Gas	Absorbent
CO_2, HCl, H_2S, HF, Cl_2, SO_3, SO_2, et al.	Aqueous KOH or NaOH solution
Cl_2, Br_2, I_2	KI solution
NH_3, volatile amines	Dil. sulfuric acid
CO	Cu^+ in HCl or NH_3 solution; I_2O_5 suspended in oleum
NO	H_2SO_4 + HNO_3; $KMnO_4$ solution
N_2O	Ethanol
H_2	Dinitroresorcinol suspension with Ni catalyst; Pd sol in sodium picrate solution.
O_2	Alkaline pyrogallol solution; $Na_2S_2O_4$ solution; $CrCl_2$ solution; ammoniacal Cu^+ solution.
O_3	Alkaline KI solution
Unsaturated hydrocarbons	Oleum (20–25% SO_3); Bromine water
Volatile organ. compounds	H_2SO_4 + $K_2Cr_2O_7$; $KMnO_4$ solution

Static headspace sampling. This sampling technique allows the separation of highly volatile substances from low volatile matrices due to the different partial pressure.

The partial pressure of a highly volatile substance in a mixture of low volatility liquids is described by Raoult's law, in simplified form by Henry's law. The first applications of this technique date back to Harger et al. (1939). The combination of vapor space sampling with gas chromatography is now standard in many food, environmental and clinical analyses.

A distinction is made between two techniques: The *static headspace* and the *dynamic headspace* technique. In the former, the liquid sample is brought into the gas/liquid headspace in a thermostat in an only partially filled glass vessel. The analyte vapor pressure in the headspace is directly related to the analyte in the sample by Henry's constant. By taking a small gas sample from the vapor space and transferring it to the gas chromatograph, multicomponent analysis of volatile contaminants can be performed. This process can be carried out reproducibly in automated form. Derivatization can be used to further separate selected analytes. The addition of additives facilitates the transition to the gas phase. *Solid phase microextraction (SPME)* is now widely used. A typical embodiment is shown in Fig. 8.5. A fused silica fiber coated with a thermally stable stationary phase is introduced into the vapor space through a septum for gas sampling, equilibrated, and then transferred to the injector block of the GC. Suitable stationary phases for this purpose are ionic liquids due to their negligible vapor pressure.

In the meantime, numerous applications for headspace sampling have been published. (Tab. 8.2).

Difficulties can result from non-reproducible gas sampling from the pressurized vapor space. The use of internal standards circumvents this.

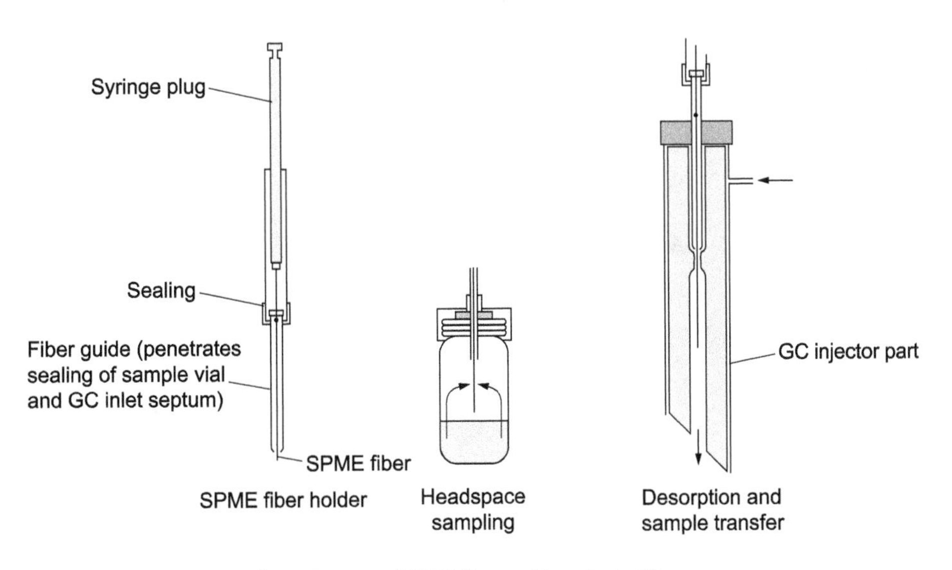

Fig. 8.5: Vapor space sampling using coated SPME fiber and transfer to GC.

Tab. 8.2: Headspace sampling and coupling with gas chromatography (examples).

Analyte	Matrix	Headspace temperature [°C]	Additive	Detection limit
Ethanol	Blood	60	NaF	3 µg/mL
HCHO	Blood	60	n-Propanol	200 µg/mL
Methyl methacrylate	Blood	70	NaCl	20 ng/mL
CCl_4	Blood	90	H_3PO_4	10 ng/mL
Amphetamines	Urine	RT	K_2CO_3	1 µg/mL

8.2.2 Continuous mode of operation

For the continuous separation of individual components from a gas stream, washing bottles are usually used, which are commercially available in numerous designs. The efficiency is increased above all by breaking up the gas bubbles as finely as possible, which can be achieved by incorporating (e.g. glass frits or layers of glass beads,) stirring devices (cf. Fig. 8.6) or also by adding surface-active substances to the absorption liquid.

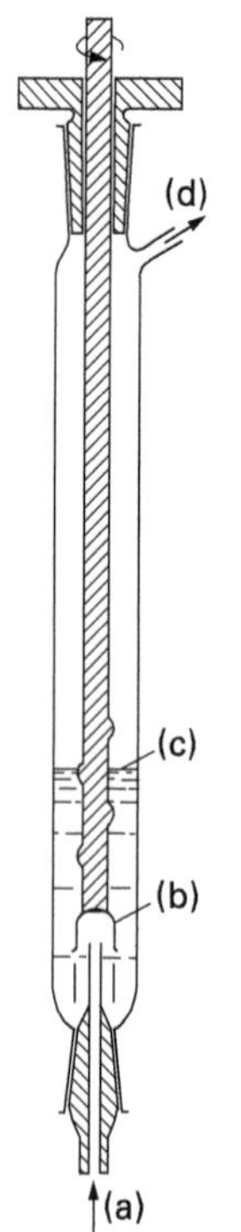

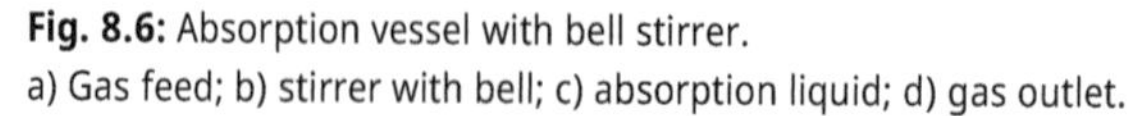

Fig. 8.6: Absorption vessel with bell stirrer.
a) Gas feed; b) stirrer with bell; c) absorption liquid; d) gas outlet.

Dynamic vapor phase sampling. This technique, known as *purge-and-trap* in Anglo-Saxon, is also used for the efficient separation of highly volatile analytes from low-volatility liquids such as water, blood or ionic liquids. An inert gas is usually used as an auxiliary phase. The expelled analyte is enriched in a downstream separation (e.g., by cryosampling or sorptive sampling of solids).

8.2.3 Circulation method

To reduce the amount of absorbing liquid, the solution can be circulated; an example of this method of operation is given in Fig. 8.7.

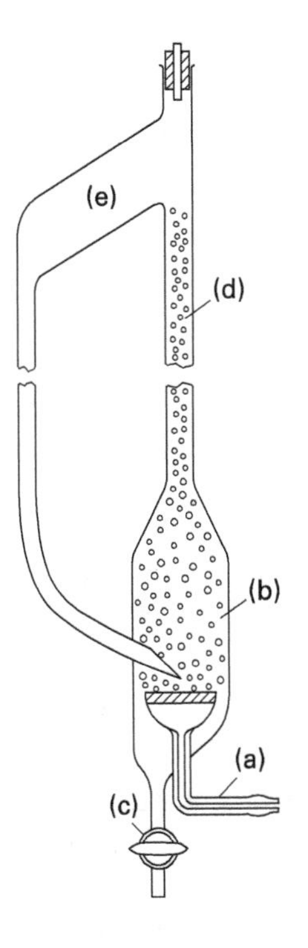

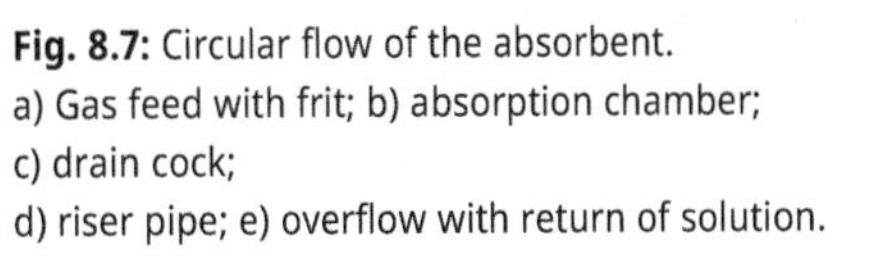

Fig. 8.7: Circular flow of the absorbent.
a) Gas feed with frit; b) absorption chamber;
c) drain cock;
d) riser pipe; e) overflow with return of solution.

8.3 Separations by one-sided repetition

The method of one-sided repetition results most simply in absorption processes by connecting several wash bottles in series. A frequently used template, in which the gas stream is divided into constrictions of the elongated absorption vessel and mixed with the solution in several chambers, is reproduced in Fig. 8.8.

The method of working with one-sided repetition of the separation step also includes enrichment methods for trace components in gases (e.g. in air), in which the

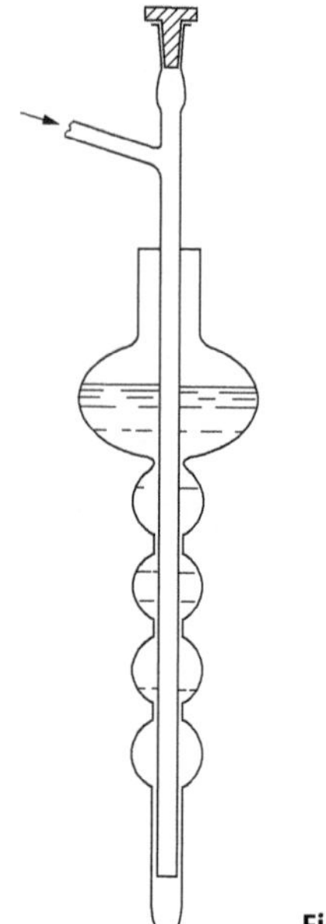

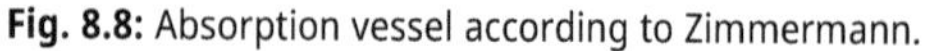

Fig. 8.8: Absorption vessel according to Zimmermann.

analysis sample is passed through a short tube filled with absorbent. The filling can be thought of as being broken down into individual sections in which the constituent to be separated is gradually held and removed from the gas stream. The enrichment is terminated before the separated component has broken through the column. Fillings of such tubes include polyethylene glycol or silicones coated on an inert support material. The absorbed compounds can then be recovered by rinsing under elevated temperature.

In a second method of enrichment, the method of incomplete separation is used in conjunction with the determination of the distribution coefficient. The gas mixture to be investigated is allowed to flow through a tube containing absorbent until distribution equilibrium has been established in the entire bed of the stationary phase; twice to three times the breakthrough volume is generally sufficient for this. For quantitative determination, the absorbed component is also removed from the column; in contrast to the previous method, the amount of stationary phase and the distribution isotherm must be known for the evaluation. Since the concentration of the separated component on the stationary phase is determined by the partition coefficient and the concentration in the gas phase, the latter quantity can now be calculated.

The advantages of the method are that no weighing is required and that stronger enrichments can be obtained than with the previous method. On the other hand, a larger sample quantity must be available and the temperature must be kept constant.

These two procedures are often performed for pre-enrichment in conjunction with additional chromatographic separation.

8.4 Separations by systematic repetition: Column method (gas chromatography)

8.4.1 Principle

If analytical separations are to be carried out by dissolving gases in liquids without chemical reaction, the separation operations must be repeated very often because of the usually only slight differences in the Henry coefficients. This is achieved by using separation columns in which a stationary liquid phase is deposited on an inert solid support material. The moving phase consists of an inert gas stream that carries the substance mixture applied to the top of the column through the separation bed. The various components of the analytical sample are eluted from the column more or less rapidly, depending on the Henry coefficients.

Gas chromatography[2] works with two auxiliary phases, the carrier gas and the stationary liquid.

8.4.2 Inert support material for the stationary phase

The inert carrier material for the stationary phase usually consists of a porous substance that absorbs the absorption liquid well. Inorganic carrier materials such as silicon dioxide (historically: diatomaceous earth and brick dust) have proven most effective, but PTFE powder, porous organic polymers of styrene/divinylbenzene (*Porapak*®) or small glass beads coated with stationary liquid are also recommended.

The support material should not be too fine-grained, otherwise the flow resistance to the moving phase becomes too great; on the other hand, the finer the grain size, the greater the effectiveness of the column. As a compromise between these mutually exclusive requirements, a particle size of about 0.3 mm diameter is often used. Furthermore, the particle size should be as uniform as possible so that non-uniform flows and thus broadening of the elution bands do not occur in the column.

A very disturbing effect, especially in the separation of polar substances, is that the support material is often not completely inert, but adsorbs the analytical substances to a greater or lesser extent. This causes a broadening of the elution bands and thus a deterioration of the separations. This disturbance is eliminated by acid or alkali treatment of the support, by superficial coating with detergents or with silver metal or – most effectively – by siliconization with $(CH_3)_2\,SiCl_2$ after acid treatment.

2 In Anglo-Saxon literature, a distinction is made between *gas-liquid chromatography* (discussed in this section) and *gas-solid chromatography* (see Section 4). Usually, both methods are referred to as *gas chromatography*.

A special case are the so-called *Golay columns* or *capillary columns*, where no carrier material is used, but the stationary liquid adheres as a thin film to the inner wall of a narrow tube (<1 mm diameter).

8.4.3 Column material and dimensions

The columns used in gas chromatography are usually made of copper or steel tubes. For the separation of sensitive organic compounds at higher temperatures, glass and quartz capillary columns are preferable, since decomposition reactions can occur in metal tubes and catalytically accelerated processes can take place.

Probably the most commonly used columns are those with 4–8 mm inner diameter at lengths of about 1–6 m. An extension beyond about 6–8 m usually does not bring any significant improvement in the separation effect, since it becomes increasingly difficult to achieve uniform filling; moreover, because of the pressure drop in the column, the most favorable flow velocity can only be achieved in part of the separation bed (see below).

For preparative separation of larger amounts of substances, columns of about 1–2 cm diameter (maximum up to about 10 cm) with lengths of about 2–3 m are used.

An exception are again the already mentioned capillary columns, which consist of thin copper or steel tubes of 0.25–0.50 mm inner diameter and 25–150 m length.

8.4.4 Sample transfer

Gaseous analysis samples are placed in glass or metal loops with calibrated contents and, after pressure and temperature measurement, are purged into the column by turning a valve with the carrier gas flow (Fig. 8.9).

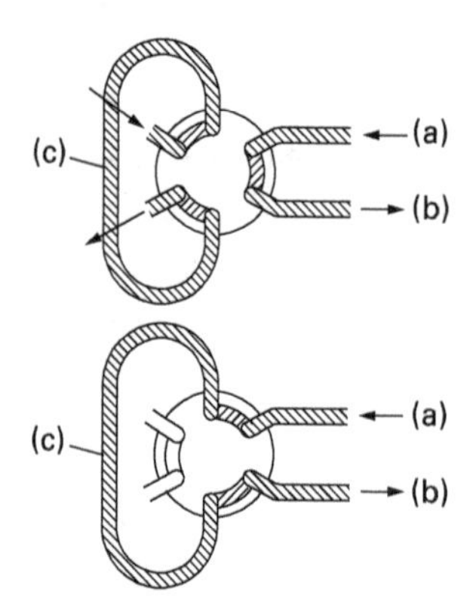

Fig. 8.9: Sample transfer with gas loop.
a) and b) Carrier gas supply and discharge; c) gas loop.
Upper picture: Filling the gas loop; the carrier gas stream is fed directly to the column.
Lower picture: Purging the gas loop with the carrier gas.

Another option is to use gas-tight injection syringes; insert the needle through a PTFE/silicone polymer membrane (*septum*) at the head of the column and then quickly introduce the sample into the carrier gas stream (Fig. 8.10). The withdrawal of the gas volume from a larger reservoir is done accordingly.

The quantities dispensed are in the order of 0.5–5 ml when using columns of the usual dimensions.

Samples that are liquid at normal temperature are also dispensed using syringes. For the normally used small quantities of approx. 1–10 µl, special so-called *microliter syringes* have been developed. The liquids are vaporized in the head of the heated column or – less frequently – in a separate prechamber.

The mentioned larger columns with diameters of 1–2 cm allow substance tasks of maximum about 100 mg, preparative columns with 10 cm diameter allow tasks up to about 100 g.

Capillary columns can separate only a few micrograms of substance. Since direct measurement of such small amounts is not possible with sufficient accuracy, a much larger amount is applied and the carrier gas stream is then divided into two unequal parts (*splitting*), the larger of which is discarded and only the smaller enters the column (Fig. 8.11). Normally, splitting ratios of about 1: 100 to 1: 2000 are selected.

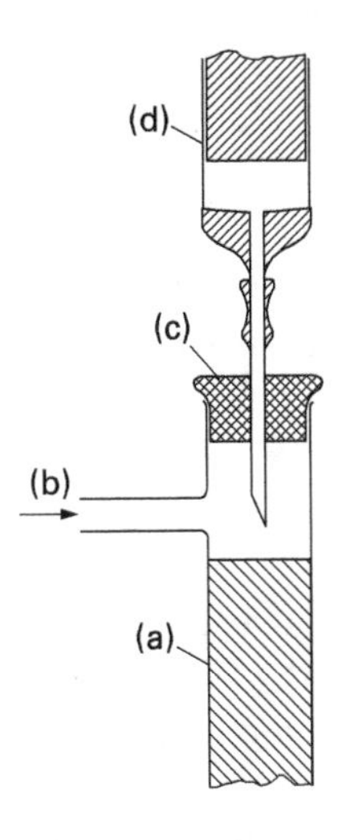

Fig. 8.10: Sample transfer with injection syringe.
a) Separation column; b) carrier gas feed;
c) septum; d) syringe.

Solids can be brought into the column with movable plungers or by turning a cock plug provided with a recess. Usually, however, they are dissolved and the solution is sprayed onto the column head in the manner described. It is advisable to use a high-boiling liquid that elutes later than the sample, so that no interference occurs from the elution band of the solvent.

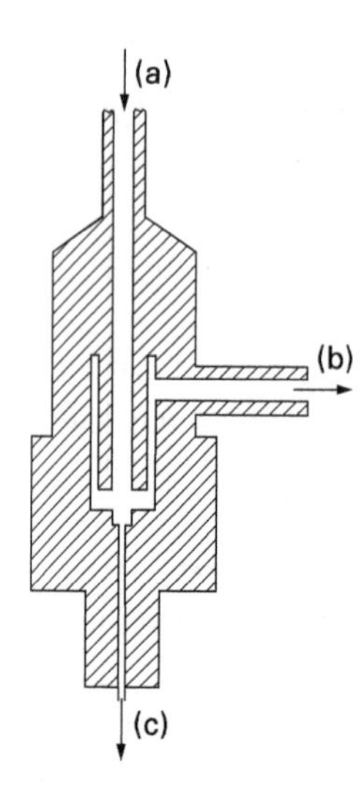

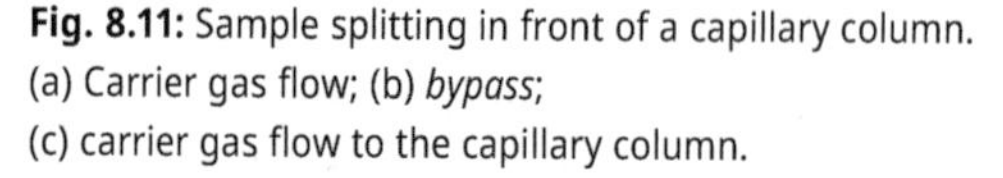
Fig. 8.11: Sample splitting in front of a capillary column.
(a) Carrier gas flow; (b) *bypass*;
(c) carrier gas flow to the capillary column.

8.4.5 Stationary phase

The type of *stationary liquid phase* is of decisive importance for the separation performance of a gas chromatographic column. The success or failure of a separation essentially depends on its selection; liquids that give the highest possible separation factor for the substances to be separated are favorable.

A distinction is made between non-polar liquids which separate the individual components according to their vapor pressures without significant interaction with the constituents of the analysis sample, and those which are more or less selective for certain substances or substance classes as a result of specific interactions. Stationary liquids suitable for separations within different substance groups have been determined empirically to a large extent; some examples are given in Tab. 8.3.

The vapor pressure of the stationary phase limits the temperature range in which a separation column can be used. If the liquid in question already has a noticeable vapor pressure at the experimental temperature, the column will quickly *bleed out*. This can be prevented to a certain extent by loading the carrier gas upstream of the column with the vapor of the separating liquid.

Recently, *ionic liquids* (salts that are liquid at room temperature without solvents) have become particularly important. They are characterized by high thermal stability and hardly measurable vapor pressure.

For gas chromatography at high temperatures, apiezone oil and some silicone oils are mainly used as stationary phases; one can work with them at about 300 °C to a maximum of about 350 °C. Even higher temperatures can be achieved using molten salts and ionic liquids.

The gas chromatographic investigation of aqueous solutions presents particular difficulties, since the water, which is present in large excess, gives very broad bands

Tab. 8.3: Stationary phases for gas chromatographic columns (examples).

Substances to be separated	Stationary liquid	Maximum temp. (°C)
Aliphatic hydrocarbons	Kerosene oil	150
	Silicone polymers	approx. 300
	Dioctyl phthalate	150
	Polyglycol ether	220
Olefins	Dimethylsulfolane	50
	β, β-Oxydipropionitrile	100
	Dinonyl phthalate	150
Aromatic hydrocarbons	Tricresyl phosphate	125
	Dinonyl phthalate	150
	Silicone oil	220
	Apiezon oil	300
Chlorinated aliphatic hydrocarbons	Tricresyl phosphate	125
	Silicone grease	250
	Di-i-decyl phthalate	175
Aldehydes	Polyglycol ether	250
Ketones	Polyglycol ether	250
Alcohols	Polyglycol ether	250
	Dinonyl phthalate	150
Ester	Polyglycol ether	250
	Silicone grease	250
	Kerosene oils	150
	Ethylene glycol succinate	200
Ether	Dinonyl phthalate	150
	Apiezon oil	300
Amine	Polyglycol ether + KOH	250
Nitrile	Silicone oil	220
	Apiezon oil	300
Phenols	Polyglycol ether	250
	Trimethylolpropane tripelargonate	160
Alkaloids	Silicone polymer	250
Fatty acids	Silicone polymer	250
Mercaptans	Dioctyl phthalate	150

due to the strong association at most stationary phases, which overlap other compounds. Only a few compounds, mainly polyethylene glycols and, more recently, ionic liquids, have proved to be suitable as stationary phases.

The solid support is generally impregnated with about 20–30% (maximum about 40%) of its weight of stationary liquid. The separation efficiency of the column increases as the loading of the support is reduced, since equilibration is faster due to the reduced film thickness of the liquid. Therefore, loadings down to a few tenths of a percent have been proposed. As a disadvantage, however, there is the risk of an increased influence of adsorption phenomena, so that the carrier material must be inactivated with particular care. In addition, the capacity of the column is reduced and thus the amount of sample that can be loaded for separation.

Capillary columns contain only a thin film of separating liquid, which is produced by blowing through a few drops of the stationary phase.

8.4.6 Mobile phase – circulation method

The *mobile phase* should react neither with the components of the analysis sample nor with the stationary liquid or the solid support material. The type of carrier gas has an influence on the separation efficiency of the column in that the random diffusion of the molecules of the analysis substances in the gas phase leads to a broadening of the elution bands; this diffusion decreases with increasing molecular weight of the carrier gas, and accordingly an improvement in separations has been observed with N_2, CO_2 or Ar as the mobile phase compared with H_2 or He. However, the effect is quite small, so that other considerations usually determine the choice of carrier gas.

If the separated components are to be recovered by freezing out after leaving the column, a low-boiling gas, e.g. H_2, He or N_2, must be used. One can further elute with condensable carrier gases, e.g., ethanol vapor, benzene, and others, and precipitate the separated substances together with the carrier gas. For the isolation of gaseous components, CO_2 is often used as a carrier gas, which is absorbed in KOH solution after leaving the column (see below).

As a rule, however, the individual components are not isolated, but the elution bands are recorded automatically with suitable detectors. In this mode of operation, the carrier gas is determined by the type of detector; for thermal conductivity detectors, H_2 and He are most favorable, for ionization detectors, He, Ar or N_2 are mainly used.

The carrier gas can be returned to the top of the column after it has left the column (recirculation method). This prevents the column from bleeding, but care must be taken to remove eluted substances beforehand.

An interesting development is gas chromatography with carrier gases at extremely high pressures, *supercritical fluid chromatography: SFC* (up to approx. 2000 bar). Here, the gas density approaches the density of liquids (*supercritical fluid*), and numerous compounds whose boiling points are above the decomposition temperatures, e.g. oligopeptides, sugars, nucleosides and various polymers, dissolve in such compressed gases, so that these can also be separated. The method represents an intermediate stage between gas and partition chromatography. CO_2, NH_3 or CCl_2F_2 are used as carrier gases;

the main difficulty is to find suitable stationary phases that are not also soluble in the supercritical gas.

8.4.7 Flow rate of the mobile phase

The flow rate of the carrier gas has a considerable influence on the separation efficiency of a gas chromatographic column. At high flow rates, the distribution equilibria cannot be fully established and the efficiency decreases. At very low velocity, the equilibria are fully established, but due to the diffusion of the individual components in the gas phase, the bands broaden more and more, and the efficiency also decreases. Therefore, there is a certain flow rate at which the column has a maximum of separation stages (or the height of a separation plate (HETP) has a minimum). These ratios are represented by an equation named after van Deemter (1956), which is in a simplified form after Keulemans (1957) (see also Fig. 8.12):

$$H = \mathrm{A} + \frac{\mathrm{B}}{u} + \mathrm{C} \cdot u$$

$$(H = \text{height of a separation plate}; \mathrm{A}, \mathrm{B}\,\text{u.}\,\mathrm{C} = \text{constants}; u = \text{carrier gas flow rate}). \quad (2)$$

The *van Deemter equation* does not always agree with experimentally determined values (cf. Fig. 8.13), so that modified equations have repeatedly been developed for more precise investigations. However, it is generally sufficient for practical purposes, where the determination of the position of the minimum of the height of a separation plate is most important.

The flow rate of the carrier gas should, if possible, be the same over the entire length of the separation column so that optimum conditions can prevail everywhere. In practice, however, this cannot be achieved because a pressure drop occurs along the column due to the flow resistance of the packing, which increases the flow velocity towards the end of the column. The resulting deterioration of the separations is, however, insignificant unless the pressure drop is very large (> approx. 2: 1). In capillary columns which have no filling, the flow resistance is relatively small.

With normal columns, one works with a gas passage of about 30–50 ml/min, with capillary columns with about 0.5–1.5 ml/min.

The carrier gas velocity is kept exactly constant during a test by fine control valves (if one does not work with pressure programming; see below).

8.4.8 Influence of temperature on the separations

In gas chromatographic separations, the column temperature is essentially determined by the vapor pressures of the substances to be analyzed. One can operate the

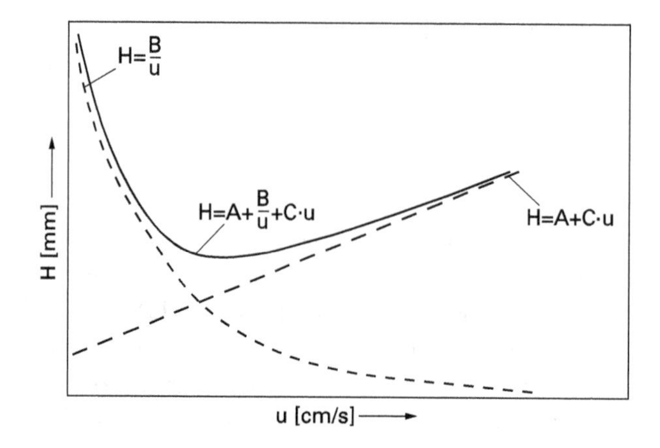

Fig. 8.12: Graphical representation of the van Deemter equation.

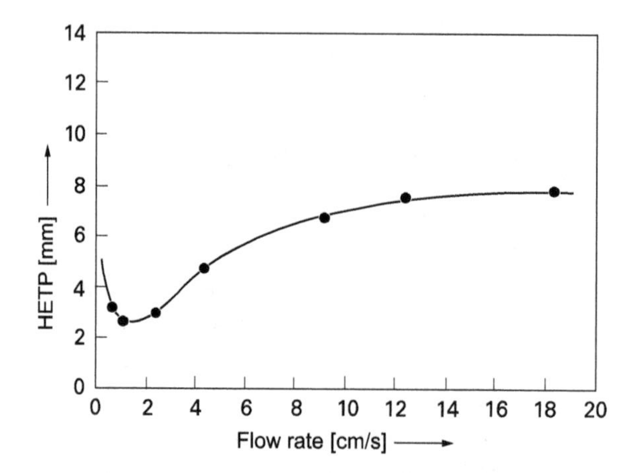

Fig. 8.13: Example of deviations from the van Deemter equation.

column about 50 °C below the boiling point of the highest boiling component of the mixture to be analyzed. Upwards, the temperature is limited by the volatility of the stationary phase, as already mentioned.

In general, the lower the column temperature, the more effective the separation, but at very low temperatures, an increase in the separation stage height may occur again (cf. Fig. 8.14).

The temperature should be kept as constant as possible during a separation (if you do not work with temperature programming, see below); this allows the baseline of the chromatogram to be kept uniform.

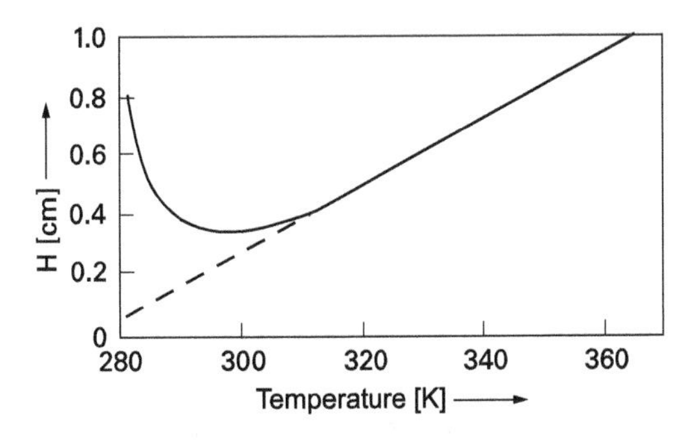

Fig. 8.14: Influence of column temperature on the separation plate height *H*.

8.4.9 Gradient elution and reverse gas chromatography

When investigating mixtures whose components have widely differing boiling points, one can either work at a high column temperature, which, however, results in poor separation of the low-boiling components, or at a relatively low temperature, which, however, results in a large time requirement for eluting the high-boiling components. In such cases, it is advisable to increase the temperature during elution (*temperature programming* or application of a *temperature gradient*).

A variant of this method is called *reversed gas chromatography*; the column is kept at a relatively low temperature so that the components of the analytical sample do not migrate or migrate only very slowly. Then a short temperature gradient in the form of a small oven, heated at the rear end and cooled at the front end, is drawn across the column at a constant speed (Zhukovitskii, 1951)[3]. For each substance of the analytical sample there is a certain temperature at which it travels as fast as the stove; thus, narrow bands of the individual substances are formed within the temperature gradient (Fig. 8.15).

The advantage of the method is that the effect of the gradient results in a tightening of the bands, which leads to a significant increase in the lower detection limit as a result of the increased concentration at the band maximum. However, due to the short length of the separation path in the temperature gradient, only a small number of substances can be separated simultaneously.

In addition to temperature gradients, pressure gradients (i.e. variable flow rates of the mobile phase) and concentration gradients of the stationary phase are also used – but relatively rarely. The former are difficult to realize experimentally. In the latter, one fills the column section by section with portions of the carrier material, progressively reducing the liquid phase loading. Finally, combinations of several gradients have also been proposed.

3 The idea of drawing a temperature field over a separation column probably originated with E. Jantzen and H. Witgert (1939).

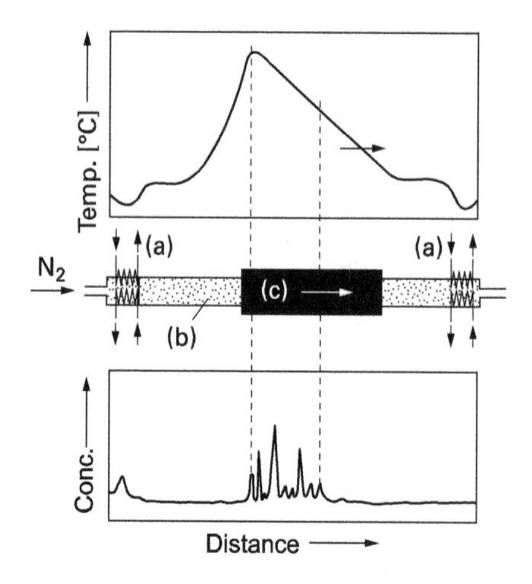

Fig. 8.15: Reversion gas chromatography.
Top: Temperature profile along the column.
Center: Column and sliding oven arrangement.
a) Cooling; b) column; c) oven.
Below: Arrangement of the components to be separated in the temperature gradient.

8.4.10 Isolation of individual components from the eluate

Gaseous components can be recovered from the eluate by using CO_2 as a carrier gas. The gas flow is directed from below into a burette filled with KOH solution; the appearance of elution bands is indicated by the fact that the gas bubbles are no longer completely absorbed by the potassium hydroxide solution. In a device for changing the template (Fig. 8.16), a new burette can be brought over the outlet of the gas stream by turning a ground joint. Losses are only caused by the solubility of the investigated gases in the KOH solution.

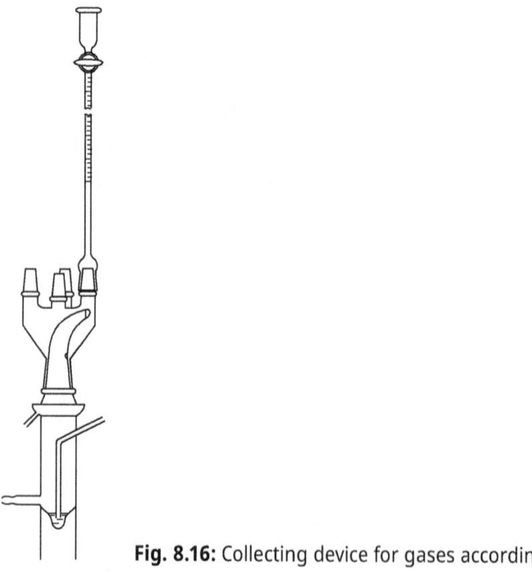

Fig. 8.16: Collecting device for gases according to Ehrenberger.

A problem that has not yet been solved satisfactorily is the isolation of components that are liquid or solid at ordinary temperature. It is possible to pass the eluate through cold traps; however, these always retain only part of the condensed components, since aerosols are formed whose particles are easily carried along by the gas flow. Even by filling the freezing-out device with inserts or adsorbents, yields of hardly more than 90–95% can be achieved.

The most favorable method is probably the complete condensation of the entire eluate, which is easily possible, for example, with the use of ethanol vapor as a carrier gas, but can also be carried out with Ar, CO_2 and other gases without particular difficulty.

Good results are also described with a similar technique in which a so-called *auxiliary phase collector* (e.g. water vapor, acetone vapor, etc.) is introduced into the eluate and this is then precipitated together with the components of the analysis sample in a cooler.

8.4.11 Elution curves and detectors

Isolation of the components of a gas chromatogram from the eluate is performed relatively rarely; predominantly, elution curves are recorded digitally using appropriate detectors and computer analysis.

Of the numerous detectors specifically designed for gas chromatography, mention should be made of thermal conductivity cells, thermistors, flame ionization detectors, photoionization detectors, and electron capture detectors. Particularly powerful is the coupling with a mass spectrometer. These devices, some of which have extremely high sensitivities down to the lower fg range, have played a major role in the widespread use of gas chromatographic methods.

8.4.12 Identifying the bands of an elution diagram

Retention data are particularly suitable for identifying individual substances separated by gas chromatography, since the reproducibility of these values is very good. One usually uses the corrected retention data, which are obtained by subtracting the amount of carrier gas contained in the column from the total retention. The value for this correction is obtained automatically, since a small amount of air inevitably enters the carrier gas stream during substance addition. This air is practically not absorbed by the common stationary phases and thus migrates through the column without delay. In the chromatogram, this contamination becomes noticeable by a so-called *air peak*; subtract the eluate volume from the substance feed to the maximum of the air peak from the retention volume V_i of substance i and obtain the corrected retention volume V_i^o (cf. Fig. 8.17).

Instead of the retention volume V_i, the retention time t_i or the corrected retention time t_i^o can be used.

The corrected retention data depend, among other things, on the carrier gas velocity and the column length. These variables are eliminated by referring to a standard substance that is added to the analytical sample prior to loading onto the column. If one designates its corrected retention volume with V_{St}^o the ratio is $V_i^o/V_{St}^o = V_{i,St}^o$ is the so-called *relative corrected retention volume of substance i, referred to standard St*, and the following relationships apply (see also Fig. 8.17):

$$V_{i,St}^o = \frac{V_i^o}{V_{St}^o} = \frac{V_i - V_{Air}}{V_{St} - V_{Air}} = \frac{t_i - t_{Air}}{t_{St} - t_{Air}} = \frac{x_i - x_{Air}}{x_{St} - x_{Air}} \tag{3}$$

Since these are ratios, it does not matter whether the retention volume or the retention time is used to determine $V_{i,St}^o$ is used. The relative corrected retention volume of a given substance is independent of the column length and the carrier gas velocity; it only depends on the type of stationary phase and the column temperature.

The relative retention data are less reproducible if the retention values of the analytical substance and the standard are far apart. In order to obtain accurate values even in such cases, it was suggested by Kováts that not only *one* standard substance should be added, but *several* which conveniently belong to a homologous series. For example, butane – hexane – octane – decane etc. can be used as standard substances; the retention volumes (or times) of these compounds are given certain indices, e.g. 400 – 600 – 800 – 1000 etc. The elution bands of the constituents of the analysis sample are classified according to their meaning and given the corresponding intermediate values as indices (*Kováts retention index*, 1958).

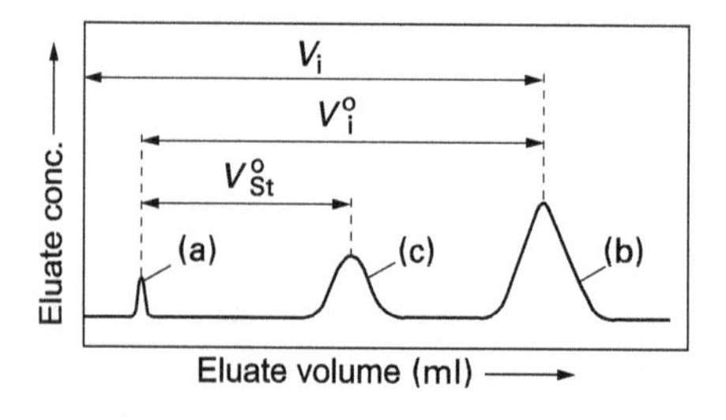

Fig. 8.17: Retention volume V_i, corrected retention volume V_i^o of substance i and corrected retention volume V_{St}^o of the standard St.
(a) Air peak; (b) elution band of substance i;
(c) elution band of the standard St.

The identification of unknown components from retention data is only possible if the values of the substances concerned are already known. For this purpose, extensive tables have been prepared for various substance classes (at specific stationary phase and column temperature). However, since a chromatogram can show only a relatively small number of components separately for purely spatial reasons, the possibility of coincidences must always be expected, and the identification of a substance from a retention value alone is not reliable. It is customary to increase the certainty of assignment by determining a second retention value on a column with a different stationary phase; if both data agree with the known values for the analytical substance, the assignment is generally regarded as certain.

A further possibility for identifying unknown elution bands results from a regularity already mentioned in distribution chromatography: If the logarithms of the (corrected) retention volumes are plotted against the number of C atoms of the compounds concerned, straight lines are obtained for the compounds of a homologous series, although small deviations often occur in the first members (cf. Fig. 8.18). The assignment is made by checking with which of the retention straight lines an experimentally found retention value gives a whole C number (cf. Fig. 8.19).

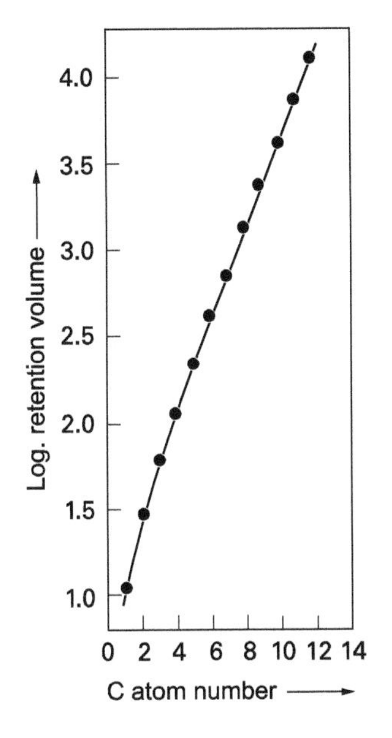

Fig. 8.18: Dependence of the logarithms of the retention volumes on the number of C atoms in the homologous series of fatty acids.

Furthermore, the retention volumes or the retention times of the members of different homologous series can be plotted against each other on two columns; for each homologous series, a straight line is obtained whose slope is characteristic for the series in question (cf. Fig. 8.20).

Finally, the temperature dependence of the Kováts indices also gives an indication of the substance class to which an unknown elution band is assigned.

Reaction gas chromatography can also be used to obtain information on the qualitative composition of the analysis sample. This involves chemical reactions that are carried out immediately upstream or downstream of the column or in the column itself. For example, using H_2 as the carrier gas, unsaturated compounds can be catalytically hydrogenated upstream of the column and, by recording one chromatogram

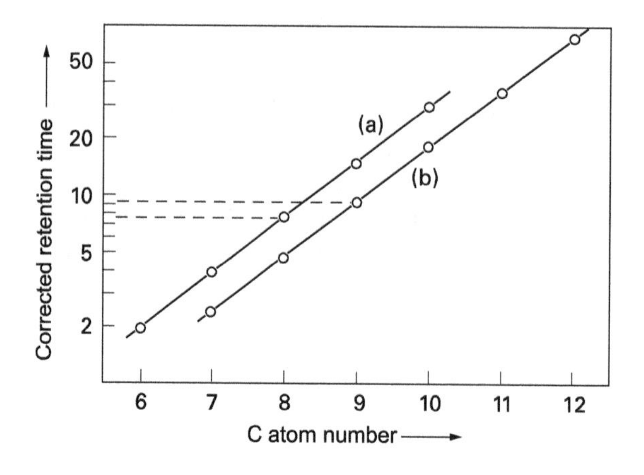

Fig. 8.19: Assignment of elution bands of fatty acid methyl esters.
(a) n-Fatty acid methyl ester; (b) 3-methyl fatty acid methyl ester.

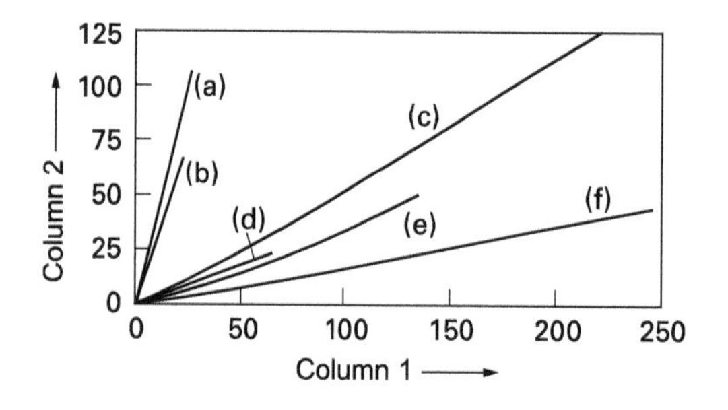

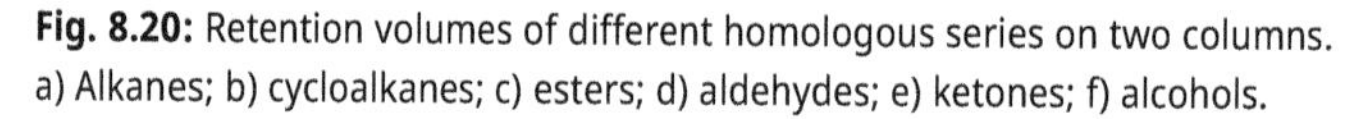

Fig. 8.20: Retention volumes of different homologous series on two columns.
a) Alkanes; b) cycloalkanes; c) esters; d) aldehydes; e) ketones; f) alcohols.

with and one without hydrogenation, it is possible to determine which elution bands belong to unsaturated compounds. The method can be extended to numerous other substance classes with the aid of other reactions (see Tab. 8.4).

Despite these aids, the interpretation of complicated gas chromatograms in particular, where knowledge of the compound classes present is not already given, presents considerable difficulties. In such cases, further methods of identification must be used; in particular, the combination of the gas chromatograph with a mass spectrometer directly connected to the column outlet has proven to be very powerful. However, final clarification is often only achieved by chemical and physical methods of structure elucidation after isolation of pure components from the eluate.

Quantitative evaluation of gas chromatograms is performed by digital integration of peak areas.

Tab. 8.4: Reaction gas chromatography (examples).

Reagent	Removed substances
H_2 + catalyst	Olefins
H_2 + catalyst	Aliphatic halogenated compounds
ZnO	Carboxylic acids
KOH solution	CO_2, H_2S, SO_2
Hg acetate + $HgCl_2$	Olefins
$LiBH_4$, $LiAlH_4$	Oxygen containing compounds
H_3BO_3	Alcohols
$NaHSO_3$	Aldehydes, ketones
HNO_2	Amines
P_2O_5, $Mg(ClO_4)_2$	H_2O
I_2O_5	CO
$HgCl_2$	CH_3SH
Maleic anhydride	Butadiene

8.4.13 Efficacy and scope of the method

As a result of the straightness of the distribution isotherms, the elution bands in gas chromatographic separations are usually symmetrical. With ordinary columns of 4–6 mm diameter, separation plate heights of about 0.5–2 mm are achieved, thus with column lengths of 4–6 m 2000–12000 separation plates. The separation plate heights of capillary columns are about 0.2–0.5 mm, so that with the very long columns possible here, separation plate numbers of several hundred thousand to over one million can be achieved.

Gas chromatography is thus one of the most effective separation methods known. Also noteworthy are the extremely low lower limits of quantification achievable with some detectors, the good reproducibility of the results and the short duration of the separations, usually only about 10–30 min. In addition, the method can be largely automated.

The range of application is limited by the fact that compounds must be present which are volatile at temperatures which are not too high (maximum approx. 300 °C). While this condition is fulfilled by only relatively few inorganic compounds, numerous organic substances with sufficiently high vapor pressures in the temperature range in question are known, and many others can be converted into volatile derivatives by chemical reactions (cf. Tab. 8.5).

Due to the advantages mentioned, gas chromatography has improved, accelerated or even made possible the analysis of organic substance mixtures to an extent that can hardly be overestimated.

The relatively low capacity of the columns in use must not always, but in many cases, be regarded as a disadvantage, since it makes the identification of unknown components more difficult. Another disadvantage is the difficulty in quantitatively recovering the separated substances from the eluate.

Tab. 8.5: Derivatization of low-volatile substances before gas chromatographic separations (examples).

Substance class	Reaction
Carboxylic acids	Esterification with CH_3OH/BF_3 or CH_3OH/HCl
Alcohols	Silylation with hexamethyldisilazane
Polyalcohols	Acetylation with acetic anhydride
Monosaccharides	Silylation with trimethylsilyl chloride
Amino acids	Esterification with diazomethane, then reaction with trifluoroacetic anhydride.
Steroids	Silylation with trimethylsilyl chloride
Polymeric natural and synthetic materials	Pyrolysis
Be^{2+}, Al^{3+} et al.	Reaction with acetylacetone or trifluoroacetylacetone

8.5 Countercurrent method

The countercurrent principle has also been used for distribution between gases and liquid phases, e.g. in annular arrangements; however, such processes have not yet gained any importance.

8.6 Cross flow method

The cross-flow method with continuous substance feed can be realized in gas chromatography in such a way that a broad liquid film flows down a vertical surface and a gas flow is passed across it. However, the method is not very effective because of the hardly avoidable vortex formation in the gas phase and has therefore not attained any significance; it is only mentioned for reasons of completeness.

General literature

P. Henrichon, Gas Chromatography: History, Methods and Applications, Elsevier, Amsterdam (2020).
H. McNair, J. Miller & N. Snow, Basic Gas Chromatography, John Wiley & Sons, New York (2019).
C. Poole, Gas Chromatography, Elsevier, Amsterdam (2021).
O. Sparkman, Z. Penton & F. Kitson, Gas Chromatography and Mass Spectrometry: A Practical Guide, Elsevier, Amsterdam (2011).

Literature to the text

Headspace – Gas chromatography

B. Kolb & L. Ettre, Static Headspace-Gas Chromatography: Theory and Practice, Wiley & Sons, New York (2006).

N. Snow &G. Bullock, Novel techniques for enhancing sensitivity in static headspace extraction-gas chromatography, Journal of Chromatography A *1217*, 2726–2735 (2012).

N. Snow & G. Slack, Head-space analysis in modern gas chromatography, Trends in Analytical Chemistry *21*, 608–617 (2002).

Separation phases

E. Ghanem & S. Al-Hariri, Chromatographia, Separation of isomers on nematic liquid crystal stationary phases in gas chromatography: A review *77*, 653–662 (2014).

X. Liang, X. Hou, J. Chan, Y. Guo & E. Hilder, The application of graphene-based materials as chromatographic stationary phase, TrAC – Trends in Analytical Chemistry *98*, 149–160 (2018).

C. Meinert & U. Meierhenrich, A new dimension in separation science: Comprehensive two-dimensional gas chromatograph, Angewandte Chemie – International Edition 51(42), pp. 10460–10470 (2012).

H. Nan & J. Anderson, Ionic liquid stationary phases for multidimensional gas chromatography, TrAC-Trends in Analytical Chemistry *105*, 367–379 (2018).

C. Poole & S. Poole, Ionic liquid stationary phases for gas chromatography, Journal of Separation Science, *34*, 888–900 (2011).

L. Vidal, M.-L. Riekkola, and A. Canals, Ionic liquid-modified materials for solid-phase extraction and separation: A review, Analytica Chimica Acta *715*, 19–41 (2012).

Y. Yu, Y. Ren, W. Shen, H. Deng & Z. Gao, Applications of metal-organic frameworks as stationary phases in chromatograph, TrAC – Trends in Analytical Chemistry *50*, 33–41 (2013).

9 Adsorption and absorption of gases on solids

9.1 General

9.1.1 Historical development

The separation of gas mixtures by adsorption and desorption on or from solids was first worked on for analytical purposes by Berl (1921) and Peters (1937). The adsorption chromatography of gases was already indicated by Schuftan (1931), but the method came to greater importance only after the work of Hesse (1941), Cremer (1951) and others.

Absorption methods were already used during the beginnings of analytical chemistry in the late 18th and early 19th centuries.

9.1.2 Definitions

The retention of gases on the surface of solids is referred to as *sorption* or *adsorption*. Attempts have been made to characterize the different binding strengths occurring in this process by the expressions *physisorption* (weak binding) and *chemisorption* (strong binding), but no clear distinction can be drawn because the transitions between the two groups are fluid.

In the following, the term *adsorption* will be used in the sense of a relatively weak bond that can be easily dissolved again, while *absorption* proceeds with the formation of chemical compounds.

9.1.3 Auxiliary phases – distribution isotherms – speed of equilibrium adjustment

In separations by adsorption of gases on solids, one works either with only one auxiliary phase, the adsorbent, or with the adsorbent and an additional gas that acts as a transport or carrier gas to drive the analysis sample through a column filled with the adsorbent.

Adsorption isotherms. The distribution isotherms known as *adsorption isotherms* occur in various forms: In the simplest – though rarely observed – case, the concentration in the adsorbent (i.e., the amount of gas taken up by a given amount of adsorbent) is proportional to the pressure or partial pressure p of the gas. Thus, a straight-line isotherm corresponding to Henry's law is present.

The more common and thus for practical purposes more important isotherms are named after Freundlich (1907) and Langmuir (1916). The *Freundlich isotherm* follows the equation

https://doi.org/10.1515/9783111181417-009

$$c_{Ads} = a \cdot p^n, \tag{1}$$

where $n < 1$ (usually between 0.4 and 1). eq. (1) is often expressed in the form

$$c_{Ads} = a \cdot p^{1/m} \tag{1a}$$

with $m > 1$ written.

For experimental verification, the logarithmic form is conveniently used:

$$\log c_{Ads} = \log a + n \cdot \log p. \tag{1b}$$

If we plot $\log c_{Ads}$ against $\log p$, we obtain a straight line with slope n and $\log \alpha$ as the ordinate intercept.

The curve corresponding to eq. (1) does not have a straight part; furthermore, it does not give a constant final value at high gas pressures (cf. Fig. 9.1).

The equation of the *Langmuir isotherm* is

$$c_{Ads} = \frac{k_1 \cdot p}{1 + k_2 \cdot p}. \tag{2}$$

The associated curve (cf. Fig. 9.1) approximates a straight line with slope k_1 at low partial pressure p as it enters the coordinate starting point ($k_2 \cdot p \ll 1$); at high pressure p, a saturation value for the concentration in the adsorbent is obtained, the magnitude of which is given by the ratio k_1 / k_2 ($c_{Ads} \to \frac{k_1 \cdot p}{k_2 \cdot p}$). If chemical bonding and thus absorption is present, the process tends to be irreversible.

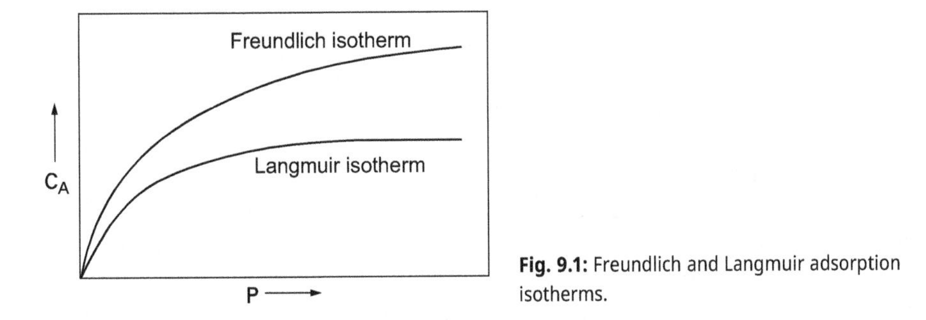

Fig. 9.1: Freundlich and Langmuir adsorption isotherms.

Among other isotherms, those according to Brunauer, Emmett and Teller (*BET isotherm*, 1938) and Temkin (1940) should be mentioned. The BET isotherm has a turning point with a renewed increase at higher pressures (Fig. 9.2); it is frequently used to determine the surface area of adsorbents.

Rate of equilibrium adjustment. When gases are adsorbed on solids, equilibrium is usually reached within a few seconds. It can be delayed up to several minutes if the

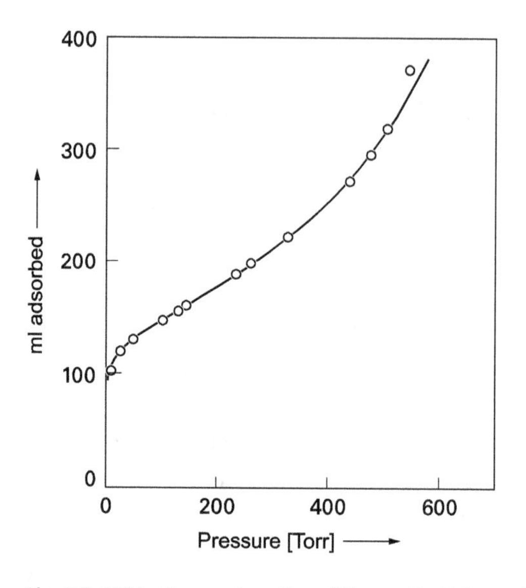

Fig. 9.2: BET isotherm; adsorption of N_2 on a Fe-Al_2O_3 catalyst at −195.8 °C.

gas has to diffuse into narrow pores of the adsorbent, which is especially the case with coarse-grained, porous materials with a large inner surface.

9.1.4 Desorption

Adsorption is always accompanied by a positive heat tone; therefore, as the temperature decreases, the amount of gas retained by a given amount of adsorbent increases, and as the temperature increases, it decreases. One uses this behavior to desorb the adsorbed substances by heating the entire system. The temperatures required depend on the strength of the bonds formed during the adsorption process; if the bonds are very weak, the adsorbed substances can be desorbed at 0 °C, while stronger bonds may require several hundred degrees Celsius. If real chemical bonds are present, the process of absorption can be practically irreversible.

Desorption can also be achieved by reducing the pressure (*pumping down*) and by displacement with other, more adsorbed gases.

9.1.5 Adsorbents

The adsorbents used for gases in analytical chemistry are mainly molecular sieves, silica gel, activated carbon and various organic polymers, and to a lesser extent aluminum oxide and some rarely used compounds, such as copper phthalocyanine or nickel complexes. Chemically selective sorbent materials are used for gas sensing.

Apart from the particle size, important characteristics of adsorbents are the specific internal surface area (in m^2/g), the average pore size (in Å or nm) and the pore size distribution. These quantities are obtained from empirically determined adsorption isotherms.

Molecular sieves are hydrous multilayer silicates (zeolites) produced by hydrothermal synthesis. This class of substances was first described by Weigl (1924) and McBain (1926).

The molecular sieves can be dehydrated without changing the basic structure of the lattice. Instead of the water molecules, cavities of completely uniform size are then present (cf. Figs. 9.3 and 9.4), which can be filled again with water, but also with other substances, releasing mostly considerable adsorption heats.

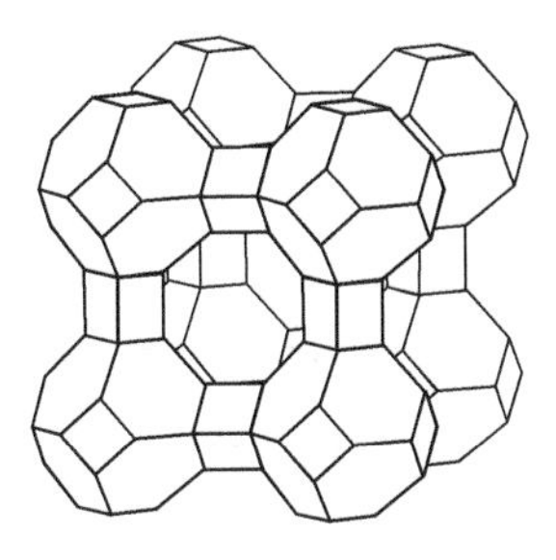

Fig. 9.3: Crystal structure of molecular sieves.

The decisive advantage of molecular sieves over other adsorbents is that only molecules whose diameter is not larger than that of the pores can diffuse into the interior of the crystals. Since the pore size is uniform, this results in sharp demarcations from larger particles, which remain excluded. In addition to adsorption, therefore, a sieving effect occurs, often bringing about separations of chemically similar compounds that are almost impossible to achieve by other methods.

Irrespective of the sieve effect, polar substances are bound more tightly than non-polar ones and unsaturated compounds more tightly than saturated ones – as far as the size ratios of the molecules allow.

In principle, the sieving effect is also present in other adsorbents with a porous structure, but it has only a minor effect on the separation efficiency in these because of the usually broad pore size distribution.

In addition to adsorption inside the crystal lattice, an adsorption effect with lower selectivity occurs on the outer surface of the individual particles in molecular sieves. However, since this effect accounts for only about 1% of the total capacity of the adsorbent, no significant interference occurs as a result.

The adsorption isotherms often (but not always) follow the formula given by Langmuir, whereby the initial part can be very steep (cf. Fig. 9.5). In the presence of such isotherms, the substances in question are well adsorbed even in ranges of small concentrations.

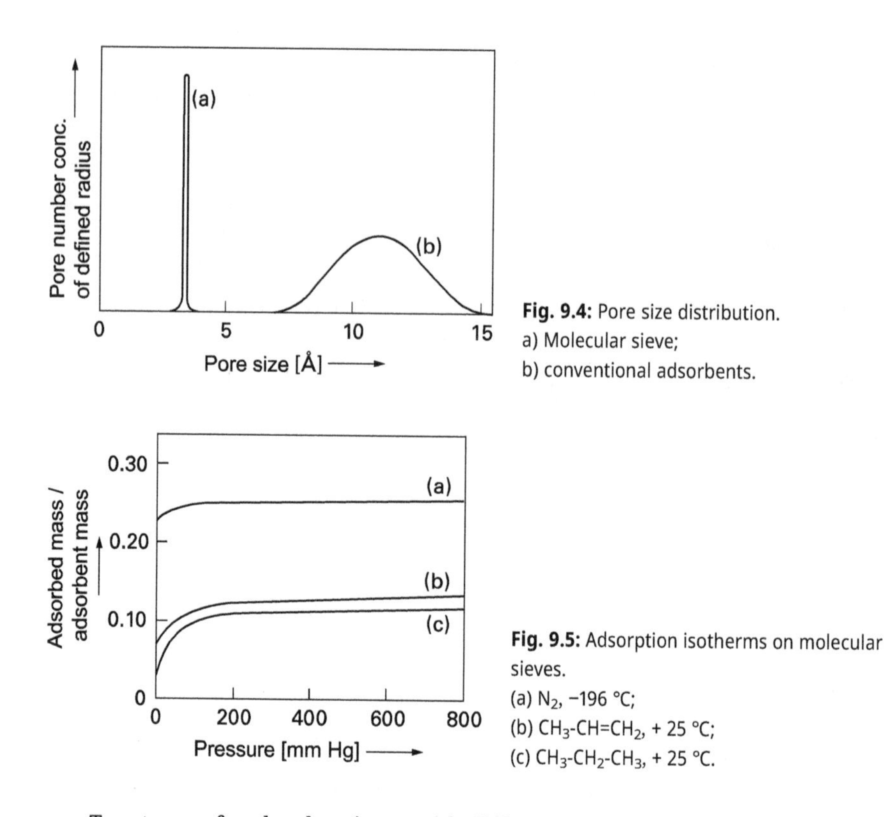

Fig. 9.4: Pore size distribution.
a) Molecular sieve;
b) conventional adsorbents.

Fig. 9.5: Adsorption isotherms on molecular sieves.
(a) N_2, −196 °C;
(b) CH_3-CH=CH_2, + 25 °C;
(c) CH_3-CH_2-CH_3, + 25 °C.

Two types of molecular sieves with different lattice structures are commercially available.

Type A corresponds to the formula $Me_{12/n}[(AlO_2)_{12}\ (SiO_2)_{12}] \cdot 27\ H_2O$, where n *is* the valence of the cation Me. The compound is prepared in the sodium form; other ions, e.g., K^+ or Ca^{2+}, can replace the Na^+ ion by changing the pore size (see Tab. 9.1).

The composition of the second type of molecular sieve (type X) is given by the formula $Me_{86/n}[(AlO_2)_{86}\ (SiO_2)_{100}] \cdot 276\ H_2O$. Here, too, the Na^+ ion introduced during synthesis can be exchanged for other cations; the calcium compound has gained practical importance (cf. Tab. 9.1).

The molecular sieves are produced as extremely fine crystalline powders; they are processed with an inert binder to form larger beads.

The thermal resistance is excellent: Types 4 A, 5 A and 13 X can be heated up to about 700 °C for a short time without damage, type 10 X is resistant to about 500 °C.

Activation, i.e. removal of the water of crystallization, is carried out by heating in a vacuum to 300–350 °C or by rinsing with a low adsorbable gas at the same temperature. Molecular sieves are regenerated in the same way after use.

When using molecular sieves, it should be noted that water is adsorbed with strong preference over all other compounds. On the one hand, this behavior can be used for the effective drying of gases and of organic liquids, but on the other hand it

Tab. 9.1: Molecular sieves.

Type	Pore diameter (Å)	Cation
3 A	3	K^+ (70%)*)
4 A	4	Na^+
5 A	5	Ca^{2+} (70%)*)
10 X	9	Ca^{2+} (70%)*)
13 X	10	Na^+

*)Only 70% of Na^+ exchanged for K^+ or Ca^{2+}.

interferes with the separation of mixtures of organic compounds containing water. In such cases, the sample must first be dried by other means.

Acidic gases in combination with moisture destroy the lattice of the molecular sieves.

Silica gel (silica gel) is prepared by precipitating sodium silicate solutions with acid, washing the precipitate and activating at about 200–300 °C. During activation, the main amount of water is expelled; the residual water content affects the adsorption capacity. Regeneration of used gels should be done at slightly lower temperatures (about 150–200 °C).

The pore size of the individual particles can be influenced by the pH values during precipitation and washout; on average, it ranges from about 15 to 60 Å; pore size distributions for some gels are given in Fig. 9.6.

Wide-pored gels have specific surface areas of about 200–400 m^2 /g, narrow-pored ones of 600–800 m^2 /g. The smaller the pore diameter, the greater the adsorption enthalpies and the more strongly adsorbed substances are retained.

The surface area of silica gels is reduced by contact with weakly alkaline solutions; in strongly alkaline liquids the gels dissolve.

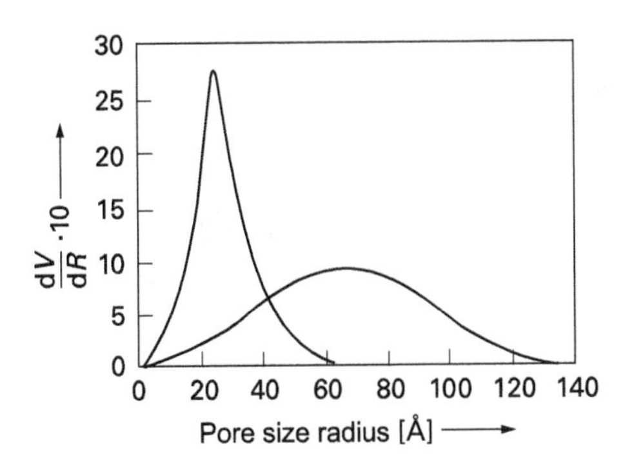

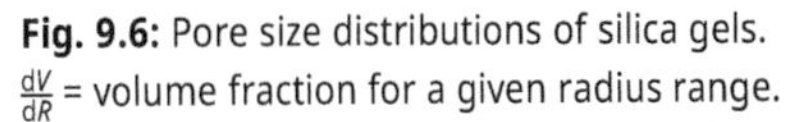
Fig. 9.6: Pore size distributions of silica gels. $\frac{dV}{dR}$ = volume fraction for a given radius range.

Activated carbon is obtained by calcination of wood, various types of coal, etc. and subsequent activation with steam at 700–900 °C. The material can be produced with different porosity; the specific internal surface varies from about 300 to 1200 m^2 /g.

A special type of activated carbon is formed by pyrolysis of polyvinylidene chloride under certain conditions (HCl splitting). The product is characterized by a relatively narrow pore size distribution (mean pore radius 12.4 Å, specific internal surface area 1072 m^2/g), it consequently exhibits molecular sieving properties and is suitable for the separation of gases.

Finally, mention should be made of the use of graphite powder, which was occasionally used for separations.

Porous polymers. The porous high-polymer plastics already mentioned in Chap. 8 (*Gas chromatography*) play an important role as adsorbents in dry form without liquid phase. These substances are produced by a special polymerization technique (polymerization in dilute solutions with relatively high additions of bifunctional compounds as crosslinkers, see also chap. 11, *Ion exchange*). When the solvent is removed, the porous structure of the compound is retained. The starting materials are mainly styrene or ethylstyrene with divinylbenzene as crosslinker, but corresponding polymers have also been synthesized from polyphenylene oxides, polyvinyl acetate or polyacrylonitrile.

For commercially available products, specific surface areas of 500–600 m^2/g, mean pore diameters of 75–3500 Å and pore volumes of 1.0–1.2 ml/g are given. The pore size can be varied by selecting the polymerization conditions; it is by no means uniform, but lies in such a wide range that the pore sizes of the different grades overlap to a certain extent.

Before use, the polymers are conditioned for a while by heating them in an inert gas stream for 1–2 h to about 220 °C. With few exceptions, they are stable up to about 250 °C. The capacity is relatively low.

Alumina is used primarily for adsorption from solutions, rather than for separation of gases (but separations of hydrocarbons on Al_2O_3 have been reported).

Metal-organic frameworks (MOFs). These represent recent attempts to use porous crystalline materials, which consist of defined docking sites for gas molecules at structure-forming regions of a two- or three-dimensional metal-organic polymer, for selective gas adsorption. A typical structure is shown in Fig. 9.7.

The achievable pore size of < 5 Å leads to enormous specific surface areas (e.g., for MOF-177: 5600 $m^2\ g^{-1}$), and is adjustable by the choice of ligands used for synthesis (cf. Fig. 9.8) in the polymer backbone. Due to the polymeric structure, MOFs possess a flexible structure. So far, only a few applications are known in analytical chemistry (for example, as a gas-sensitive color-forming recognition structure in chemical sensors). However, there is great potential especially for catalysis and gas storage (H_2, CH_4, CO_2 etc.).

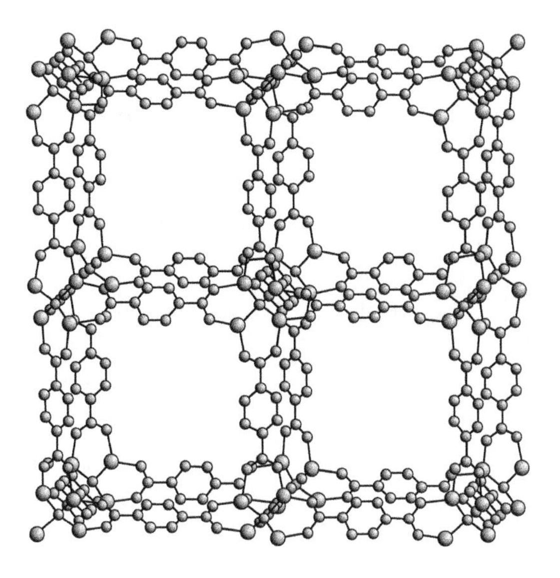

Fig. 9.7: Nanoscopic cross-section of a metal-organic framework structure.

9.1.6 Absorbents

A number of the more commonly used solid absorbents are listed in Tab. 9.2. The most important applications are the absorption of water vapor and of *acidic* gases such as HCl, SO_2, CO_2 and others.

9.2 Separations by one-time equilibrium adjustment

Separations of gases by one-time adsorption of one component or some components of a mixture are carried out mainly at low temperatures. As an example of the method of operation, consider the determination of traces of helium in argon according to Melhuish (1966): The gas mixture is placed in a vessel with activated carbon which has been cooled to −190 °C and, after the adsorption equilibrium has been set, the non-adsorbed helium (and possibly traces of hydrogen) are pumped into a second vessel with the aid of a Toepler pump.

Mention should also be made here of older methods for separating gas mixtures by fractional desorption, which generally give better results than the adsorption methods because desorption equilibria are established more quickly. In this process, the entire mixture is first adsorbed at low temperature and then pumped off one gas after the other with gradual heating.

The one-time equilibration method is also occasionally used in absorption processes, e.g., to remove H_2O vapor, CO_2 and others, but by and large plays only a minor role.

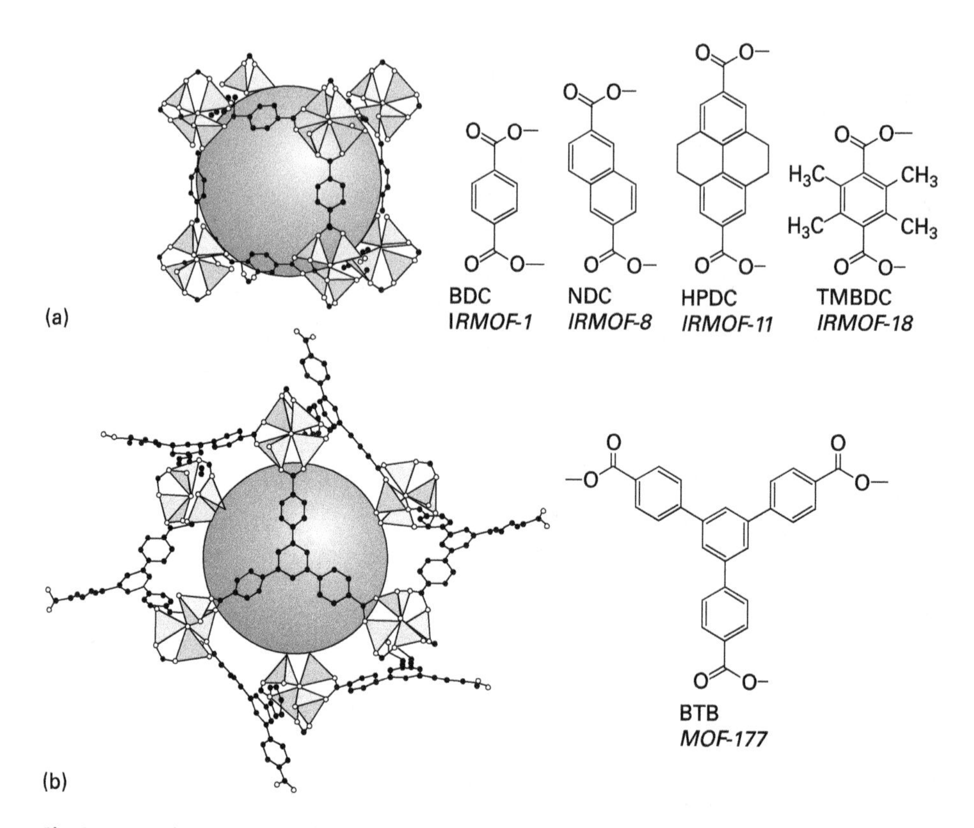

Fig. 9.8: Examples of ligands in MOF synthesis.

Tab. 9.2: Solid absorbents for gases (examples).

Absorbent	**Absorbed gases**
$CaCl_2$	H_2O
$Mg(ClO_4)_2$	H_2O
P_2O_5	H_2O
Soda lime (NaOH + CaO)	CO_2
Sodium asbestos (NaOH on asbestos)	CO_2
NaOH, KOH	CO_2, HF, HCl, HBr, SO_2, SO_3 and others.
$PbCrO_4$	SO_2, SO_3
MnO_2, PbO_2	Nitrogen oxides
Yellow phosphorus	O_2

9.3 Separations by one-sided repetition

9.3.1 Adsorption process

When repeating the separation step on one side, the gas mixture is passed through a straight or U-shaped tube containing the adsorbent. The quantity of gas must be such that the adsorbed component does not yet break through, but is completely retained. The substance in question is then expelled again for quantitative determination by heating or by displacement with a more strongly adsorbed compound, or is balanced directly together with the tube.

In addition, the method of incomplete separation from a gas sample passed in excess over the adsorbent and evaluation using the partition coefficient described in section 8.3 can be used.

Both methods are recommended primarily for the preconcentration of impurity traces in gases in conjunction with a subsequent gas chromatographic determination (cf. Fig. 9.9).

9.3.2 Absorption method

The technique of one-sided repetition is also frequently used in absorption processes, in which a single contact of the gas mixture with the solid phase is sufficient for quantitative conversion. Here, too, straight or U-shaped absorption tubes are used, through which the gas mixture is allowed to flow.

The method is used to separate impurities from gases or for quantitative determination of individual components (usually by weighing out the tube, e.g. for H_2O, CO_2 and others). Organic elemental analysis is also an example.

9.3.3 Circulation method

A recirculation method, in which the gas mixture to be analyzed is passed repeatedly over the solid absorbent in a closed apparatus, has been devised by Spence (1940) for micro-gas analysis, but has probably not become more important.

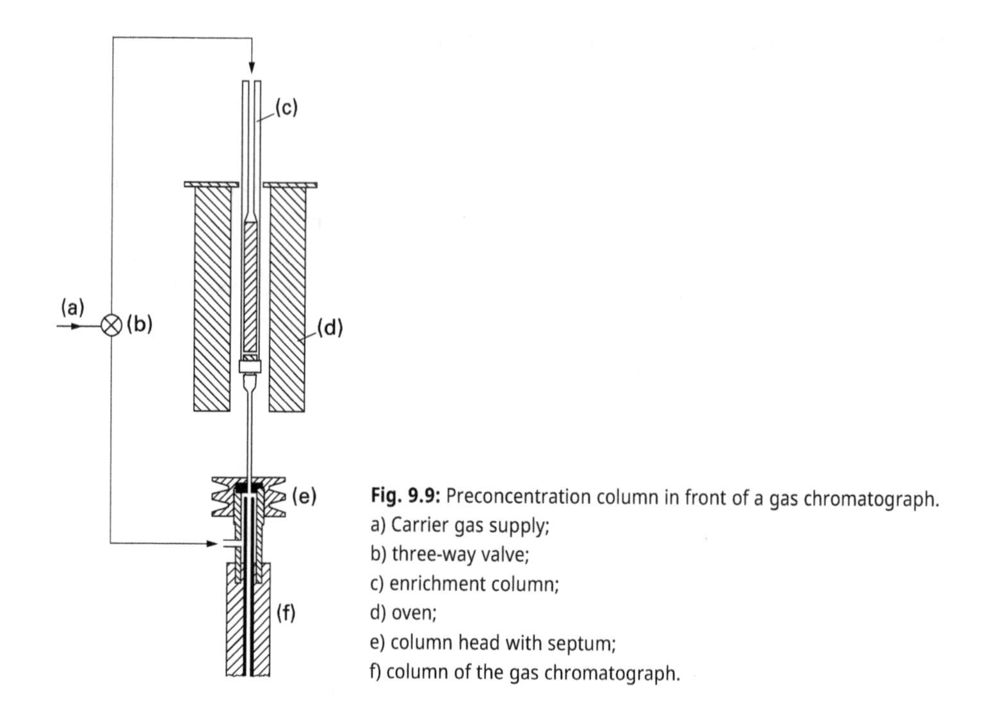

Fig. 9.9: Preconcentration column in front of a gas chromatograph.
a) Carrier gas supply;
b) three-way valve;
c) enrichment column;
d) oven;
e) column head with septum;
f) column of the gas chromatograph.

9.4 Separations by systematic repetition: Column method (gas adsorption chromatography)

9.4.1 Principle

Gas chromatography on solid adsorbents basically uses the same working technique as when liquid stationary phases are used; the adsorbent is filled into a column, the analytical sample is applied and a carrier gas is allowed to flow through as the moving phase.

9.4.2 Column material and dimensions

Column materials are usually glass or metal (copper or steel), occasionally PTFE, but the type of material generally does not matter.

The columns usually consist of tubes of about 1–6 m length with 4–6 mm clear width, furthermore capillary columns up to 15 m length have been produced.

9.4.3 Stationary and mobile phase

The stationary phases used in gas chromatography on solid adsorbents are almost exclusively the substances mentioned in section 9.1.5 (molecular sieves, silica gel, activated carbon, porous polymers, Al_2O_3), and glass powder, which can be important for special purposes.

Particle sizes are usually between 0.2 and 0.8 mm, with grit sizes of about 0.5 mm likely to be used predominantly.

It is often recommended to eliminate particularly active centers on the surface of the adsorbents by coating them with small amounts of a readily adsorbable liquid (e.g. 1.5% squalane), a procedure that leads to gas chromatography with liquid stationary phase. As a result, the adsorption capacity of the surface becomes more uniform, the adsorption isotherms of the substances to be separated become less curved, and the elution bands become more symmetrical. A low water content and siliconization of silica gel have the same effect.

Before commissioning the columns, a pretreatment of the stationary phase is usually prescribed in order to remove water residues, adsorbed gases and impurities left over from the synthesis. This pretreatment generally consists of bakeout in the carrier gas or in an inert gas stream (cf. Tab 9.3).

Tab. 9.3: Pretreatment of adsorbents in gas adsorption chromatography.

Adsorbent	Pretreatment
Molecular sieves	He or H_2 flow, 4–8 h, 250–300 °C
Silica gel	Depending on the desired activity in the carrier gas stream, bake out at 100–250 °C.
Activated carbon	H_2 or Ar flow, some hrs, 600–700 °C
Porous polymers	He, Ar, or N_2 flow, 12 h, 200–250 °C
Al_2O_3	Bake out for a few hrs under carrier gas flow at 600–700 °C.

H_2, He, Ar or N_2 are usually used as mobile phases. It should be noted that the chemical nature of the carrier gas influences the retention of the compounds to be separated to a much greater extent than is the case in gas chromatography with liquid stationary phases. The cause is the adsorption, which cannot be eliminated, of even the molecules of the carrier gas, resulting in a more or less pronounced displacement effect (cf. Tab 9.4). To a certain extent, this effect can also be exploited to eliminate active centers on the surface of the adsorbent and thus to improve separations.

The velocity of the carrier gas flow influences the separation efficiency of a column in the same way as it does in gas chromatography with liquid stationary phase; in principle, the van Deemter equation discussed in the previous section also applies here (cf. Fig. 9.10).

Tab. 9.4: Influence of the carrier gas on the retention time of CH_4; activated carbon column.

Carrier gas	Retention time (min)
He	34
Ar	22
N_2	16
Air	15
Ethin	5

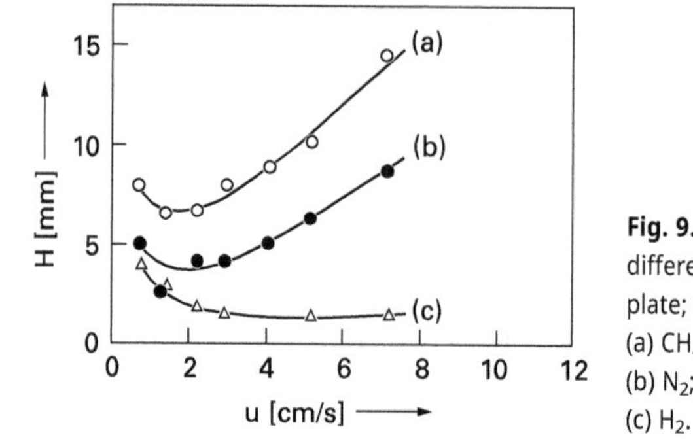

Fig. 9.10: Influence of the carrier gas velocity *u* of different gases on the height *H* of a separation plate; activated carbon column, + 109 °C.
(a) CH_4;
(b) N_2;
(c) H_2.

9.4.4 Temperature influence

For the separation of substances in the gas phase at room temperature, work is usually carried out at room temperature or lower temperatures (cf. also Tab. 9.5). Since the straight-line part of the adsorption isotherms usually extends toward larger concentrations at elevated temperature, improved separations are occasionally obtained at about 100–150 °C.

9.4.5 Gradient elution – reverse gas chromatography – displacement technique

Gradient elution is used almost exclusively in gas adsorption chromatography in the form of temperature programming. A gradient in the flow rate of the carrier gas can be achieved by gradually narrowing the column, but this procedure has not become important.

To achieve particularly high sensitivities, the reversion technique described in the previous section is recommended.

Desorption by displacement with a more adsorbable compound has been used relatively little. The carrier gas stream is passed through a suitable liquid, e.g. ethyl acetate, bromobenzene, dioxane, etc., before the column, and the vapor of this liquid then pushes the components of the analysis sample in front of it on the column.

9.4.6 Sampling of components – elution curves – detectors

When isolating individual components after separations by gas adsorption chromatography, the same difficulties arise as in gas chromatography with liquid stationary phase. The most effective method is that of Janák, already described in Chap. 8.4.10, in which CO_2 is used as a carrier gas, which one absorbs in KOH solution.

In general, the elution curves are recorded automatically using one of the detectors also listed in the previous section (especially the thermal conductivity detector).

9.4.7 Efficacy and scope of the method

In gas adsorption chromatography, the elution bands are usually approximately symmetrical despite the curved isotherms, since one tends to work with low loading and thus in the initial part of the adsorption isotherms. Furthermore, the rapid equilibration resulting from the fact that diffusion occurs only in the gas phase and not in a liquid has a favorable effect. Another advantage of the method is that one can work at high temperatures without having to fear bleeding of the column.

The separation efficiency of the process is very good; for example, separation stage heights of several mm were determined for activated carbon columns, approx. 1 mm for an Al_2O_3 column, 0.4 mm and 1.3–3.5 mm for columns with porous polymers, respectively; values of even approx. 0.1 mm were achieved with extremely carefully packed silica gel columns and with capillary columns.

Gas adsorption chromatography is particularly suitable for separating low-boiling gases for which gas chromatography with liquid stationary phase largely fails because of the only slight differences in solubility of the gases. Activated carbon, silica gel and, above all, molecular sieves are used for such separations (see Tab. 9.5). Molecular sieve 5 A is particularly suitable for separating straight-chain from branched hydrocarbons; only the former can penetrate into the interior of the pores due to their small diameter.

Porous polymers are mainly used for the separation of strongly polar compounds, especially those containing hydroxyl groups. The polymers retain water only very weakly, so that on the one hand aqueous solutions (e.g. of lower alcohols) can be analyzed and on the other hand the determination of water in organic solvents is possible. Al_2O_3 is mainly used for the separation of lower hydrocarbons.

Reverse gas chromatography, in conjunction with pre-enrichment, can be designed into an extremely sensitive method that can still detect impurities in gases in the pptv range.

Tab. 9.5: Separation of gases by gas adsorption chromatography (examples).

Column	Temp. (°C)	Gas mixture
Molecular sieve 5 A, 3 m	25–400	O_2-N_2, CO-C_2H_6, N_2O-CO_2
Molecular sieve 5 A, 2 m	approx. 25	He-O_2, Ar-N_2
Molecular sieve 5 A, 0.75 m	−72	Ar-O_2
Molecular sieve 5 A, 5 m	100	H_2-O_2-N-CH_4-CO
Molecular sieve 5 A	−78	He-Ne-N_2
Activated carbon, 0.35 m	40	H_2-CH_4-CO_2-C_2H_4-C_2H_6
Activated carbon, 2.25 m	20	He, Ne-Ar-Kr-Xe
Silica gel, 2 m	−70 to + 25	N_2-NO-CO-N_2O-CO_2

9.5 Countercurrent method

A countercurrent method, in which the adsorbent flows down a column, entraining a component of the gaseous mixture of substances supplied in the middle, has been described, but is probably less suitable for analytical chemistry.

General literature

Adsorption, General

E. Bottani & J. Tascón, Adsorption by Carbons, Elsevier, Amsterdam (2008).

S. Brunauer, The Adsorption of Gases and Vapors, Vol. 1 – Physical, Lightning Source Inc., La Vergne (2008).

J. Condon, Surface Area and Porosity Determinations by Physisorption, Elsevier, Amsterdam (2006).

J. Keller & R. Staudt, Gas Adsorption Equilibria: Experimental Methods and Adsorptive Isotherms, Springer, Heidelberg (2005).

F. Kerry, Industrial Gas Handbook: Gas Separation and Purification, CRC Press, Boca Raton (2007).

N. Quirke, Adsorption and Transport at the Nanoscale, CRC Press, Boca Raton (2005).

R. Roque-Malherbe, Adsorption and Diffusion in Nanoporous Materials, CRC Press, Boca Raton (2018).

Literature to the text

Adsorbents

E. Baltussen, C.A. Cramers, and P. Sandra, Sorptive sample preparation – a review, Analytical and Bioanalytical Chemistry *373*, 3–22 (2002).

J. Chia & E. Zellers, Multi-adsorbent preconcentration/focusing module for portable-GC/microsensor-array analysis of complex vapor mixtures, Analyst *127*, 1061–1068 (2002).

T. Cserhati, Carbon-based sorbents in chromatography. New achievements, Biomedical Chromatography *23*, 111–118 (2009).

S. Firooz & D. Armstrong, Metal-organic frameworks in separations: A review, Analytica Chimica Acta 1234, 340208 (2022).
G. Kuvshinov, V. Maizlish, S. Kuvshinova, V. Burmistrov & O. Koifman, Copper and nickel complexes of tert-Butyl substituted phthalocyanines as modifiers for films based on polyvinyl chloride and adsorbents for gas chromatography, Macroheterocycles *9*, 244–249 (2016).
Q. Li & D. Yuan, Evaluation of multi-walled carbon nanotubes as gas chromatographic column packing, Journal of Chromatography A, *1003*, 203–209 (2003).
J. de Zeeuw & J. Luong, Developments in stationary phase technology for gas chromatography, TrAC, Trends in Analytical Chemistry *21*, 594–607 (2002).
J. de Zeeuw, The development and applications of PLOT columns in gas-solid chromatography, LC-GC Europe *24*, 38–45 (2011).
V.I. Zheivot, Gas chromatography on carbon adsorbents: characterization, systematization, and practical applications to catalytic studies, Journal of Analytical Chemistry *61*, 832–852 (2006).

Gas adsorption chromatography

V. Berezkin & J. de Zeeuw, Capillary Gas Adsorption Chromatography, Wiley-VCV, Heidelberg (2002).
R. Sitko, B. Zawisza & E. Malicka, Graphene as a new sorbent in analytical chemistry, TrAC – Trends in Analytical Chemistry *51*, 33–43 (2013).

Metal-organic framework structures

R. Fischer & C. Wöll, Functionalized coordination space in metal-organic frameworks, Angewandte Chemie Int. Edition *47*, 8164–8168 (2008).
Z.-Y. Gu, C.-X. Yang, N. Chang & X.-P. Yan, Metal-organic frameworks for analytical chemistry: from sample collection to chromatographc separation, Accounts of Chemical Research *45*, 734–745 (2012).
L. Kreno, J. Hupp, & R. Van Duyne, Metal-organic framework thin film for enhanced localized SPR gas sensing, Analytical Chemistry *82*, 8042–8046 (2010).
J.-R. Li, R. Kuppler, & H.-C. Zhou, Selective gas adsorption in metal-organic frameworks, Chem. Soc. Rev. *38*, 1477–1504 (2009).
S. Ma, Gas adsorption applications of porous metal-organic frameworks, Pure and Applied Chemistry *81*, 2235–2251 (2009).

Molecular sieves

M. Aroon, A. Ismail, T. Matsuura & M. Montazer-Rahmati, Performance studies of mixed matrix membranes for gas separation: A review, Separation and Purification Technology *75*, 229–242 (2010).
P. Goh, A. Ismail, S. Sanip, B. Ng & M. Aziz, M., Recent advances of inorganic fillers in mixed matrix membrane for gas separation, Separation and Purification Technology, 243–264 (2011).
C. Lin, K. Dambrowitz, and S. Kuznicki, Evolving applications of zeolite molecular sieves, Canadian Journal of Chemical Engineering *90*, 207–216 (2012).
L. McCusker, D. Olson & C. Baerlocher, Atlas of Zeolite Framework Types, Elsevier, Amsterdam (2007).
T. Wong, Handbook of Zeolites: Structure, Properties and Applications, Nova Science Publishers, Hauppauge (2009).

Reverse gas chromatography

L. Farmakis, A. Koliadima, G. Karaiskakis & J. Kapolos, Reversed-flow gas chromatography as a tool for studying the interaction between aroma compounds and starch. J Agric Food Chem. 66, 12111–12121 (2018).

J. Kapolos, Environmental applications of reversed-flow GC, Encyclopedia of Chromatography *1*, 776–782 (2010).

A. Koliadima, Reversed-flow GC, Encyclopedia of Chromatography (3rd Edition) 3, 2037–2043 (2010).

V. Warren, Gas Chromatography: Analysis, Methods and Practices. Nova Science Publishers Inc., Hauppauge (2017).

Separation plate heights

F. Gritti & G. Guiochon, Mass transfer kinetics, band broadening and column efficiency, Journal of Chromatography A, *1221*, 2–40 (2012).

F. Gritti, I. Leonardis, J. Abia & G. Guiochon, Physical properties and structure of fine core-shell particles used as packing materials for chromatography. Relationships between particle characteristics and column performance, Journal of Chromatography A, *1217*, 3819–3843.

A.V. Kozin, A.A. Korolev, V.E. Shiryaeva, T.P. Popova, and A.A. Kurganov, The influence of the natures of the carrier gas and the stationary phase on the separating properties of monolithic capillary columns in gas adsorption chromatography, Russian Journal of Physical Chemistry A *82*, 276–281 (2008).

C. Poole, Gas Chromatography, Elsevier, Amsterdam (2021).

10 Adsorption of dissolved substances on solids

10.1 General

10.1.1 Historical development

The oldest of the separation processes based on adsorption effects is probably paper chromatography, which was already developed by Goppelsroeder (1861) on the basis of a suggestion by Schönbein (1861). The name *chromatography* derives from the appearance of colored zones when separating colored substances, the first field of application; of course, colorless substances can also be separated.

The method came into wide use after improvements by Brown (1939) and Consden et al. (1944). An extension and generalization of the method, thin-layer chromatography, can be traced back to Ismailov and Schraiber (1938), but only gained greater importance after work by Stahl et al. (1956).

The technique of adsorption chromatography in columns was essentially worked out by Tswett (1903), whose main merit is the introduction of the developing technique. However, the technique fell into oblivion and was rediscovered only by Kuhn et al. (1931).

The development to chemically selective sorbents has been significantly advanced by the use of biological receptors (Lerman, 1953), on the one hand, and by the synthesis of molecularly imprinted polymer structures (Mosbach, Shea & Wulff, 1980), on the other.

10.1.2 Auxiliary phases – distribution isotherms – speed of equilibration – irreversible adsorption

Auxiliary phases. In the processes that use adsorption from solutions for separation, one works with two auxiliary phases, a liquid solvent and the solid adsorbent.

As in the case of adsorption of gases on solids, the *adsorption isotherms* are usually nonlinear; isotherm shapes according to Freundlich or Langmuir or BET isotherms are frequently observed (cf. Figs. 9.1 and 9.2), but very irregular shapes also occur (cf. Fig. 10.1).

By loading the adsorbents weakly, it is possible in many cases to work in the practically straight-line part of the isotherms, and by inactivating the adsorbent (e.g., by wetting it) it is often possible to extend the straight-line part toward higher concentrations.

In the graphical representation of adsorption isotherms, the concentration in the adsorbent is generally chosen as a quantitative ratio (e.g. mg of adsorbed compound per g of adsorbent) and the concentration in the solution is chosen in the dimension *g/v* (e.g. g of solute per 100 ml of solution).

https://doi.org/10.1515/9783111181417-010

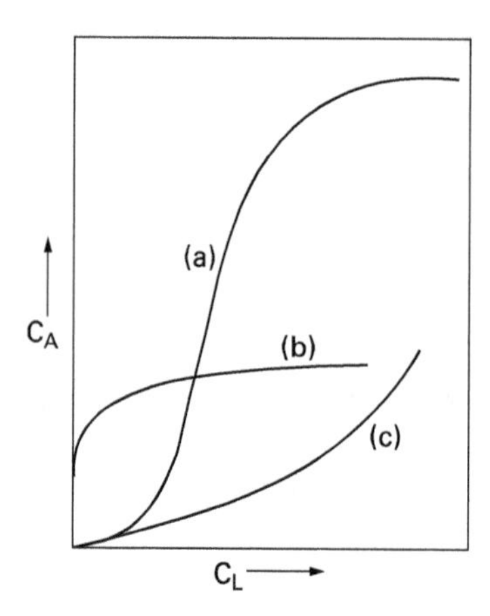

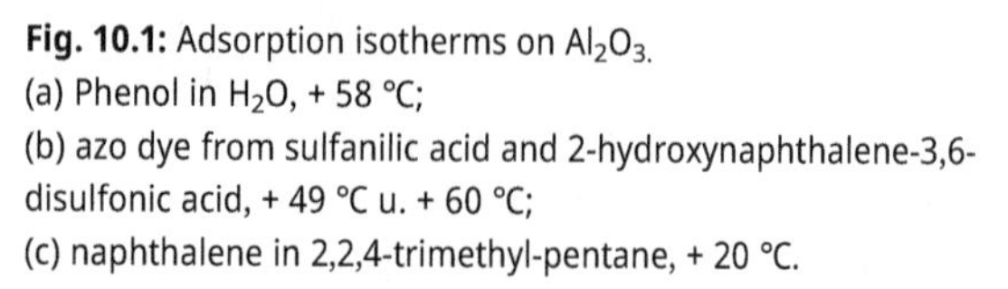
Fig. 10.1: Adsorption isotherms on Al_2O_3.
(a) Phenol in H_2O, + 58 °C;
(b) azo dye from sulfanilic acid and 2-hydroxynaphthalene-3,6-disulfonic acid, + 49 °C u. + 60 °C;
(c) naphthalene in 2,2,4-trimethyl-pentane, + 20 °C.

Rate of equilibrium setting. Adsorption equilibria are usually established within a few seconds, unless delays occur due to larger diffusion distances in the solution. However, much longer times (minutes) are also reported. Furthermore, in some adsorption processes, the bulk is adsorbed rapidly, and only when the solution and adsorbent are in contact for a longer time is a small further portion of the solute taken up by the solid phase.

Irreversible adsorption. Often, a small amount of adsorbed substance cannot be readily desorbed again, and the adsorption is then practically irreversible for this fraction. For compounds present in larger quantities, this interference can usually be neglected, but when separating trace components, the losses incurred in this way become intolerably large. The effect is associated with a high adsorption enthalpy for the first adsorbed moieties (> 25 kJ/mol). One can reduce this disturbance by partially inactivating the adsorbent.

10.1.3 Adsorbents

If the compounds to be separated are present in aqueous solution, Al_2O_3 (for polar compounds) or activated carbon (for non-polar substances) are usually used as adsorbents.

If the substances to be analyzed are dissolved in organic liquids, separations can be performed with a whole number of oxygen-containing hydrophilic compounds, which can be arranged in the following series (*Zechmeister's series*) according to decreasing adsorption capacity:

$$Al_2O_3 - Al(OH)_3 - MgO - CaO - Ca(OH)_2 - CaCO_3 - CaSO_4 -$$
$$Ca_3(PO_4)_2 - \text{talcum} - \text{sugar} - \text{inulin}.$$

This order is not completely fixed, but depends on the nature of the adsorbed substances; therefore, slightly different arrangements can be found in the literature.

In addition to these adsorbents, various types of silica gel, cellulose and cellulose derivatives, and gel-like organic polymers are used extensively. Mention should also be made of hydroxyapatite, calcium phosphates, magnesium phosphate, magnesium silicate, glass and quartz powders and several organic polymers such as some polyamides, polyacrylamide and polyacrylonitrile. Molecular sieves can also be used for separations of dissolved compounds (in non-aqueous solvents).

Some of the adsorbents listed have already been discussed in the previous section; the properties of some others are described below.

Alumina is probably the most widely used adsorbent in analytical chemistry, as it has excellent adsorption capacity for almost all inorganic and organic compounds. It can be applied to both aqueous and non-aqueous solutions.

The most active product is obtained by heating precipitated aluminum hydroxide to about 350 °C; it is hygroscopic; water absorption reduces the *activity* (i.e. adsorption capacity). This behavior is exploited to obtain oxides of different activities by adding defined amounts of water (Tab. 10.1).

Tab. 10.1: Activity levels of alumina.

Water addition (%)	Activity level (a. Brockmann)
0	I
3,0	II
4,5	III
9,5	IV
15	V

The activity can be controlled with the aid of various organic dyes whose adsorption behavior is determined. The aluminum oxides available for sale usually still contain small amounts of alkali (Na_2CO_3) or calcium compounds, which are important in the separation of acids or bases. Grades standardized with respect to alkali content are also available, which give a specific pH setting with water (measurement in 10% slurry).

The specific surface area of alumina preparations is relatively low, reported at about 200–250 m^2/g. The particle size is generally about 70 µm.

The adsorption isotherms on Al_2O_3 are often straight, especially when working with somewhat water-containing preparations and at low loading; on the other hand, very irregular isotherms have often been found (cf. Fig. 10.1). The behavior of aluminum oxide during adsorption has been studied in great detail, and some rules about the retention of different classes of substances can be established.

Sensitive organic compounds can be altered when adsorbed on Al_2O_3 as a result of the catalytic efficiency of the surface.

Silica gel is of great importance not only for the separation of gases (cf. Chap. 9), but also of dissolved substances. Linear isotherms – at least in small concentration ranges – are often observed with this adsorbent as well. The pore widths are of little influence on the separations, since normally the molecular diameters are small enough so that the particles can penetrate unhindered into the interior of the gel. Only in separations on very narrow-pored gels do noticeable sieving effects occasionally occur. The activity falls with increasing water content.

Attempts have been made to some extent with success to produce silica gels with specific separation effects by carrying out the precipitation in the presence of the organic compound to be adsorbed, which is intended to impart a certain structure to the inner surface. However, such gels have not yet attained greater importance.

Magnesium silicate. By combining magnesium sulfate and sodium silicate solutions, washing out and strongly heating the precipitates, products are obtained which have proved useful as adsorbents for the separation of numerous organic compounds. The activity can be influenced by the addition of water (about 7–35%); activation usually occurs at about 650°, resulting in a virtually anhydrous product. When activated at lower temperatures (e.g. 110–250°), small amounts of water remain in the material.

The specific surface area is about 200 m^2/g; it depends strongly on the residual Na_2SO_4 content resulting from the production of the magnesium silicate. The commercially available products contain about 0.2–0.5% Na_2SO_4.

Magnesium oxide is to some extent comparable to Al_2O_3 in adsorption properties, but has not attained its importance, since interferences can occur due to the easy water absorption and release. It is best to use preparations with a specific surface area of about 90 m^2/g and a water content of 3–7%.

Porous glasses; porous quartz spheres. Leaching of soluble fractions from biphasically solidified borosilicate glasses yields porous structures whose pore diameters can be influenced by the manufacturing conditions and the concentration of the soluble fractions. The pores of these materials are of different sizes, but the distribution can be kept within relatively narrow limits. The separating capacity is therefore significantly poorer than that of molecular sieves, but the acid resistance is much better.

Porous quartz beads have also been made and used for separations; they are obtained by annealing silica gels.

Cellulose is widely used as powder or paper strips for separations. The effectiveness is not based on adsorption alone, but – at least in some applications – additionally on a distribution of the substances to be separated between the organic solvent and the water skin on the surface of the cellulose particles. Furthermore, ion exchange reactions with carboxyl groups of the adsorbent can occur. However, the various effects can hardly be distinguished from each other experimentally, and cellulose will therefore be dealt with in this section under *Adsorbents*.

In addition to ordinary cellulose, chemically modified products, e.g. acetyl cellulose, are also used; some of these are hydrophobic and suitable for separating water-insoluble compounds (see also section 11, *Ion exchange resins*).

High-polymer gels. The term *gels* is used to describe high-polymer organic substances that can absorb solvents under swelling and then form a porous three-dimensional network. Dissolved molecules can penetrate from the solution into the interior of the gels, and in addition to adsorption, a sieving effect occurs; this excludes molecules with diameters larger than the pore width. The sieving effect of such compounds was first observed in separation experiments with swollen starch, and later more effective synthetic compounds were prepared.

There is a relationship between these gels and the porous polymers mentioned above in that the swelling capacity of the gels can be reduced almost at will by increasing additions of crosslinking compounds. If acidic or basic functional groups are incorporated into the polymers, the so-called *ion exchangers* result (cf. section 11).

A distinction is made between hydrophilic and hydrophobic gels.

The important group of hydrophilic *dextran gels* is produced from the bacterial metabolite dextran, a polysaccharide, by crosslinking with epichlorohydrin

O CH Cl
CH_2 CH_2

produced (Fig. 10.2).

Several types are available under the name *Sephadex®*, which differ in the degree of crosslinking and thus in the average pore sizes. They are used for fractionation in different molecular weight ranges (Tab. 10.2).

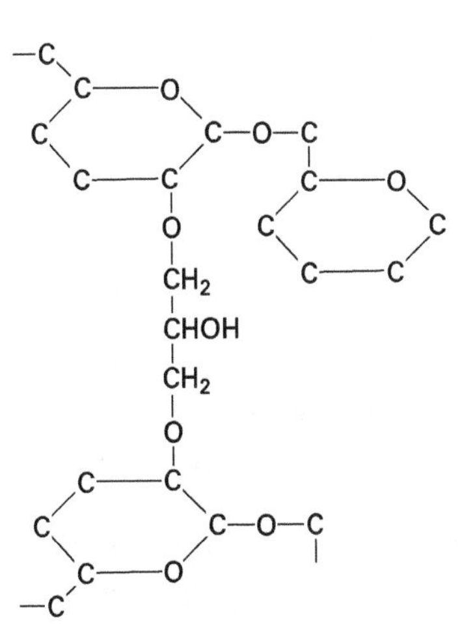

Fig. 10.2: Structure of dextran gels (schematic).

Of other hydrophilic gels, crosslinked agarose and crosslinked acrylamide should be mentioned.

Cross-linked polyethylene oxides can be used for separations in both aqueous and organic solutions, they transfer to the organophilic gels. These are mainly produced from styrene, vinyl acetate or methyl methacrylate.

Polyamides represent a somewhat different class of adsorbents; they are obtained either by polycondensation of ω-aminocarboxylic acids or from aliphatic diamines and dicarboxylic acids. Thus, caprolactam gives rise to

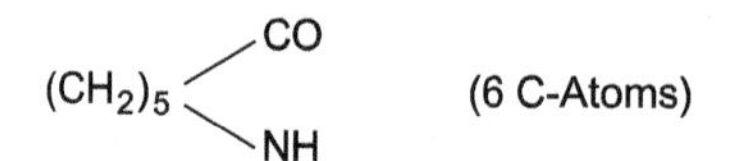

Tab. 10.2: Sephadex® types.

Type	Approximate fractionation range (compounds of higher molecular weight are excluded)
G - 10	Up to 700
G - 15	Till 1500
G - 25	100–5 000
G - 50	500–30 000
G - 75	3000–70 000
G - 100	4000–150 000
G - 150	5000–400 000
G - 200	5000–800 000

the so-called 6-polyamide (*Perlon*®)

$$-\mathrm{CO}-\mathrm{NH}-(\mathrm{CH_2})_5-\mathrm{CO}-\mathrm{NH}-(\mathrm{CH_2})_5-\mathrm{CO}-\mathrm{NH}-$$

and from hexamethylenediamine (6 C atoms) and adipic acid (6 C atoms) the 6,6-polyamide (*Nylon*®)

$$-\mathrm{CO}-\mathrm{NH}-(\mathrm{CH_2})_6-\mathrm{NH}-\mathrm{CO}-(\mathrm{CH_2})_4-\mathrm{CO}-\mathrm{NH}-\,.$$

The polyamides can adsorb various organic compounds in solid solution, therefore they have a much higher capacity than the other adsorbents. However, the capacity strongly depends on the conditions during the production of these substances; it is greatest when the individual chains are stored irregularly and smallest in stretched material with regular storage of the molecules.

The sorption isotherm of phenol on Perlon® is linear up to medium concentrations and then increases until the polyamide softens. Langmuir isotherms were found in other systems.

Adsorbents for affinity chromatography. In the so-called *affinity chromatography* indicated by Lerman (1953), the high selectivity of enzyme reactions is exploited to improve chromatographic separations. As is well known, the effect of enzymes consists in the fact that a certain site of the molecule combines with the *substrate* (i.e. the reaction partner). Furthermore, there are so-called *inhibitors* which block the normal enzyme reaction by being more tightly bound by the enzyme than the substrate.

If either a substrate or an inhibitor is bound to a solid, insoluble framework substance in such a way that the reaction with the enzyme can still take place, this material represents a selective or even specific adsorbent for the enzyme in question. Such adsorbents can be prepared, for example, by ether-like binding of substrates and of inhibitors to cellulose or other polysaccharides. Conversely, enzymes can also be anchored to scaffold substances and inhibitors selectively isolated with the adsorbent obtained.

The application of the also very selective antigen-antibody reaction (cf. Chap. 12) for the isolation of antibodies should also be mentioned here. For example, albumin is bound as antigen to cellulose (Lerman) or ion exchangers containing -COOH groups (cf. chap. 11) are esterified with antigens (Isliker, 1954). The resulting material then selectively retains the associated antibodies (the second example is not an ion exchange reaction).

These reactions between adsorbent and solute can be reversed, i.e. the enzymes, inhibitors or antibodies can be detached again by changing the experimental conditions. Especially the use of highly selective antibodies nowadays allows the enrichment of trace substances from complex liquid matrices (e.g. urine, wastewater, milk etc.). Recent developments use porous sol-gel glass as a confining medium for freely mobile antibodies for selective pre-purification and enrichment.

Molecularly imprinted polymers. These *molecularly imprinted polymers (MIPs)* represent an attempt to introduce biomimetic receptors into separation technology. This technique, introduced by Wulff (1980) and Mosbach (1981), uses a molecular imprinting process in which, according to the lock-and-key principle, the analyte to be separated serves as a template for a polymer synthesis. After polymerization, the imprinting molecule is washed out. The resulting cavity serves as the receiving cavity in the separation experiment.

10.1.4 Solvents in adsorption from solutions

In the adsorption processes, the solvent plays a decisive role, since its molecules are also adsorbed and thus very strongly influence the partition coefficients of the dissolved compounds. For hydrophilic adsorbents, e.g. Al_2O_3, the solvents can be arranged in a series according to their effectiveness in eluting (*eluotropic series* according to Trappe, 1940):

Petrolether – CCl_4 – Trichlorethylene – Benzene – CH_2Cl_2 – $CHCl_3$ –

Diethylether – Ethylacetate – Acetone – n – Propanol – Ethanol –

Methanol – Water – Pyridine.

The solvents at the beginning of this row are adsorbed most weakly; compounds dissolved in them can therefore be held particularly well by the adsorbent. The liquids at the end of the series occupy even the active centers of the adsorbent very firmly; dissolved substances can therefore hardly be adsorbed. On the other hand, substances that have already been adsorbed can be displaced from the adsorbent particularly well with these solvents. By mixing several solvents, the desorbing properties can be tuned almost at will.

The above series changes slightly when moving to other polar adsorbents, strong changes occur when using non-polar adsorbents, e.g. activated carbon.

When using such series, however, one must be aware that specific interaction forces between solute and adsorbent or solvent are not taken into account, so that irregularities often occur. For the same reason, attempts to measure the eluting effect of solvents by ratios have only been partially successful.

10.1.5 Distribution coefficients and separation factors

In adsorption, the partition coefficients generally depend on the concentration of the solute because of the mostly odd isotherms, furthermore they are influenced by fellow solvents; this influence is especially noticeable at high concentrations, it can practically disappear at low concentrations.

The distribution coefficients are therefore of less importance in adsorption processes than, for example, in distribution between two liquid phases.

Advantageously, high partition coefficients can be reduced to almost any extent by partially inactivating the adsorbent and improving the solvent's solubilizing effect.

In the enrichment of trace constituents from highly dilute solutions, the fact that the distribution is often strongly in favor of the solid phase, especially at small concentrations of the solute, is important. For example, partition coefficients $> 10^3$ and separation factors for the separation of Mo and Re or of W and Re of about 10^4 have been observed in the adsorption of tungsten (VI) and molybdenum (VI) on Al_2O_3.

The situation is different when separations are carried out on gels with the aid of sieve effects (size exclusion). In the simplest case, adsorption is negligible; if the partition coefficient is defined as the ratio of the concentration in the liquid phase inside the gel to the concentration in the external solution, its value is either 0 (complete exclusion of the solute) or 1 (unhindered diffusion into the gel). In reality, however, one also observes values between 0 and 1, which are interpreted as partial exclusion, as well as distribution coefficients that are considerably higher than 1; in these, adsorption effects apparently play a significant role.

10.2 Separations by one-time equilibrium adjustment

For adsorption separations by one-time equilibration, different working methods have to be distinguished: Either the adsorbent is introduced into the solution of the substances to be separated (either as granules or as a porous, sorbent surface on a carrier) and separated after the adsorption process has been carried out (*batch process*), or the adsorbent is produced in the solution itself by a precipitation reaction, in which case the substances to be adsorbed are entrained (*precipitation from homogeneous solution*). However, this so-called *entrainment effect* will be discussed later in the treatment of precipitation methods (see chap. 12), since other phenomena are of importance in addition to adsorption.

The stirring of adsorbents into solutions is used relatively rarely in analytical chemistry – in contrast to preparative and technical chemistry. This process is mainly used to remove natural substances from highly dilute aqueous solutions (e.g. reducing components can be removed from sugar solutions with activated carbon); it plays a considerable role in the enrichment of enzymes. Also worth mentioning here are processes for concentrating aqueous protein solutions; dry Sephadex® is added, which absorbs water and other low-molecular compounds, while the high-molecular protein bodies remain excluded.

An example from inorganic analysis is the separation of phosphate; this is adsorbed as barium phosphate on a precipitate stirred into the solution.

Solid phase microextraction. *Solid phase microextraction (SPME)* was introduced 1990 by Pawliszyn to enable rapid enrichments from aqueous solutions without large organic solvent volumes, both in the laboratory and in the field. In this process, the adsorbing phase is usually on a fiber surface, which is then placed in the solution to be measured for equilibration. The process consists of two steps: Adsorption and desorption.

Often, it is a combination of low-volatile and/or solid phase that is used for enrichment. In the meantime, coatings of ionic liquids are already being used. Especially the combination of enrichment in an injection syringe needle and GC is widely used. The micro method is particularly suitable for the determination of volatile analytes (cf. Tab. 10.3).

Tab. 10.3: Solid phase microextraction with fibers for preconcentration.

SPME fiber type	Analytes	Determination procedure
100 µm PDMS	Volatile org. molecules (60–275 Da) VOC, BTEX, PAH	GC/HPLC
65 µm PDMS/DVB	Volatile amines, nitroaromatics	GC
80 µm Carboxes/PDMS	Chlorobenzenes, metal comp.	GC-ICP
85 µm polyacrylate	Pesticides, herbicides Phenols	GC/HPLC

10.3 Separations by one-sided repetition

Separations by one-sided repetition of the adsorption step can be performed by stirring several portions of the adsorbent into the solution in question. More common than this somewhat cumbersome method is to filter the solution through a more or less thick layer of the adsorbent, repeatedly bringing the analytical sample into contact with fresh layers of the solid phase. This is usually done with short column arrays; in a similar procedure, the analytical solution is filtered through a thin layer of adsorbent placed in a filter crucible or a groove (*adsorptive filtration* a. Fink, 1939). The adsorbed compounds are then recovered by suitable solvents.

The technique of one-sided repetition of adsorption steps is mainly used for enrichment of trace components of inorganic or organic nature from dilute solutions, e.g., adsorption of Fe^{3+} and Al^{3+} on silica gel, of Na^+ on $Sb_2O_5 \cdot aq$ or selective separation of HSO_4^- with Al_2O_3. Furthermore, the purification of organic solvents should be mentioned; for example, H_2O traces can be removed with molecular sieves, and ethanol additions can be separated from $CHCl_3$ with Al_2O_3.

Finally, it is worth mentioning the separation of n-paraffins from mixtures of hydrocarbons with molecular sieve 5 A, working also in ranges of higher concentrations of the compounds to be separated.

Immunoaffinity chromatography. *Immunoaffinity chromatography* columns with immobilized antibodies for the pre-separation of traces from difficult matrices, such as liquid clinical samples or foodstuffs, have become somewhat popular. Sample volumes of about 100 ml are aspirated through small columns filled with a few mg of receptor material. In many cases, antibodies or molecularly imprinted polymers (as antibody replacement) are used as receptors. Subsequently, a few milliliters are used for washing and, after changing the eluent, desorption is initiated. This achieves gentle enrichment of the trace and removal of an interfering matrix, such as fat. An advantage is the multiple (n < 100) repeatability of this separation step, resulting in acceptable costs overall despite the high production costs for antibody production.

Monolithic columns. While solid-phase preconcentration has also conquered commercially wide application areas of liquid-liquid extraction in recent years, one disadvantage remained. Cartridges had to be filled and the sorption sites remained on the filled spherical material. A reduction of the sphere diameters leads to an increasing pressure drop during the aspiration of the sample.

An elegant solution is the production of so-called *monolithic columns*, consisting of a piece of porous polymer material with gusset-free channels and high sorption site density. The first monoliths were produced from acrylamide, silica, styrene or methacrylate. Typically, the starting mixture consists of several monomers, porogens, cross linker substances and initiator to start the polymerization. The mixture can be polymerized directly in a suitable tubular empty cartridge made of steel, glass or plastic. In the case of SiO_2 monoliths, C_{18} phases or ion exchange functions can be incorporated on the inner pore surfaces. Table 10.4 shows some examples of monoliths. Alcohols or acetonitrile are used as porogens.

Tab. 10.4: Monolithic separation columns.

Outer casing	Polymer	Specific surface area [m^2/g]
PEEK tube	Divinylbenzene, ethylstyrene	329
	2-Hydroxyl methacrylate	367
Stainless steel tube (10 mm × 4.6 mm i. d.)	Glycidyl methacrylate, ethylene dimethacrylate	44.3
Stirring rod (31 mm × 6 mm)	2-acrylamido-2-methyl-1-1-propanesulfonic acid	60
Capillary (20 mm × 75 µm i. d.)	Glycidyl methacrylate	4,8
Quartz rod (40 mm long)	Sol gel derived from tetramethylsilane & polyethylene glycol	317,5
Quartz rod in cartridge (20 mm long)	Sol gel derived from tetramethylsilane & polyethylene glycol	745

Applications for monoliths are manifold. Especially the enrichment of biomaterials (peptides, proteins, bacteria, viruses) is successful. Monoliths with built-in receptor functions are extremely interesting. For example, antibiotic molecules directed against bacterial cell walls have been successfully applied to enrich *E. coli* from surface water in monoliths. All the typical advantages of repeated regenerability as in solid phase extraction are retained. Disadvantages, such as swelling of the polymer phase, temperature sensitivity and microbial contamination are not applicable to SiO_2-based monoliths.

Some successful applications of monoliths, partly in automated form, are summarized in Tab. 10.5.

Tab. 10.5: Applications of monolithic separation columns.

Monolith material	Analytes	Matrix	Detection limits
Poly(MAA-EMDA)	Antidepressants	Urine	11–50 μg l^{-1}
	Opiates	Urine	7–20 ng l^{-1}
Poly(GMA-EDMA)	Antibody (IgG)	Blood serum	nmol l^{-1}
C_{18}-SiO_2	Anesthetics	Urine	7–37 μg l^{-1}
SiO_2	Methionine enkephaline	Cerebrospinal fluid	1 μg l^{-1}

10.4 Separations by systematic repetition: Separation series

For repeated adsorption in a separation series, one portion of the adsorbent is added to each of a number of vessels, the solution containing the substances to be separated is poured into the first vessel and its contents are stirred until equilibrium is reached. After the solid phase has settled, the liquid is decanted and brought into contact with the second portion of adsorbent in the next vessel. The liquid is allowed to pass through the entire series of vessels in this manner and the process is repeated with fresh additions of solvent until the desired separation is achieved.

In this way, a mixture of anthracene with chrysene in cyclohexane solution at Al_2O_3 was largely separated (Fig. 10.3).

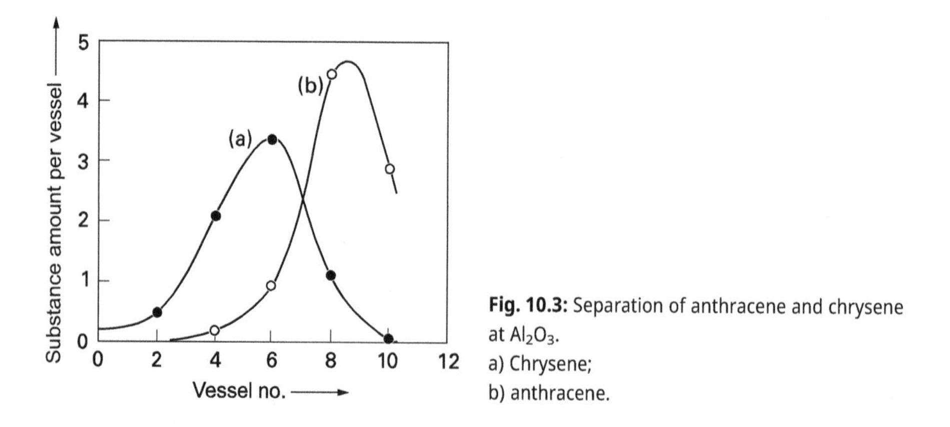

Fig. 10.3: Separation of anthracene and chrysene at Al_2O_3.
a) Chrysene;
b) anthracene.

The advantage of the method is that larger approaches can be put to work; however, it is cumbersome, especially when a larger number of separation stages are required, and it has therefore not gained importance.

10.5 Separations by systematic repetition: Column methods (adsorption chromatography)

10.5.1 Principle

As with the column methods discussed previously, a tube is filled with the stationary phase (the adsorbent), the analysis sample is applied to the head of the column in the form of a solution, and then conveyed through the column as the moving phase using a suitable solvent. Either the chromatogram is developed, with the substances remaining as separate zones in the column, or the components of the analytical sample are eluted one after the other and collected with a fraction collector or detected individually by a detector.

10.5.2 Column material and dimensions – filling the columns

The columns used in adsorption chromatography are usually made of glass. Very long columns, where the moving phase must be fed under pressure, are made of steel tubes.

The diameter of the usual columns is about 0.5–2 cm with lengths of 30–200 cm. These can be used to separate substance quantities in the order of a few milligrams. For weights of several hundred milligrams, column diameters of about 5–7 cm are required, and the length is then suitably about 100–250 cm. If cellulose is used as the stationary phase, packs of the usual round filters can be stacked on top of each other and columns of about 2.5–6 cm ø can be produced in this way. For the separation of very small amounts of substance, micro-columns with 0.1 cm ø have been used.

The column diameter has no significant influence on the separation level of a column as long as it is not considerably larger than about 1 cm. For columns with diameters greater than 2 cm, uniform filling with stationary phase is difficult to achieve, and the efficiency tends to decrease significantly.

High-performance columns with lengths of 10 m and more are conveniently divided into individual sections of about 1 m in length, which are connected to each other by capillaries. The shorter pieces can be filled better than a single long column.

Some column designs are shown in Fig. 10.4.

The column filling rests on a glass frit or on a glass wool pad at the bottom of the column. The adsorbent must be filled with great care, since the quality of the column depends essentially on the uniformity of the separating bed. The stationary phase is either added to the column in dry portions while being continuously shaken and rotated, or it is slowly and uniformly slurried with an inert liquid so that the formation of channels in the separation bed is avoided as far as possible.

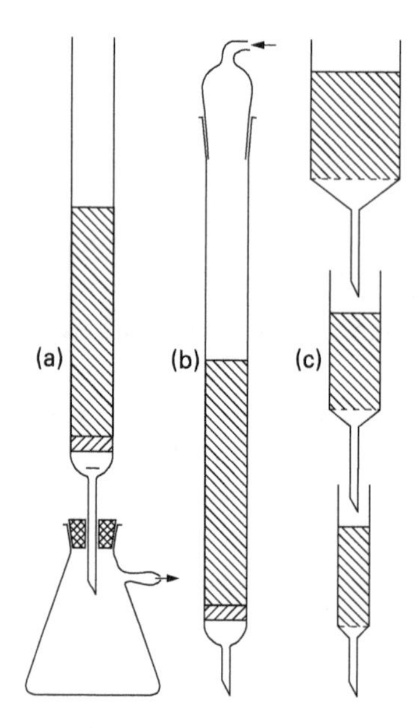

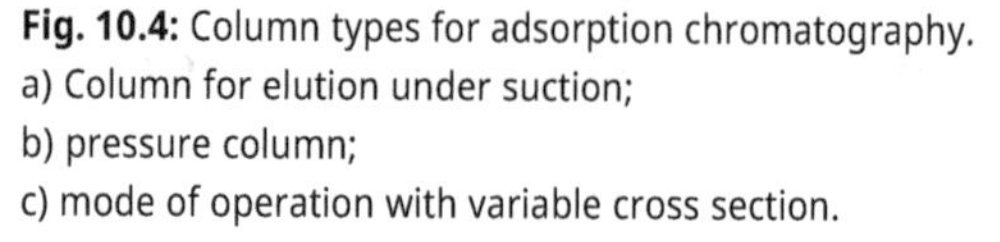
Fig. 10.4: Column types for adsorption chromatography.
a) Column for elution under suction;
b) pressure column;
c) mode of operation with variable cross section.

10.5.3 Stationary and mobile phase

When selecting the stationary phase, not only the adsorbing effect but also the particle size must be taken into account. In general, it is also true for adsorption methods that the separation plate height becomes smaller as the particle size of the stationary phase decreases. On the other hand, the flow resistance increases, so that elution can be greatly delayed. This disadvantage can occasionally be eliminated by selecting a liquid with lower viscosity.

Deactivating adsorbents, e.g. by adding water, generally reduces the retention volumes; this effect affects different types of molecules in approximately the same way, so that the order of elution of a given number of compounds is generally maintained.

10.5.4 Substance task and loading of the columns

In adsorption chromatography, the substances to be analyzed are applied in liquid form. To ensure that the separation effect is not impaired, the exit zone at the head of the column must remain as narrow as possible.

For this purpose, the column must not be overloaded; the sample quantity must be based on the diameter of the stationary phase and the capacity of the adsorbent. The influence of the loading on the separation level is shown in Fig. 10.5.

Furthermore, the substance should be applied carefully and in the middle of the surface of the separation bed so that the top layer of the adsorbent is not stirred up.

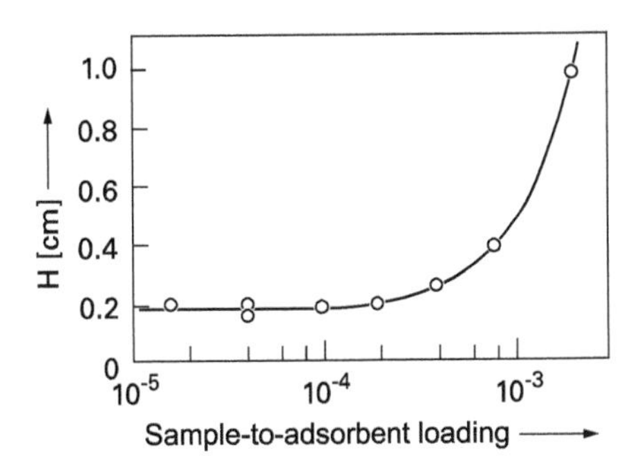

Fig. 10.5: Dependence of the separation plate height on the loading of a column.

To prevent this, the stationary phase is usually covered with some glass wool, a filter paper disc or a layer of glass beads.

10.5.5 Influence of the flow velocity of the mobile phase

When considering the influence of the flow velocity of the eluent on the separation plate height, it is expedient to start from the *van Deemter equation* (cf. Chap. 8. Eq. 2):

$$H = \mathrm{A} + \frac{B}{u} + \mathrm{C} \cdot u \qquad (1)$$

However, the second member of this equation can usually be neglected here; it reflects the contribution of diffusion to the band broadening, but this is small because of the low diffusion velocity in liquids. (Only at extremely low flow velocities, which are of no practical significance, would diffusion become noticeable.) Therefore, a simplified equation (2) applies, which gives a linear dependence of the separation plate height H on the flow velocity u:

$$H = \mathrm{A} + \mathrm{C} \cdot u \qquad (2)$$

$$(\text{A and C} = \text{constants}).$$

This equation was also confirmed experimentally (Fig. 10.6).

A somewhat different behavior was observed for separations with dextran gels: In wide velocity ranges, the separation plate height was independent of the flow velocity, so that it was possible to elute relatively quickly.

10.5.6 Influence of the temperature

The influence of temperature changes in separations by adsorption is complex: With an increase in temperature, the equilibrium setting is accelerated, so that an improvement

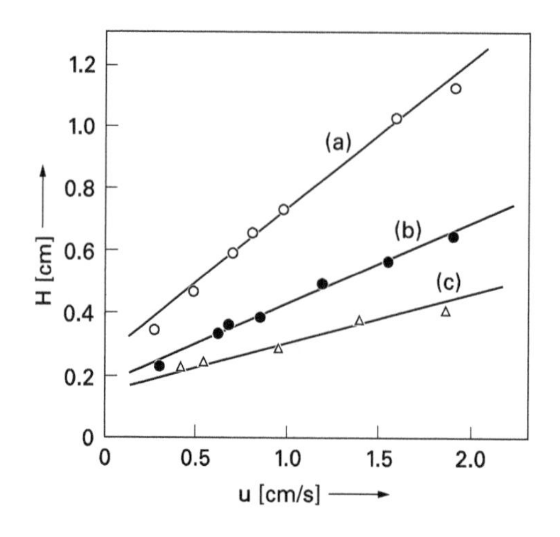

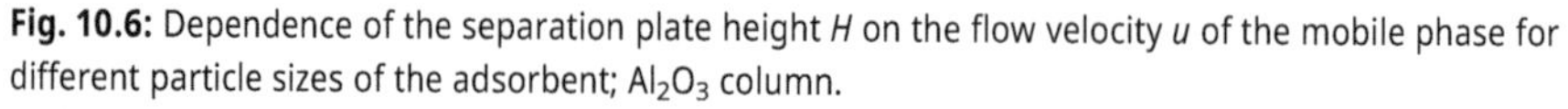

Fig. 10.6: Dependence of the separation plate height *H* on the flow velocity *u* of the mobile phase for different particle sizes of the adsorbent; Al_2O_3 column.
a) 181 µm grain diameter;
b) 128 µm;
c) 57 µm.

in the efficiency of the column can be expected. At the same time, however, the viscosity of the eluent decreases, and in the usual mode of operation, in which the liquid is driven through the separation bed by gravity, an increase in the flow velocity results. This may compensate or even overcompensate for the favorable effect of the temperature increase. But under favorable conditions, the separation efficiency of a column can be considerably improved by increasing the temperature.

A second temperature effect is the change in partition coefficients acting in the same sense for all adsorbates. Although this does not result in any significant improvements in separations, it is possible to achieve favorable partition coefficients (and retention times) for weakly adsorbable substances at low,[1] and for overly adsorbable compounds at high temperatures.

10.5.7 Gradient elution

If the analysis sample contains components that are adsorbed very differently, the elution of the components that migrate most slowly can be accelerated by changing the eluent. More common than the stepwise change of the properties of the moving phase is the continuous admixture of a second, better eluting liquid, i.e. gradient elution. The best results are usually obtained with linear concentration gradients, but occasionally a

1 E.g., the separation of hydrocarbons at −78 °C.

weakly convex gradient or a ternary mixture of liquids may be somewhat more favorable. The components of the mixture should not differ too much in eluting effect, otherwise interfering displacement effects may occur with degraded resolution of the bands. pH gradients are also used for the separation of acidic or basic compounds.

Apart from concentration gradients, gradients of flow velocity can be achieved by stepwise (cf. Fig. 10.4c) or continuous change of the column cross-section, but these, like temperature gradients, have gained little importance for adsorption chromatography.

10.5.8 Closed loop recirculation procedure

The amount of eluent required for separation in a column can be reduced by recirculation. The experimental arrangement consists of the column, a pump, the detector and a valve by which separated compounds can be removed from the circuit (Fig. 10.7).

If, in addition to the eluent, incompletely separated components also pass through the column several times, this corresponds to the use of a correspondingly extended separation bed, whereby, in addition to the material savings, the advantage of more uniform filling of the short column also comes into play (cf. Fig. 10.8). With this method of working, continuous monitoring of the eluate must prevent slow-moving substances from being overtaken by faster ones.

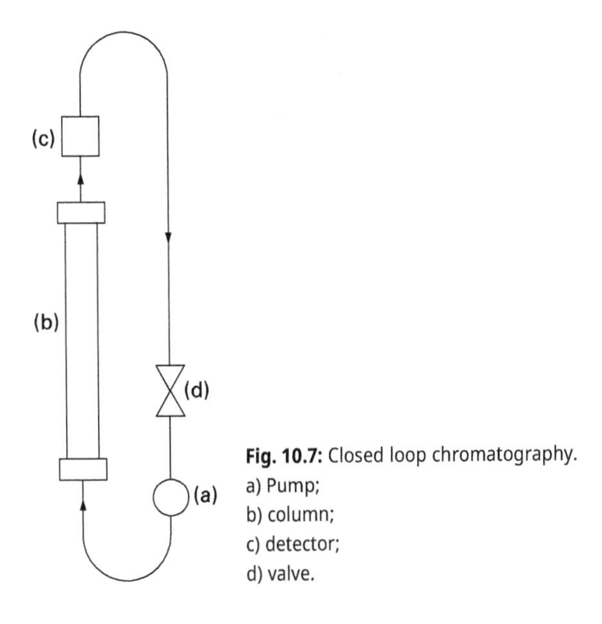

Fig. 10.7: Closed loop chromatography.
a) Pump;
b) column;
c) detector;
d) valve.

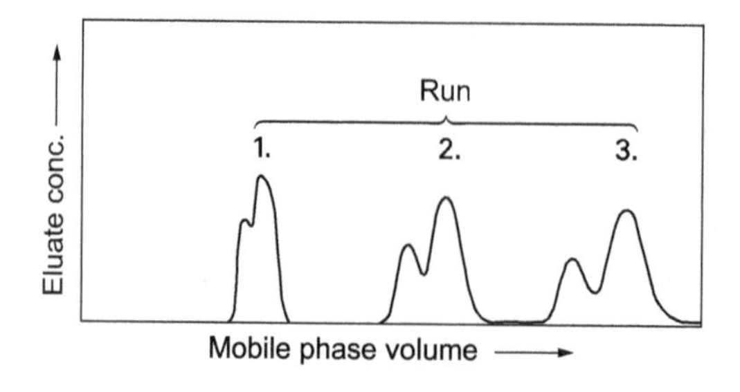

Fig. 10.8: Improvement of the separation of two compounds after multiple passes through the column.

10.5.9 Displacement technique – interposition of auxiliary substances

The normally undesirable mutual influence of the partition coefficients of several substances, which occurs above all at high concentrations, is exploited in the so-called *displacement method.* In this method, a compound is added to the eluent that is adsorbed more strongly than any component of the mixture to be separated. During elution, this compound pushes the other compounds ahead of it, and after equilibration in a sufficiently long column, each substance has a constant migration rate and a constant concentration on the adsorbent. The individual zones follow each other closely with no spaces in between, and the length of a zone is a measure of the initial concentration. Although no noticeable overlapping of the zones occurs, exact separation is difficult, on the other hand, no tail formation occurs and very long columns can be used. A variation of this method consists in the choice of an auxiliary substance which does not push all the compounds of the analysis sample in front of it, but lies in the adsorption behavior between two compounds which are pushed apart and thus better separated. This avoids the disadvantage of the direct succession of zones, but the added compound must be removed again. The process has been used, for example, to separate amino acids with the addition of various alcohols and palmitic and stearic acids with the addition of methyl palmitate.

10.5.10 Detection

To identify the zones of colorless substances on the separation column, fluorescent components can be made visible with UV radiation after the chromatogram has been developed (the column must be made of quartz). If the adsorbent is removed from the column, the zones can be stained by coating with color reagents; finally, the substances of the analysis sample can also be converted to colored compounds before separation.

In the more common method of working with elution, conductivity detectors, pH sensors or – for the separation of inorganic substances – coupling methods with element spectroscopy or ICP mass spectrometry are often used for aqueous solutions.

For organic liquids, the continuous measurement of light absorption (also in the UV range), the continuous measurement of light refraction with sensitive differential refractometers and the measurement of the radioactivity of labeled substances are particularly suitable after elution. Coupling with mass spectrometers is also very common. Fluorescent substances can be detected in the ultra-trace range when UV laser light is applied.

10.5.11 Efficacy and scope of the method

For Al_2O_3 – and silica gel columns, separation plate heights of about 2–4 mm are reported. Efficiency increases with decreasing particle size of the adsorbent; one can expect the separation plate height to be about twenty times the particle diameter. Particularly carefully packed high-performance columns of 4 mm diameter yielded heights of about 1 mm, and the optimum particle size of the silica gel was about 50 µm.

In gel chromatography, similar values are generally obtained (separation plate height approx. 2–5 mm, rarely 1 mm and less).

From these figures it can be seen that the usual columns of about 30–100 cm in length have only 150–500 separation plates in the order of magnitude, even with careful manufacture, and are therefore by no means particularly effective. If several thousand separation plates are required for a separation problem, one has to work with columns of about 10 m length and elute under high pressures with eluents of low viscosity.

The range of applications of adsorption chromatography is extraordinarily wide, and the method has probably been applied to all classes of substances. Some examples of applications are given in Tab. 10.6.

Tab. 10.6: Separations by adsorption chromatography (application examples).

Adsorbent	Separated analytes
Al_2O_3	Alkaloids, terpenes, carotenoids, steroids, polycyclic aromatics, synth. dyestuffs, inorg. compounds
Silica gel	Hydrocarbons, chlorinated hydrocarbons, org. sulfur compounds, alkyl phenols, polycyclic aromatics, hormones, metal complexes
Magnesium silicate	Steroids, hormones, vitamins
$CaCO_3$	Carotenoids
MgO	Aromatics; olefins, drugs
Polyamide	Carboxylic acids, phenols, sulfonic acids

Gel chromatography has brought significant progress in the separation of water-soluble organic compounds. Above all, procedures for the investigation of biological materials should be mentioned here; as a further example, the separation of polyglycols, which can hardly be achieved by other methods, should be mentioned (Fig. 10.9).

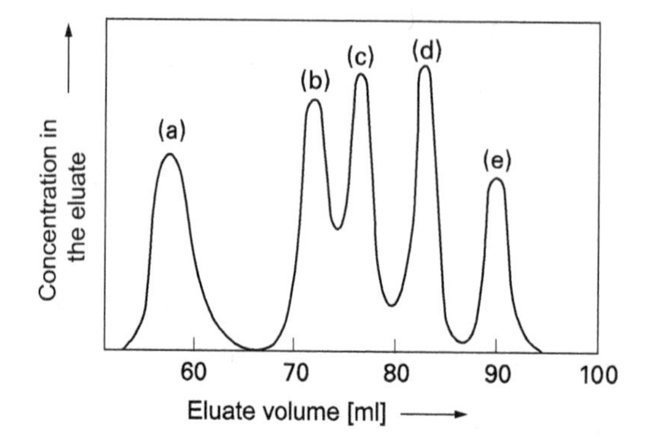

Fig. 10.9: Separation of polyglycols on a dextran gel column.
a) Polyethylene glycol, mol. wt. 600;
b) tetraethylene glycol;
c) triethylene glycol;
d) diethylene glycol;
e) ethylene glycol.

With gels of suitable pore sizes, even very high molecular weight substances can be separated; a simple relationship between the molecular weight M and the retention volume V is obtained:

$$V = \mathrm{A} - \mathrm{B} \cdot \log M \qquad (\mathrm{A\ and\ B} = \mathrm{constants}). \tag{3}$$

Plotting log M against V yields curves with a straight part for which the above relationship holds (cf. Fig. 10.10). The bending into a parallel to the abscissa at very high molecular weights is caused by complete exclusion of the compounds concerned, i.e. lack of retention. The constants A and B are dependent on the substance class, therefore the molecular weight of a substance can only be determined when the position of the calibration line in the coordinate network has been determined with the aid of compounds of known molecular weight.

Finally, an important application of gel chromatography is the study of polymers; individual members of a polymer homologous series cannot be separated, but the molecular weight distribution can be determined quickly and with little substance amount.

In general, adsorption chromatography is a distinctly micromethod in which the substance requirement is of the order of a few micrograms and less. The advantages of the method are its great versatility and efficiency; since the choice of adsorbent and eluent can be varied widely, the conditions can be adapted to the separation

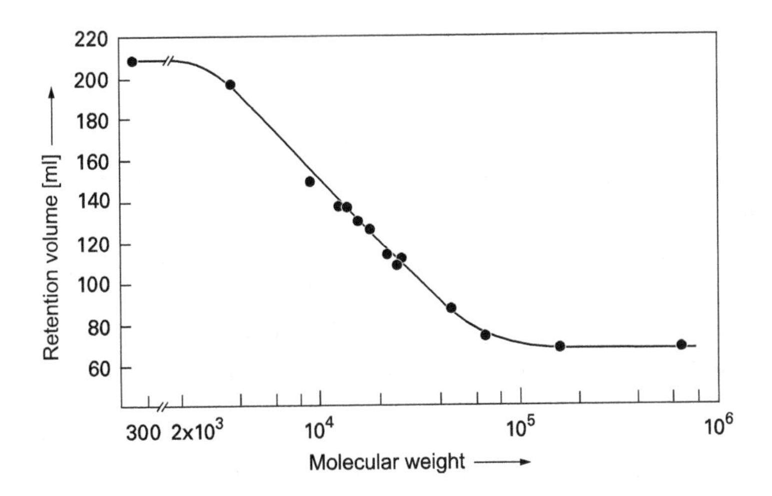

Fig. 10.10: Dependence of protein retention volumes on molecular weight.

problem at hand, and compounds which are chemically very similar can often be separated. Experimental simplicity has also contributed to the widespread use of the method, although the aforementioned high-performance columns require considerably greater effort.

The main disadvantage is the occurrence of irreversible adsorption, which considerably limits the range of application in trace analyses, as well as the often observed tail formation and the *sagging* of the zones caused by irregular flow of the moving phase.

Another major disadvantage is the difficulty in keeping the properties of the adsorbents constant; it is almost impossible to avoid the fact that different batches of the same adsorbent occasionally exhibit different properties. Apparently incidental phenomena, e.g. fluctuations in humidity, also play a role here, which can lead to different water absorption and activity.

Finally, it should be noted that sensitive organic substances are occasionally altered during adsorption, e.g., by isomerization, hydrolysis, or decomposition.

10.6 Separations by systematic repetition: Planar methods (thin-layer chromatography, paper chromatography)

10.6.1 Principle and adsorbents

In the planar methods, a more or less thin layer is used instead of a columnar arrangement of the adsorbent. The same compounds are used as adsorbents as in column chromatography. The difference between the two methods is essentially in the experimental procedure.

10.6.2 Performing thin-layer chromatography

Whereas in the past *thin-layer chromatography plates* had to be largely manufactured in-house, a variety of commercial products are now available. The plates consist of strips or have square dimensions (usually 20 × 20 cm); in the latter case, several separations can be performed side by side.

Layer thicknesses and substance amounts. The thickness of the adsorbent layer is usually 0.15–0.25 mm; about 5–20 µg, at most about 50 µg of substance is applied to such plates. For somewhat larger amounts (a few milligrams), plates of 1 mm layer thickness can be used; even thicker plates crack easily during drying, but can be made up to 2 mm thickness using special adsorbents.

Extremely small sample amounts (< 1 µg) can be separated with very thin layers prepared by immersing the plates in a suspension of the adsorbent; for example, 0.05 mm thick layers of silica gel, cellulose or polyamides on polyester films are obtained. Another method consists in precipitation of the adsorbent on the support, e.g. 6 µm thick layers of silica gel on cellophane were prepared in this way.

However, the lower limit of the feed amount is given not only by the layer thickness, but also by the lower limits of the detection reactions for the separated components.

Compound application. The compounds to be separated are applied as a dilute solution (approx. 1%) with micropipettes or microliter syringes, as far as possible in a punctiform manner, about 1.5 cm from the edge of the plate. Larger amounts are applied to the layer by repeated dripping while drying the stain after each application or by drawing a slightly longer narrow transverse line. Volatile substances can be vaporized onto the layer from a capillary.

Developing and eluting. In *thin-layer chromatography*[2], the developing technique is usually used, and elution only rarely. A suitable solvent is allowed to flow through the layer by placing the plate in a closed vessel (to avoid evaporative loss of flow agent) approximately vertically in a trough of solvent about 0.5 cm deep; the liquid rises upward due to capillary forces in this method, which is referred to as *ascending development*. In the *descending method*, the liquid is fed from a vessel at the top using a strip of paper as a wick. Correspondingly, development can also be carried out on a horizontally positioned plate.

Circular technique has also been proposed in thin layer chromatography.

Elution, carried out less frequently, can be realized in the descending method by placing a trough with flow medium on the upper edge of the thin film plate. In ascending

2 The correct term would be *thin-layer adsorption chromatography*, as there are other types of thin-layer chromatography. But the shorter term has become common.

chromatography, for example, the adsorbent can be placed on the outside of a glass capillary and the flow medium allowed to flow down through the capillary.

Stepwise development – gradient method. If the substances to be analyzed are only incompletely separated with a particular flow agent, the separation can be completed with a second liquid; in this case, two-dimensional development on square plates is expedient.

Instead of stepwise development, the gradient method can be used; concentration and pH gradients are obtained by continuously adding a second liquid to the contents of the flow agent trough. Less common are temperature gradients, which can be produced by heating the thin film plate on one side.

Often an improvement of the chromatograms is achieved by wedge-shaped thin film plates in which a gradient of the flow velocity occurs. Finally, layer gradients have also been described in which two adsorbents of different activities or different pH values are applied along the plate in mixtures with continuously changed composition.

10.6.3 Performing paper chromatography

In terms of method, *paper chromatography* represents a variety of thin-layer chromatography; the mode of operation is practically the same, except that paper strips are used instead of layers of powdered adsorbents. For a few years now, paper chromatography has been gaining increasing importance as a rapid measurement technique in combination with antibodies or other biological receptors. Combined with this is a simple visual readout of the analyte's reaction in a color reaction. Such *dip stick techniques* allow rapid and straightforward detection of single analytes in real matrices. The capillary action of the paper replaces the pump.

Paper chromatography is carried out on paper strips of various types whose thickness is somewhat increased compared to the usual filter papers and which consist of pure cellulose. Modified papers are obtained by acetylation or by impregnation with formamide, etc., as well as by incorporating various other adsorbents into the mass. Some of these papers are also suitable for separating nonpolar substances with organic flow agents[3]. A special case are the so-called *glass fiber papers*, which are particularly chemically and thermally resistant.

For the separations, apply substance amounts of about 10–50 µg about 1 cm from the edge of a rectangular 30–40 cm long paper strip in the form of dilute solutions and develop the chromatogram in ascending or descending order in closed vessels.

A variant in which somewhat larger substance quantities can be separated is the circular technique, which can also be used in paper chromatography; the sample

3 Referred to as *reversed phase chromatography* in Anglo-Saxon literature.

solution is fed into the center of a round sheet of paper lying between two glass plates and developed radially outward (cf. Fig. 6.32). By rotating the paper around the center, the flow velocity can be increased and the development accelerated (*centrifugal chromatography*).

Two-dimensional development is often used to separate more complicated mixtures.

10.6.4 Detection and identification of the separated substances

Probably the most important detection methods for colorless substances on thin film plates or paper strips are color reactions. The chromatograms are sprayed with suitable reagent solutions, whereby the position of the separated compounds can be recognized by the formation of *spots*. Usually, amounts of order-of-magnitude 1 μg can still be detected in this way. Even more sensitive are methods that use fluorescent phenomena for detection: Either fluorescent substances are made directly visible by UV radiation after developing the chromatogram, or a fluorescent compound is added to the adsorbent and the fluorescence quenching by the analytical substances is observed; the individual components stand out as dark spots on a lighter background when the thin-layer plate is viewed in UV light. Furthermore, radioactive substances can be detected by their radiation.

If the location of the stains is known, the compounds in question can occasionally be removed from the adsorbent by sublimation. Non-volatile substances are isolated by extraction after the layer has been mechanically scraped off or after the paper spot has been cut out, and then identified by chemical or physical methods.

An important tool for characterizing individual compounds are the R_f and R_{St} values. These values have been determined and tabulated for various classes of substances (e.g. amino acids, steroids, etc.) under uniform conditions. However, the R_f values are influenced by numerous variables (activity and layer thickness of the adsorbent; humidity; chamber shape; presence of other components, etc.), so that their reproducibility leaves much to be desired.

10.6.5 Efficacy and scope of thin film methods

Since the adsorbents in thin-layer chromatography may have very small particle diameters (favorable are about 2–6 μm), this method achieves lower separation plate heights than chromatography in columns; thus, values of 0.07–0.3 mm are given, and values of 0.0025–0.01 mm for layers of about 1 μm thickness.

On the other hand, the running distance is at most only about 20–30 cm, so that with the usual thin-film plates one can expect no more than an order-of-magnitude 1000 separation plates, which, moreover, are only partially utilized when the development technique is applied.

Both thin-layer chromatography and paper chromatography are used extensively for the separation of organic mixtures, and there is hardly a class of substances that has not been studied more or less successfully in this way; of the extremely numerous applications, only a few examples can be given (cf. Tab. 10.7). The separation of inorganic ions by these methods plays a relatively minor role.

Tab. 10.7: Application examples of thin-layer chromatography.

Mixture to be separated	Adsorbent	Mobile phase
Alkaloids	Silica gel	$CHCl_3$ + 5-15%, CH_3OH
Alkaloids	Al_2O_3	Cyclohexane + $CHCl_3$, 3:7
Amine	Silica gel	C_2H_5OH + 25% NH_3 solution, 4:1
Amino acids	Silica gel	50% n-propanol; phenol-H_2O, 10:4
Amino acids	Silica gel	n-Butanol – CH_3COOH – H_2O, 60:20:20
Fatty acids, glycerides	Silica gel	Xylene; toluene; benzene; $CHCl_3$; acetone etc.
Cholesterol ester	Silica gel	$CHCl_3$; $CHCl_3$ + CH_3OH, 95:5
Phenols	Silica gel	Benzene; cyclohexane + $CHCl_3$ + diethylamine, 5:4:1
Pyrones, anthraquinones, tannins	Silica gel	Benzene + methyl acetate + HCOOH, 2:2:1
Steroids	Silica gel	$CHCl_3$; 1,2-dichloroethane; $CHCl_3$ + acetone, 9:1 and others.
Vitamins	Silica gel	Hexane; $CHCl_3$ et al.
Sugar	Silica gel	Acetic ester + i-propanol-H_2O (2:1), 65:35

The molecular weight determination described for column chromatography with gel stationary phases can also be carried out using appropriate thin-film plates. The logarithms of the molecular weights are plotted against the R_f values (or better against the R_{St} values, which are determined against a standard running on the same plate); for homologous series, straight lines are obtained, which can then be used to determine the molecular weight of unknown compounds of the same homologous series.

The advantages of the thin-layer methods are mainly the experimental simplicity and the speed of the separations (the development of a chromatogram usually takes only about 30–60 min). Furthermore, the separated substances in the individual spots are easily accessible. When microsamples are present, it is advantageous that the feed quantities are very small.

Disadvantages, apart from the relatively low separation efficiency, are the poor reproducibility of the R_f values and the losses due to irreversible adsorption, which are also observed with these methods. In contrast to column chromatography, the flow rate of the moving phase can hardly be influenced. Finally, difficulties often arise when identifying unknown compounds, since the separated substance quantities are not sufficient for further investigations.

10.7 Cross-flow method

A cross-flow method in which the adsorbent and eluent are moved at an angle of 90° to each other is shown in Fig. 10.11: The stationary phase is located in an annular gap a between two concentric tubes, with some free space at the top to accommodate the mobile phase. The eluent flows through the annular gap from top to bottom, at the same time rotating in the direction of the arrow. The substance mixture b is continuously applied to the surface of the adsorbent at a specific point in the gap; the component c migrating more rapidly through the separation bed is discharged from the gap after a smaller angular deflection than the more slowly migrating substance d.

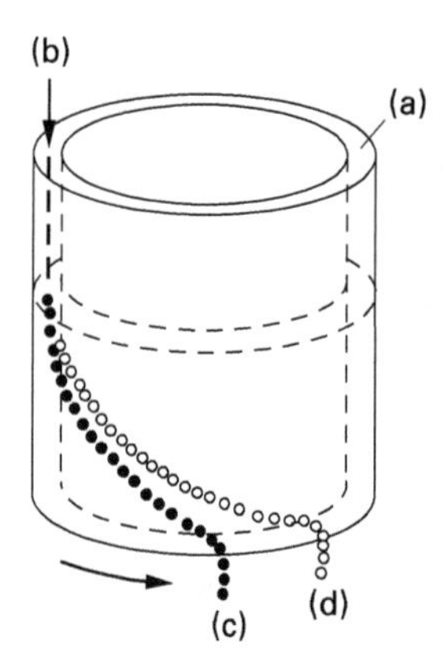

Fig. 10.11: *Continuous column chromatography* [after Fox et al. (1969)].
a) Annular gap;
b) substance supply;
c, d) discharge of the separated components.

Because of the cumbersome method of operation and the difficulty of producing a uniform separation bed, the process has not yet achieved any significance, and even somewhat modified methods do not play a major role.

General literature

General & Column Chromatography

P. Brown, E. Grushka & S. Lunte, Advances in Chromatography 43, 1–325 (2004).

W. Cheong, S. Yang & F. Ali, Molecular imprinted polymers for separation science: A review of reviews, Journal of Separation Science *36*, 609–628 (2013).

J.V. Hinshaw, Solid-phase microextraction, LCGC North America *21*, 1056, 1058–1061 (2003).

K. Kaczmarski, W. Prus, M. Sajewicz & T. Kowalska, Adsorption planar chromatography in the nonlinear range: selected drawbacks and selected guidelines, Chromatographic Science Series *95* (Preparative Layer Chromatography), 11–40 (2006).

I. Krull, Reaction Detection in Liquid Chromatography, Taylor & Francis, New York (2017)

M. Michel & B. Buszewski, Porous graphitic carbon sorbents in biomedical and environmental applications, Adsorption *15*, 193–202 (2009).

C. Poole, Liquid-Phase Extraction, Elsevier, Amsterdam (2019).

Thin layer chromatography

B. Fried & J. Sherma, Practical Thin-Layer Chromatography: A Multidisciplinary Approach, Taylor & Francis, New York (2017).
C. Poole, Instrumental Thin-Layer Chromatography, Elsevier, Amsterdam (2014).
P.E. Wall, Thin-Layer Chromatography: A Modern Practical Approach, Royal Society of Chemistry, London (2005).

Gel chromatography

E. Bouvier & S. Koza, Advances in size-exclusion separations of proteins and polymers by UHPLC, TrAC – Trends in Analytical Chemistry 63, pp. 85–94 (2014).
J.-C. Janson, Protein Purification: Principles, High Resolution Methods, and Applications: Third Edition, J. Wiley & Sons, New York (2011).
A. Striegel, W. Yau, J. Kirkland & D. Bly, Modern Size-Exclusion Liquid Chromatography: Practice of Gel Permeation and Gel Filtration Chromatography: Second Edition, J. Wiley & Sons, New York (2009).

Literature to the text

Adsorption isotherms (linear part)

F. Gritti & G. Guiochon, Systematic errors in the measurement of adsorption isotherms by frontal analysis, Journal of Chromatography A*1097*, 98–115 (2005).

Adsorption heat

I. Dekany & F. Berger, Adsorption from liquid mixtures on solid surfaces, Surfactant Science Series *107*, (Adsorption), 573–629 (2002).
A. Imre, Microcalorimetric control of liquid sorption on hydrophilic/hydrophobic surfaces in nonaqueous dispersions, Surfactant Science Series *93* (Thermal Behavior of Dispersed Systems), 357–412 (2001).
L. Roselin, M. Lin, P.-H. Lin, Y. Chang & W.-Y. Chen, Recent trends and some applications of isothermal titration calorimetry in biotechnology, Biotechnology Journal *5*, 85–98 (2010).

Affinity chromatography

F. Batista-Viera, J.-C. Janson, and J. Carlsson, Affinity chromatography, Methods of Biochemical Analysis *54* (Protein Purification), 221–258 (2011).
I. Chaiken, Analytical Affinity Chromatography, CRC Press, Boca Raton (2018).
K. Engholm-Keller & M.P. Larsen, Titanium dioxide as chemo-affinity chromatographic sorbent of biomolecular compounds – Applications in acidic modification-specific proteomics, Journal of Proteomics *75*, 317–328 (2011).

J.-C. Janson, Protein Purification: Principles, High Resolution Methods, and Applications: Third Edition, J. Wiley & Sons, New York (2011).

C. Mant & R. Hodges, High-Performance Liquid Chromatography of Peptides and Proteins: Separation, Analysis, and Conformation, CRC Press, Boca Raton (2019).

A.-C. Neumann, S. Melnik, R. Niessner, E. Stoeger & D. Knopp, Microcystin-LR enrichment from freshwater by a recombinant plant-derived antibody using Sol-Gel-Glass immunoextraction, Analytical Sciences *35*, 207–214 (2019).

J. Porath, Strategy for differential protein affinity chromatography, International Journal of Bio-Chromatography *6*, 51–78 (2001).

A.C. Roque & C.R. Lowe, Affinity chromatography. History, perspectives, limitations and prospects, Methods in Molecular Biology *421* (Affinity Chromatography), 1–21 (2008).

F. Tsagkogeorgas, M. Ochsenkühn-Petropoulou, R. Niessner & D. Knopp, Encapsulation of biomolecules for bioanalytical purposes: Preparation of diclofenac antibody-doped nanometer-sized silica particles by reverse micelle and sol-gel processing, Analytica Chimica Acta *573–574*, 133–137 (2006).

Influencing the separation plate height

B. Grimes, S. Lüdtke, K. Unger, & A. Liapis, Novel general expressions that describe the behavior of the height equivalent of a theoretical plate in chromatographic systems involving electrically-driven and pressure-driven flows, Journal of Chromatography A *979*, 447–466 (2002).

F. Gritti & G. Guiochon, General HETP Equation for the Study of Mass-Transfer Mechanisms in RPLC, Analytical Chemistry *78*, 5329–5347 (2006).

I. Schmidt, F. Lottes, M. Minceva, W. Arlt, and E. Stenby, Estimation of chromatographic columns performances using computer tomography and CFD simulations, Chemie Ingenieur Technik *83*, 130–142 (2011).

Determination of molecular weights

S. Fekete, A. Beck, J.-L. Veuthey & D. Guillarme, Theory and practice of size exclusion chromatography for the analysis of protein aggregates, Journal of Pharmaceutical and Biomedical Analysis *101*, 161–173 (2014).

H. Pasch, Analytical techniques for polymers with complex architectures, Macromolecular Symposia *178* (Polymer Characterization and Materials Science), 25–37 (2002).

S. Podzimek, Light Scattering, Size Exclusion Chromatography and Asymmetric Flow Field Flow Fractionation: Powerful Tools for the Characterization of Polymers, Proteins and Nanoparticles, J. Wiley & Sons, New York (2011).

A. Striegel, W. Yau, J. Kirkland & D. Bly, Modern Size-Exclusion Liquid Chromatography: Practice of Gel Permeation and Gel Filtration Chromatography, J. Wiley & Sons, New York (2009).

D. Sykora & F. Svec, Synthetic polymers, Journal of Chromatography Library *67* (Monolithic Materials), 457–487 (2003).

Properties of Al_2O_3

J. Nawrocki, C. Dunlap, A. McCormick & P. Carr, Part I. Chromatography using ultra-stable metal oxide-based stationary phases for HPLC, Journal of Chromatography A *1028*, 1–30 (2004).

Properties of silica gel

Z. Bayram-Hahn, B. Grimes, A. Lind, R. Skudas, K. Unger, A. Galarneau, J. Iapichella, and F. Fajula, Pore structural characteristics, size exclusion properties and column performance of two mesoporous amorphous silicas and their pseudomorphically transformed MCM-41 type derivatives, Journal of Separation Science *30*, 3089–3103 (2007).

J. Choma, M. Kloske, & M. Jaroniec, An improved methodology for adsorption characterization of unmodified and modified silica gels, Journal of Colloid and Interface Science 266, 168–174 (2003).

P. Jandera, Stationary and mobile phases in hydrophilic interaction chromatography: A review Analytica Chimica Acta *692*, 1–25 (2011).

L. Singh, S. Bhattacharyya, R. Kumar, (. . .), G. Singh & S. Ahalawat, Sol-Gel processing of silica nanoparticles and their applications Advances in Colloid and Interface Science *214*, 17–37 (2014).

H. Zou, S. Wu & J. Shen, Polymer/Silica Nanocomposites: Preparation, characterization, properties, and applications, Chemical Reviews *108*, 3893–3957 (2008).

Gradient elution

K. Broeckhoven & G. Desmet, Theory of separation performance and peak width in gradient elution liquid chromatography: A tutorial, Analytica Chimica Acta *1218*, 339962 (2022).

A. Felinger, Optimization of preparative separations, Chromatographic Science Series *88* (Scale-Up and Optimization in Preparative Chromatography), 77–121 (2003).

P. Jandera, Programmed elution in comprehensive two-dimensional liquid chromatography, Journal of Chromatography A *1255*, 112–129 (2012).

P. Nikitas & A. Pappa-Louisi, Retention models for isocratic and gradient elution in reversed-phase liquid chromatography, Journal of Chromatography A *1216*, 1737–1755 (2009).

D. Stoll, X. Li, X., Wang, (. . .), S. Porter & S. Rutan, Fast, comprehensive two-dimensional liquid chromatography, Journal of Chromatography A *1168*, 3–43 (2007).

Cross-flow method

R. Giovannini & R. Freitag, Continuous isolation of plasmid DNA by annular chromatography, Biotechnology and Bioengineering *77*, 445–454 (2002).

M. Lay, C.J. Fee, and J.E. Swan, Continuous radial flow chromatography of proteins, Food and Bioproducts Processing *84*, 78–83 (2006).

A. Uretschläger & A. Jungbauer, Preparative continuous annular chromatography (P-CAC), a review, Bioprocess and Biosystems Engineering *25*, 129–140 (2002).

Solvent during adsorption

M. Borowko & B. Oscik-Mendyk, Adsorption model for retention in normal-phase liquid chromatography with ternary mobile phases, Advances in Colloid and Interface Science *118*, 113–124 (2005).

M. Borowko & B. Oscik-Mendyk, Selectivity in normal-phase liquid chromatography with binary mobile Phase, Adsorption *16*, 397–403 (2020).

H. Oka & Y. Ito, Solvent systems: systematic selection for HSCCC, Encyclopedia of Chromatography *3*, 2192–2197 (2010).

L. Snyder, Solvent selectivity in normal-phase TLC, Journal of Planar Chromatography-Modern TLC *21*, 315–323 (2008).

Planar chromatography

H. Kalász, Planar displacement chromatography, Journal of Planar Chromatography – Modern TLC *17*, 464–446 (2004).

J. Sherma, Planar chromatography, Analytical Chemistry *82*, 4895–4910 (2010).

B. Spangenberg, C. Poole & C. Weins, Quantitative Thin-Layer Chromatography: A Practical Survey, Springer, Heidelberg (2011).

A.-M. Siouffi, From paper to planar: 60 years of thin layer chromatography, Separation and Purification Reviews *34*, 155–180 (2005).

Polymer gels

H. Chen, I. Yuan, W. Song, Z. Wu & D. Li, Biocompatible polymer materials: Role of protein-surface Interactions, Progress in Polymer Science (Oxford) *33*, 1059–1087 (2008).

Q. Wei, T. Becherer, S. Angioletti-Uberti, (. . .), M. Ballauff & M. Haag, Protein interactions with polymer coatings and biomaterials, Angewandte Chemie – International Edition *53*, 8004–8031 (2014).

11 Ion exchange

11.1 General

11.1.1 Historical development

The discovery of ion exchange on solid compounds is attributed to Thompson and Way (1850); important further investigations are by Lemberg (1870) and Gans (1905); organic ion exchangers were first synthesized by Adams & Holmes (1935).

Folin (1917) first applied ion exchangers to analytical chemistry, and eventually ion-exchange chromatography was elaborated by several research groups led by Boyd, Spedding, and Tompkins (1947).

11.1.2 Definitions – functional groups – exchange reactions – regeneration – titration curves – exchange capacity

Definitions. Ion exchangers are defined as practically insoluble, solid substances with a more or less high content of functional groups which are capable of absorbing ions from solutions while releasing an equivalent amount of other ions of the same charge sign into the solution.[1]

Functional groups. Ion exchangers can contain acidic or basic *functional groups.*[2] In the former case, they absorb cations from solutions with salt formation in the solid phase; in the latter, they absorb anions. Accordingly, a distinction is made between cation and anion exchangers.

Within these two basic types, differences arise depending on the acidity of the acidic or basicity of the alkaline reacting groups.

Acidic functional groups are present in certain aluminum silicates, polymolybdates, polytungstates and other inorganic compounds. In organic polymeric plastics, phenolic $-OH$, $-COOH$ or $-SO_3H$ (increasing acidity) are mainly introduced as exchangeable groups. The functional groups of the corresponding anion exchangers usually consist of $-NH_2$, $=NH$ or $-N^+R_3$ (increasing basicity), where R is usually a methyl or ethyl group.

Four types in particular have gained analytical importance: strongly and weakly acidic cation exchangers and strongly and weakly basic anion exchangers. Exchangers

1 Liquid compounds, e.g. organic bases, which are used to carry out exchange reactions between two immiscible liquid phases, are often included in the ion exchange resins. However, this class of substances is not covered by the definition given here.

2 The so-called *redox resins*, with which reductions and oxidations can be carried out, are not dealt with here.

https://doi.org/10.1515/9783111181417-011

that have different functional groups at the same time are less suitable for analytical purposes. Exchangers in which selectivity for certain ions is sought by introducing special functional groups are described below.

Exchange reactions. In principle, all exchange reactions on ion exchangers are double reactions, which take place between a liquid and a solid phase[3]. The following three types of reactions are distinguished: Neutralization, neutral salt splitting and double conversion.

If an exchanger with acidic functional groups is placed in an alkaline solution, *neutralization* occurs in the solid phase with salt formation, e.g.

$$\begin{aligned} &\mathrm{H-R+NaOH \longrightarrow Na-R+H_2O} \\ &\mathrm{(R = Exchanger-anion).} \end{aligned} \tag{1}$$

The exchanger changes from the *H^+ form* to the *Na^+ form.* As long as acidic groups are still present in the exchanger, reaction (1) proceeds practically completely from left to right.

For *neutral salt splitting*, a cation exchanger in the H^+ form is brought into contact with a salt solution; an exchange also occurs and the solution becomes acidic due to the H^+ ions released by the exchanger:

$$\mathrm{H-R+NaCl \rightleftharpoons Na-R+HCl} \tag{2}$$

In contrast to the conditions in neutralization, this reaction usually leads to an equilibrium which is not extremely shifted to one side.

In ion exchange, *double conversions* are those reactions in which neither H^+ – nor OH^- ions are involved, e.g.:

$$\mathrm{NH_4-R+NaCl \rightleftharpoons Na-R+NH_4Cl} \tag{3}$$

Such reactions also lead to equilibria, which are usually not very far on one side.

Corresponding equations apply to anion exchangers (cf. Tab. 11.1).

Tab. 11.1: Exchange reactions (written in ionic form).

Exchange reaction	Cation exchanger	Anion exchanger
Neutralization	$H^+-R^- + NaOH \rightarrow Na^+-R^- + H_2O$	$OH^--R^+ + HCl \rightarrow Cl^--R^+ + H_2O$
Neutral salt splitting	$H^+-R^- + NaCl \rightleftharpoons Na^+-R^- + HCl$	$OH^--R^+ + NaCl \rightleftharpoons Cl^--R^+ + NaOH$
Double conversion	$NH_4^+-R^- + NaCl \rightleftharpoons Na^+-R^- + NH_4Cl$	$Cl^--R^+ + NaBr \rightleftharpoons Br^--R^+ + NaCl$

3 Ligand exchange reactions, which follow a slightly different scheme, are discussed in Section 11.4.9.

All exchange reactions are reversible; furthermore, they proceed strictly stoichiometrically because of the electroneutrality condition.

Regeneration. The neutralization reaction (1) proceeds in the opposite direction when an exchanger present in the salt form is treated with acid, e.g.:

$$\mathrm{Na-R+HCl \rightleftharpoons H-R+NaCl} \tag{4}$$

In contrast to eq. (1), this reaction proceeds only incompletely to the right, i.e., only a portion of the H^+ ions is taken up by the exchanger. If the supernatant NaCl-containing solution is poured off after equilibration and the exchanger is repeatedly treated with portions of fresh acid solution, the alkali ion can be gradually removed from the exchanger and it can be completely restored to the H^+ form. The process is called *regeneration*. Thanks to this regeneration capability, an exchanger can be used very often.

Reactions (1) and (4) are basically identical, as shown by the notation in ionic form:

$$\mathrm{H^+ - R^- + Na^+ \rightleftharpoons Na^+ - R^- + H^+} \tag{5}$$

In reaction (1), however, the H^+ ion going into solution is intercepted by the OH^- ions present, so that the equilibrium is continuously disturbed and the reaction is forced in one direction.

The regeneration of anion exchangers is carried out accordingly by treating with aqueous sodium hydroxide solution or potassium hydroxide solution.

Titration curves. The neutralization reaction can be followed by continuous pH measurement of a slurry of the exchanger in water with the addition of NaOH solution in portions. However, the reaction is much slower than neutralization in homogeneous aqueous solution, so that the system must be stirred for some time after each addition of alkali. In this way, characteristic titration curves are obtained depending on the type of exchanger and the functional groups (Fig. 11.1).

Exchangers with COOH and SO_3H groups give curves corresponding to those expected in aqueous solutions, while the exchanger with phenolic OH shows no distinct turnover region. The steepness of the titration curves in the transition range thus gives an indication of the acidity of the functional groups.

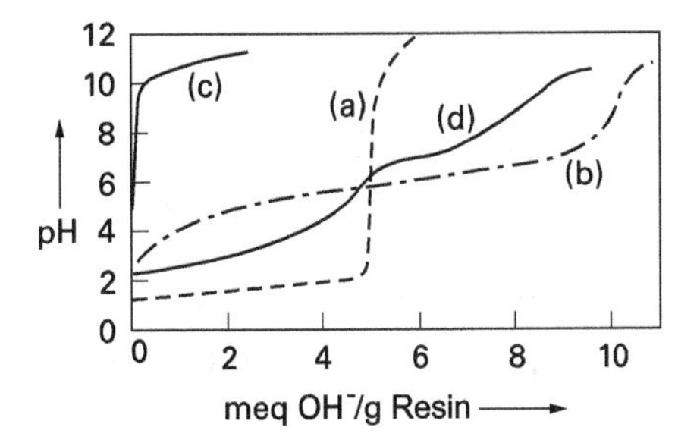

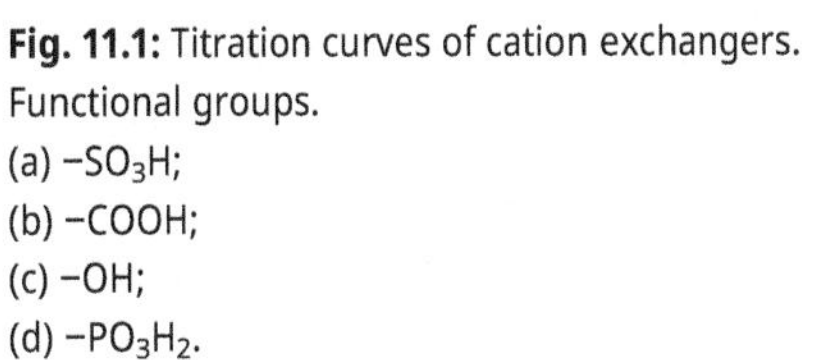

Fig. 11.1: Titration curves of cation exchangers. Functional groups.
(a) $-SO_3H$;
(b) $-COOH$;
(c) $-OH$;
(d) $-PO_3H_2$.

When treated with acid solutions, anion exchangers in the OH form provide corresponding titration curves from which conclusions can be drawn about the basicity of the functional groups.

The titration curves also show the so-called *exchange capacity* of the respective exchanger. This is the amount of alkali required to neutralize all the functional acid groups of a given amount of cation exchanger (or the amount of acid for an anion exchanger).

The exchange capacity is calculated in milliequivalents of base or acid and is usually related to 1 g of air-dry, more rarely oven-dry, or to 1 ml of exchanger. In the examples shown in Fig. 11.1, the exchanger with COOH groups has a capacity of about 10.5 meq of alkali per gram, and the SO_3H exchanger has a capacity of about 4.7 meq/g. It is appropriate to use exchangers with as high a capacity as possible.

The exchange capacity is a constant only to a first approximation; for very large ions or when exchanging ions dissolved in organic solvents, it may decrease because only part of the functional groups may react.

11.1.3 Exchanger types – porosity and swelling

The basic structure of an ion exchanger (also called *matrix* carrying the functional groups can consist of inorganic or organic material.

Various sodium aluminum silicates of the zeolite group, whose sodium ions can be replaced by other cations, act as inorganic exchangers; these compounds have found wide application in water softening technology, but do not play a role in analytical chemistry because they can only be used in a narrow pH range near the neutral point due to low chemical stability.

Some salts of heteropolyacids behave more favorably, offering advantages especially in the separation of alkali metal and alkaline earth metal ions. As an inorganic anion exchanger, hydroxyapatite is able to exchange OH^- for F^- ions; this material has also been used for separations of proteins, but adsorption effects are likely to be of decisive importance here.

Numerous other inorganic precipitates have been proposed for special separations, especially hardly soluble oxide hydrates, phosphates and tungstates of higher-value elements (e.g. zirconium and titanium), as well as porous glass. These compounds lead over to precipitates such as AgCl, CdS, CuS, $Hg(IO_3)_2$, Ag_2CrO_4 and others, on the surfaces of which inorganic ions from strongly diluted solutions are held under ion exchange; this reaction is called *precipitation exchange* or *exchange adsorption.*

Much more important than the inorganic compounds have been organic ion exchangers, which are obtained by introducing functional groups into crosslinked high-polymer synthetic resins (cf. Tab. 11.2).

The matrix of such exchangers usually consists of polystyrene crosslinked with divinylbenzene (cf. Fig. 11.2) or of a phenol-formaldehyde condensate (cf. Fig. 11.3). The resins are not completely chemically stable, but always release small amounts of organic matter into the solution; especially exchangers freshly taken into use may show considerable solubilities as a result of the presence of non-polymerized low molecular weight fractions.

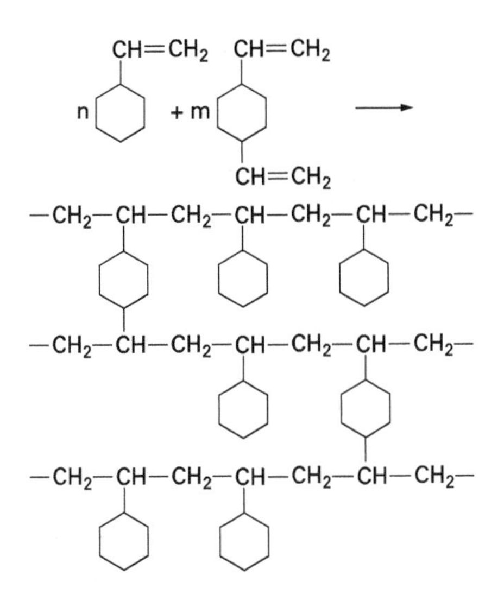

Fig. 11.2: Structure of a styrene-divinylbenzene polymer (schematic).

Furthermore, resin exchangers are attacked by oxidizing agents, which is especially true for anion exchangers and phenol-formaldehyde condensates, less so for styrene resins with acidic groups. But very strong oxidizing agents, such as $KMnO_4$, will also degrade the latter.

Various exchanger types decompose in higher temperature ranges; it is therefore recommended to carry out drying at room temperature in the air. Fresh resin exchangers should be *run in* before they are actually put into operation. To do this, they should be transferred several times in succession alternately to the H^+ (or OH^-) form and to a salt form; only then can it be assumed that they will work reversibly.

Another type of ion exchanger is formed by introducing functional groups into cellulose; the exchange capacities of such products are somewhat lower than those of the resin exchangers (cf. Tab. 11.3).

Porosity and swelling. Most ion exchangers have a porous structure, so that both solvents and dissolved compounds can penetrate into the interior of the granules. Dry organic resin exchangers swell in the process, and the extent of volume increase depends on the degree of crosslinking. In the case of inorganic exchangers with a rigid framework, this effect is very slight.

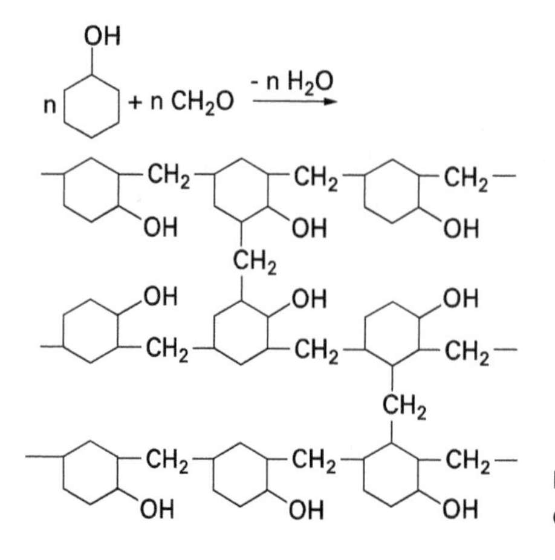

Fig. 11.3: Structure of a phenol-formaldehyde condensate (schematic).

Tab. 11.2: Synthetic *resin ion exchangers.*

Exchanger type	Trade name* and supplier		Mol. frame	Functional group	Capacity meq/g
Cation exchanger					
Strongly acidic	Amberlite IR 120	Rohm and Haas	Polystyrene	$-SO_3H$	4–5
Strongly acidic	Dowex 50**	Dow Chem. Co.	Polystyrene	$-SO_3H$	approx. 5
Strongly acidic	Lewatite S 100	Lanxessa	Polystyrene	$-SO_3H$	approx. 5
Strongly acidic	Merck I	Merck Chem.	Polystyrene	$-SO_3H$	approx. 4,5
Strongly acidic	Wofatit KPS 200	VEB Paint Factory	Polystyrene	$-SO_3H$	approx. 5
Strongly acidic	Zeocarb 225	Permutit Co.	Polystyrene	$-SO_3H$	approx. 5
weakly acidic	Amberlite IRC 50	Rohm and Haas	Polyacrylic acid	–COOH	approx. 10
weakly acidic	Lewatite CNO	Lanxess	Phenolic resin	–COOH	approx. 5
weakly acidic	Merck IV	E. Merck	Polyacrylic acid	–COOH	approx. 10
weakly acidic	Wofatit CP 300	Wolfener Farbenfabrik	Polyacrylic acid	–COOH	approx. 8
weakly acidic	Zeo carb 226	Permutit Co.	Polymethacrylic acid	–COOH	approx. 9
Anion exchanger					
Strongly alkaline	Amberlite IRA 400	Rohm and Haas	Polystyrene	$-NR_3^+$	approx. 4
Strongly alkaline	Dowex 1	Dow Chem. Co.	Polystyrene	$-NR_3^+$	approx. 3
Strongly alkaline	Lewatite MN	Lanxess	Phenolic resin	$-NR_3^+$	approx. 3
Strongly alkaline	Merck III	E. Merck	Polystyrene	$-NR_3^+$	approx. 3

Tab. 11.2 (continued)

Exchanger type	Trade name* and supplier		Mol. frame	Functional group	Capacity meq/g
Strongly alkaline	De-Acidite FF	Permutit Co.	Polystyrene	$-NR_3^+$	approx. 4
Weakly basic	Amberlite IRA 45	Rohm and Haas	Polystyrene	$-NH_2$, $=NH$	approx. 6
Weakly basic	Dowex 3	Dow Chem. Co.	Polystyrene	$-NH_2$, $=NH$	approx. 6
Weakly basic	Lewatite MIH	Lanxess	Phenolic resin	$-NH_2$, $=NH$	approx. 3,5
Weakly basic	Merck II	E. Merck	Polystyrene	$-CH_2NH_2$	approx. 5
Weakly basic	De-acidite G	Permutit Co.	Polystyrene	$-NH_2$, $=NH$	approx. 3

*Registered trademark.
**Dowex styrene resins often indicate the degree of crosslinking, e.g. Dowex 50 × 4 = Dowex 50 with 4% divinylbenzene.

Tab. 11.3: Modified celluloses as ion exchangers.

Acidity or basicity	Name	Functional group	Capacity (meq/g)
Strongly acidic	SE Cellulose	Sulfoethyl $-O-C_2H_4-SO_3H$	0,2
Strongly acidic	P Cellulose	Phosphoric acid $-O-PO_3H_2$	0,8
Weakly acidic	CM Cellulose	Carboxymethyl-$O-CH_2-COOH$	0,7
Strongly alkaline	GE Cellulose	Guanidoethyl $-O-C_2H_4-NH-C(NH_2Cl)-NH_2$	0,4
Strongly alkaline	TEAE Cellulose	Triethylamino-ethyl $-O-C_2H_4-N(C_2H_5)_3Cl$	0,5
Medium alkaline	DEAE Cellulose	Diethylamino-ethyl-$O-C_2-N(C_2H_5)_2$	0,4–0,9
Medium alkaline	PEI Cellulose	Polyethylenimine $(-NH-CH_2-CH_2)_n$	0,2
Weakly basic	AE Cellulose	Amino-ethyl $-O-C_2H_4NH_2$	0,8
Weakly basic	PAB cellulose	p-Aminobenzyl $-O-C_2H_4-C_6H_4-NH_2$	0,2
Different	ECTEOLA cellulose	Several basic groups	0,3

11.1.4 Auxiliary phases – exchange equilibria – distribution coefficients – exchange isotherm – selectivity and separation factors – exchange of ions of unequal valences – order of binding strengths of different ions

Auxiliary phases. In separations using ion exchangers, two auxiliary phases are used, the solid exchanger and the liquid solvent.

Exchange equilibria. The law of mass action applies to ion exchange reactions. For example, an exchanger in the NH_4 form is brought into contact with a NaCl solution (cf. Fig. 11.4); after equilibrium has been reached, both cations can be found in the solid as well as in the liquid phase.

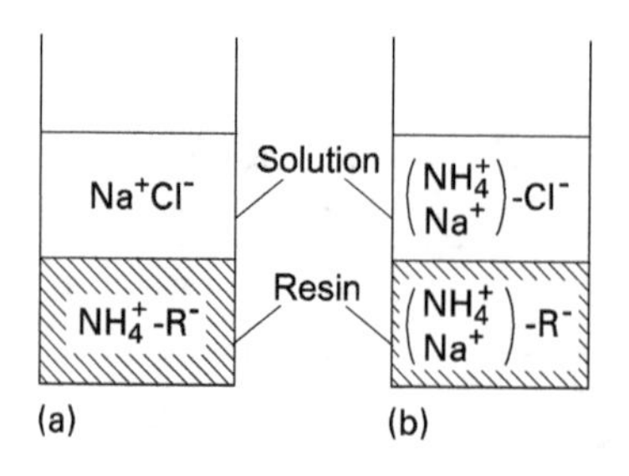

Fig. 11.4: Exchange of Na^+ vs. NH_4^+ on a cation exchanger.
a) Distribution of ions before the exchange;
b) distribution after the exchange.

Applying the law of mass action results in

$$\frac{[Na^+]_R \cdot [NH_4^+]_{Lsg}}{[NH_4^+]_R \cdot [Na^+]_{Lsg}} = K, \tag{6}$$

where the index *R* denotes the exchange phase, and the index *Lsg* denotes the solution phase.

The concentration of an ion in the exchanger phase is usually expressed as the amount by weight of ion per gram of air-dry exchanger; it can also be defined as equivalent ion per equivalent of functional groups contained in the exchanger or as the amount by weight of ion per ml of exchanger; in the latter case, attention must be paid to volume changes due to swelling.

The mass action law is generally well fulfilled in ion exchange in the range of low concentrations (cf. Tab. 11.4). At higher concentrations, deviations occur as usual.

Selectivity. The equilibrium constants of ions of the same charge sign with respect to a given exchanger are generally different, and the ions are thus taken up by the exchanger to different extents. This behavior is the basis for separations of identically charged ions with ion exchangers; it is referred to as *selectivity*. The greater the differences in the equilibrium constants of two ions, the better they are separated from each other.

Tab. 11.4: Dependence of the equilibrium coefficients on the NH_4Cl concentration when H^+ is exchanged for NH_4^+.

NH_4Cl concentration (mol/l)	$K_{H^+}^{NH_4^+}$
0,01	1,20
0,10	1,20
1,0	1,15
2,0	0,83
4,0	0,51

When various monovalent cations are exchanged for NH_4^+ (action of salt solutions on a cation exchanger in the NH_4^+-form), the equilibrium constants given in Tab. 11.5 were obtained.

Tab. 11.5: Equilibrium constants of the exchange of different monovalent cations vs. NH_4^+.

Ion	$K_{NH_4^+}^{Me^+}$	Ion	$K_{NH_4^+}^{Me^+}$
Li^+	0,40	Rb^+	1,70
H^+	0,47	Cs^+	2,40
Na^+	0,67	Ag^+	3,20
NH_4^+	1,00	Tl^+	12,7
K^+	1,05		

Partition coefficients. To describe the separation effects during ion exchange, the distribution coefficients α and the separation factors β can also be used instead of the equilibrium constants K. A distribution coefficient can be specified for each type of ion exchanged, e.g. the system shown in Fig. 11.4:

$$\alpha_{Na}^{+} = \frac{[Na^+]_R}{[Na^+]_{Lsg}} \tag{7}$$

and

$$\alpha_{NH_4}^{+} = \frac{[NH_4^+]_R}{[NH_4^+]_{Lsg}}. \tag{7a}$$

The partition coefficients are usually defined as

$$\alpha = \frac{\text{Amount of ion per g exchanger}}{\text{Amount of ion per ml solution}}$$

The partition coefficients of ions between aqueous solutions and the most commonly used synthetic resin exchangers are generally not very high; in many cases they are of the order of about 1–10. Unusually high values, in some cases up to more than 5000, are observed when complex chloro- and bromo-acids are exchanged on anion exchangers. In such systems, the equilibrium is almost quantitatively on the side of the resin even when working in strongly acidic solutions, i.e. in the presence of a large excess of other anions. On the other hand, high HCl or HBr concentrations are often required to form such complex acids (cf. Fig. 11.5).

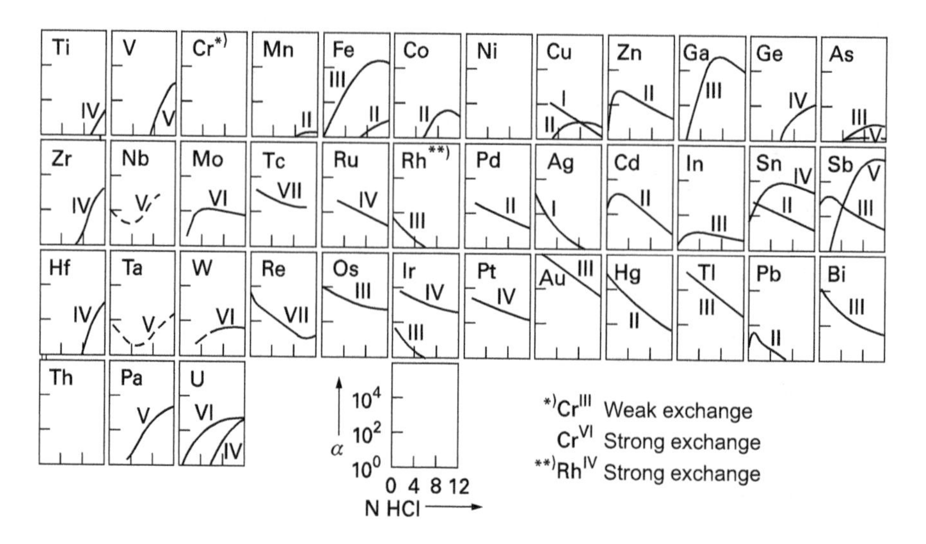

Fig. 11.5: Partition coefficients of complex chloroacids between aqueous HCl solutions and a strongly basic anion exchanger as a function of acid concentration.

Exchange isotherm. If the concentration of an ion in the exchanger is plotted against the concentration in the solution, distribution isotherms (*exchange isotherms*) of approximately the form shown in Fig. 11.6 are obtained. The initial part of the isotherm can be approximated as straight-line. At higher ion concentrations in the solution, the concentration in the exchanger approaches a limit value corresponding to saturation, i.e., the capacity of the exchanger quantity present.

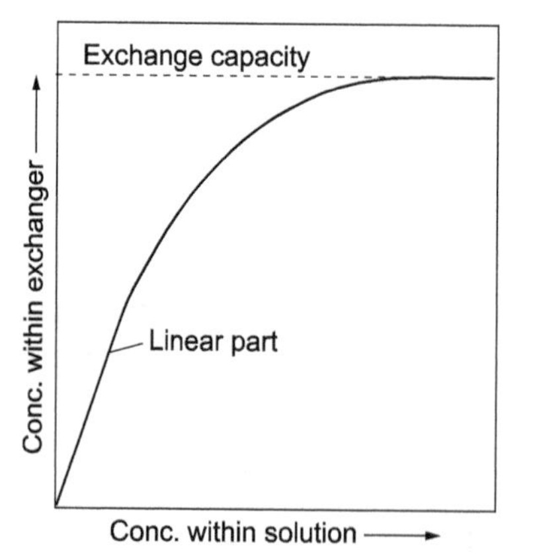

Fig. 11.6: Distribution isotherm during ion exchange.

Separation factors. If we consider the exchange of two ions (which, according to the nature of the exchange, must always be present), the separation factor β can be defined in the usual way. For the exchange of Na^+ vs. NH_4^+ (Fig. 11.4), for example:

$$\beta^{\mathrm{Na}^+}_{\mathrm{NH}_4} = \frac{\alpha^+_{\mathrm{Na}}}{\alpha^+_{\mathrm{NH}_4}} = \frac{[\mathrm{Na}^+]_{\mathrm{R}}/[\mathrm{NH}_4^+]_{\mathrm{R}}}{[\mathrm{Na}^+]_{\mathrm{Lsg}}/[\mathrm{NH}_4^+]_{\mathrm{Lsg}}}. \tag{8}$$

Since eq. (8) is identical to eq. (6), in this example the constant of the mass action equation is equal to the separation factor. However, this only applies to cases where the valences of the exchanged ions are the same (see below). The numerical values of the constant K in Tab. 11.5 are thus at the same time the separation factors for the separations $Me^+ - NH_4^+$; the separation factors for the separation of any two ions in this table are obtained from them by dividing the two values applicable to the separation of NH_4^+ are obtained by dividing the two values valid for the separation of

If one plots the concentrations of an ion in the exchange phase against the concentration in the solution using the concentration measure

$$\textit{Percentual equivalent} = \frac{\text{Equivalent Me}^+}{\text{Eq. Me}^+ + \text{Eq. NH}_4^+} \cdot 100$$

the quadratic representation is obtained, from which the separation possibilities can also be seen. In Fig. 11.7 this is done for the values of Tab. 11.5.

In this representation, the absolute concentrations in the two auxiliary phases are not considered, but the concentrations are related exclusively to the exchanged ions.

Exchange of ions of unequal valence. If the law of mass action is applied to the exchange of ions of unequal valence, then instead of eq. (6) an expression is obtained which contains concentrations in higher powers; e.g., the following applies

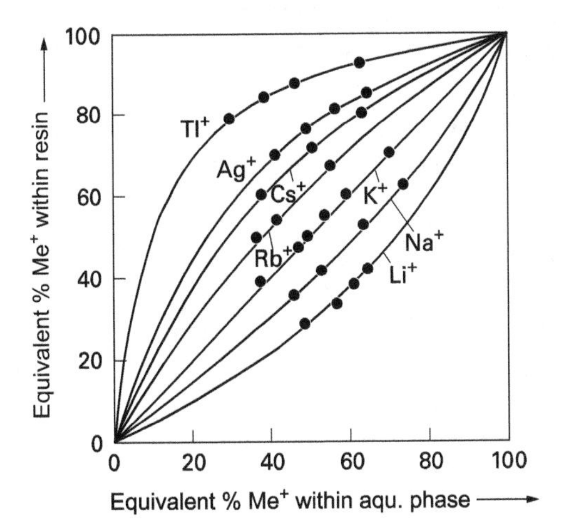

Fig. 11.7: Separations $Me^+ - NH_4^+$ on a cation exchanger in quadratic representation.

for the exchange of Na^+ against Cu^{2+} on a cation exchanger

$$2\,Na - R + Cu^{2+} \rightleftharpoons Cu - R_2 + 2\,Na^+, \tag{9}$$

and for the equilibrium constant K we obtain

$$K = \frac{[Cu^{2+}]_R \cdot [Na^0]^2_{Lsg}}{[Na^+]^2_R \cdot [Cu^{2+}]_{Lsg}}. \tag{10}$$

It follows that the separation factor

$$\beta^{Cu^{2+}}_{Na^+} = \frac{[Cu^{2+}]_R / [Na^+]_R}{[Cu^{2+}]_{Lsg} / [Na^+]_{Lsg}} \tag{11}$$

can no longer be independent of concentration, but increases with increasing dilution of the solution. Therefore, the more dilute the solution, the better multivalent ions are retained by the exchanger and separated from ions of lower valence. At high concentrations, the selectivity can be completely lost.

This has a double effect when plotted as a square: First, the curves become asymmetrical, and second, they change their position against the diagonal when the ion concentration in the solution changes (cf. Fig. 11.8).

According to eqs. (9) and (10), exchange equations can be formulated for any higher valence ions.

Order of binding strengths of various ions. An overview of the behavior of numerous ions during exchange can be obtained by arranging them in the order of their binding strengths. Such series are strictly valid only for the exchanger on which they were experimentally determined; since in analytical chemistry predominantly only a few types, mainly synthetic resin exchangers, are used whose behavior toward different ions is similar, these series have a more general significance (Tab. 11.6).

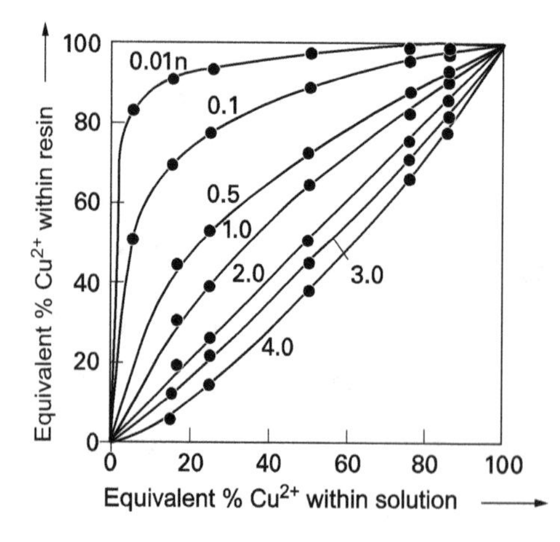

Fig. 11.8: Exchange of Na^+ for Cu^{+2+} on a cation exchanger at different Cu^{2+} concentrations in the solution.

Tab. 11.6: Order of binding strengths of cations and anions on strongly acidic and strongly basic resin exchange resins, respectively.

Cations:
$Ti^{+} > Ag^{+} > Cs^{+} > Rb^{+} > K > {}^{+}NH_4^{+} > Na^{+} > H^{+} > Li^{+}$;
$Ba^{2+} > Sr^{2+} > Ca^{2+} > Mg^{2+}$;
$La^{3+} > Ce^{3+} > Y^{3+} > Sc^{3+} > Al^{3+}$.
Anions:
Citrate > sulfate > oxalate > iodide > nitrate > chromate > bromide > thiocyanate > chloride > formate > hydroxide > fluoride > acetate.

11.1.5 Influence of the distribution coefficients

The partition coefficients and thus the separation factors are influenced by the presence of other ions competing for the functional groups of the exchanger; furthermore, in the case of exchange of unequal ions, the partition coefficients depend on the concentration, as already mentioned.

Apart from these influences, which are mostly due to the system, the partition coefficients can be influenced by the addition of complexing agents, by exchange in non-aqueous solvents and by using special exchangers with selectively acting functional groups.

Complexing agents can change the size, number of charges, or sign of charge of an ion.

For example, the separation of the Cu^{2+} ion from Na^{+} ions is significantly improved by exchange in ammoniacal solutions compared to acidic solutions, as the tetraammine complex is formed. In the case of the Fe^{2+} ion, complex formation with dipyridyl results in an increase in the partition coefficient.

Complexation can further trap ions wholly or partially from equilibria; e.g., Hg^{2+} is no longer taken up from chloride-containing solutions by formation of the virtually undissociated $HgCl_2$ from cation exchangers. Another example of this effect is the formation of citrate or tartrate complexes of rare earths, which are only slightly dissociated in weakly acidic solutions; since the degree of dissociation differs among the individual earths, the already existing (albeit small) differences in partition coefficients are further enhanced.

Furthermore, in many cases, especially for cations, it is possible to obtain ions with reversed charge sign by forming complexes. For example, the UO_2^{2+}-ion in sulfuric acid solutions gives the negative $UO_2(SO_4)_3^{4-}$-complex, and in addition numerous metal ions form anionic complexes with inorganic and organic acids such as HCl, HBr, HSCN, tartaric acid, citric acid, etc. Such complexes are no longer taken up by cation exchangers but by anion exchangers and can thus be separated from numerous other cations which do not react with the complexing agent in question or do not react in the same way.

In *non-aqueous solvents,* the exchange of ions is basically the same as in aqueous solutions, although the dissociation of the dissolved compounds is less. The capacity of the exchangers should in itself be independent of the solvent; however, one often obtains apparently smaller capacities in organic liquids than in water, since the exchange rates tend to be much lower, so that the equilibria are not reached even after a relatively long period of experimentation.

The partition coefficients and separation factors depend considerably on the solvent; often a significant improvement of the separation effects can be achieved by changing the solvent (cf. Tab. 11.7).

Tab. 11.7: Improvement of Ag^+ – Na^+ – and Na^+ – H^+ -separation during the transition from aqueous to methanolic solutions.

	H_2O	CH_3OH
$\beta^{Ag^+}_{Na^+}$	2,9	12,3
$\beta^{Na^+}_{H^+}$	1,5	3,2

If the solvent is not completely replaced by another, but only modified by adding a second liquid, the separation factor depends on the mixing ratio of the components; it may show a maximum at a certain ratio (cf. Fig. 11.9).

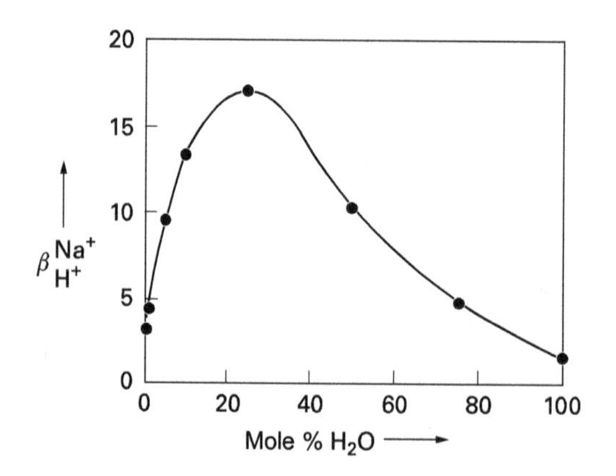

Fig. 11.9: Dependence of the separation factor β during Na^+ – H^+ exchange on the mixing ratio H_2O:CH_3–OH of the solution.

Exchangers with selectively acting functional groups were first prepared by Skogseid (1948); a resin with groups derived from dipicrylamine gave clear selectivity for potassium ions:

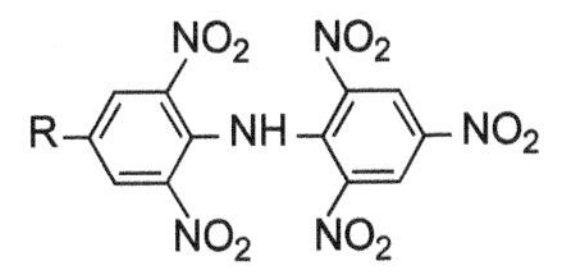

Later, numerous other exchangers with various complex-forming groups were described (cf. Tab. 11.8), but these have not found wider application in analytical chemistry because the selectivities are usually not very pronounced.

Tab. 11.8: Exchangers with selective functional groups (examples).

Functional group	Polymer frame	Selectivity given for
$-N(CH_2COOH)_2$	Polystyrene	Cu^{2+}, Fe^{3+}, Alkaline earths
8-Hydroxyquinoline	Phenolic resin	Heavy metals
–SH	Polystyrene	Hg^{2+}
$-Hg^{+}$	Phenolic resin	Mercaptans
Ethylenediaminetetraacetic acid	Phenolic resin	Ca^{2+}, Ni^{2+} et al.
Anthranilic acid	Phenolic resin	Zn^{2+}, Co^{2+}, Ni^{2+}
Hydroxamic acid	Polyacrylate	Fe^{3+}
o,o′-Dihydroxy-azo group	Phenolic resin	Cu^{2+}
–COOH + –OH	Alginic acid	Fe^{3+}

11.1.6 Rate of equilibration

In contrast to ion reactions in homogeneous solutions, the equilibria at solid ion exchangers are established relatively slowly, since the ions must first diffuse into the interior of the individual particles.

The exchange rate depends on numerous factors (type of exchanger, degree of crosslinking, grain size, temperature, type and valence of the exchanging ions); therefore, no specific values can be given for this quantity. Under favorable conditions, which are present in resin exchangers with SO_3H groups, weak to moderate crosslinking and small grain size, equilibria are approximately reached in a few minutes; exchangers with COOH groups as well as anion exchangers usually require much longer times. As already mentioned, the exchange in organic solvents is particularly slow.

In Fig. 11.10, the approximation to equilibrium as a function of time is given for different ions.

Various attempts have been made to increase the exchange rate. Leaving aside the possibility of increasing the temperature, resins can be used which carry functional groups only on the surface of the individual particles, so that diffusion of the ions into the interior is omitted. Such exchangers can be obtained, for example, by very short sulfonation of polystyrene granules. Inorganic precipitates should also be mentioned here,

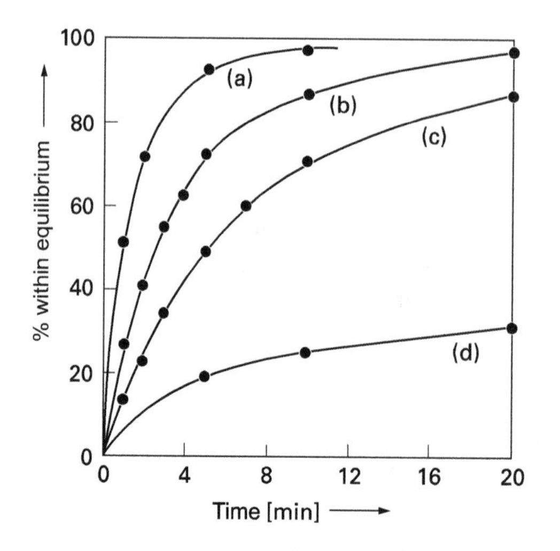

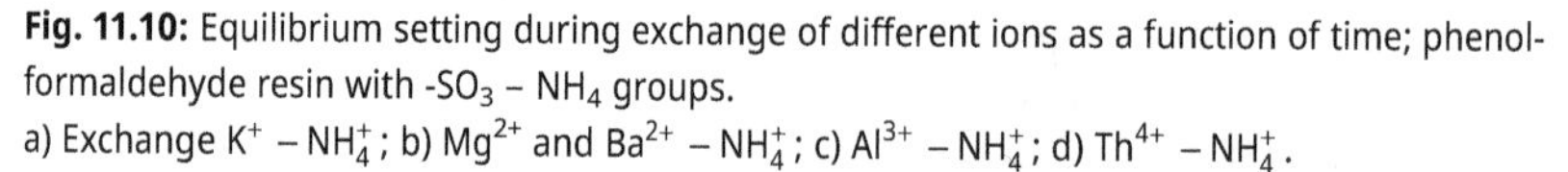
Fig. 11.10: Equilibrium setting during exchange of different ions as a function of time; phenol-formaldehyde resin with -SO_3 – NH_4 groups.
a) Exchange K^+ – NH_4^+; b) Mg^{2+} and Ba^{2+} – NH_4^+; c) Al^{3+} – NH_4^+; d) Th^{4+} – NH_4^+.

where only the surface layer is capable of exchange. In both cases, however, the increase in exchange rate is at the expense of capacity.

Another principle lies in the use of resins whose pores are much larger than those of the usual exchangers, which thus have, as it were, a foamy structure. Such *macroporous* exchangers are produced by polymerization in the presence of certain additives, and significantly improved separations have been achieved with them.

11.1.7 Side effects of ion exchange: neutral salt adsorption – uptake of nonelectrolytes – sieve effect – irreversible adsorption

Ion exchangers practically only accept ions from dilute solutions whose charge is opposite to that of their functional groups, i.e. they exclude ions with the same charge sign. However, when the salt concentration in the solution is increased more strongly, both types of ions can increasingly penetrate the exchanger. This effect, known as *neutral salt adsorption*[4], differs from ion exchange proper in that the additional amount of solute can be washed out with water. Since dilute solutions are generally used in analytical separations, it does not play a major role here.

Absorption of nonelectrolytes. If a solution contains nonelectrolytes, these enter the exchanger with the solvent during the swelling process. The concentration in the exchanger differs from that in the solution, since adsorption can still take place on the

4 One also finds the expression *neutral salt absorption*.

exchanger framework. Weak electrolytes can behave like nonelectrolytes if the undissociated portion strongly predominates.

This effect differs from neutral salt adsorption in that the nonelectrolytes are absorbed even at low concentrations in the solution. On the other hand, the nonelectrolytes, like the neutral salts, can easily be washed out of the exchanger again with pure solvent. The uptake of nonelectrolytes depends strongly on the solvent.

Sieve effect. In general, the larger the diameter of ion exchange resins, the better they absorb ions. However, if this exceeds the width of the channels inside the exchanger particles, exchange is only possible to a negligible extent at the surface of the solid phase for spatial reasons.

If several ion types of different sizes are present in the solution, those that no longer fit into the channels of the exchanger are, in a sense, sieved out and separated from the smaller ions. Such separations depend very much on the type of exchanger, in particular on its degree of crosslinking. However, the screening effect is quite blurred in the most commonly used synthetic resin exchangers, since the channels have different pore widths.

Irreversible adsorption. Occasionally, irreversible adsorption phenomena have been described on ion exchangers, e.g. of organic dyes and especially of inorganic ions in trace analyses and separations of radioactive substances in imponderable quantities. This effect is observed mainly with anion exchangers and weakly acidic cation exchangers, whereas it tends to be much less pronounced with the much-used polystyrene resins with SO_3H-groups.

11.2 Separations by one-time equilibration

The one-time equilibration method requires a high partition coefficient of the ion to be separated. The solution is stirred with a portion of the exchanger for a sufficiently long time; however, this simple procedure is only rarely applicable, e.g. for acid-base exchange and in the presence of complex chloro or bromo acids.

For example, NH_3 and thiamine can be quantitatively removed from urine with zeolites; anions and nonelectrolytes such as urea remain in the aqueous phase. Similarly, enzymes have been isolated with DEAE cellulose.

For trace detection of ions in aqueous solutions, a grain of exchanger is added to the liquid and a color reaction is carried out after the ion in question has accumulated on the exchanger; effective separations are achieved by this method when the ions in question are present as chloro or bromo acids.

Instead of exchangers in granular form, exchanger membranes can also be used. Thus, various cations were enriched on slices of cation exchange membranes in the H^+ form from very dilute solutions and then determined directly on the membrane, e.g. by X-ray fluorescence.

Furthermore, the decomposition of hardly soluble precipitates with ion exchangers should be mentioned here. Precipitates that are trace soluble are slurried in water and brought into contact with an exchanger; in the process, one of the dissolved ions is continuously intercepted so that the solubility equilibrium is disturbed and the precipitate is gradually dissolved. The exchanger must be present in large excess, and because of the usually low dissolution rate, the use of elevated temperatures is recommended.

In this way, $CaSO_4$ -, $SrSO_4$ – and $BaSO_4$-precipitates were decomposed with cation exchangers in the H^+-form, furthermore poorly soluble phosphates such as apatite, organic compounds such as calcium mandelate, and others.

11.3 Separations by one-sided repetition

In the vast majority of cases, the partition coefficients in ion exchange are not sufficiently large to remove an ion virtually completely from a solution in a single partition step; the exchange reaction must then be repeated. In this case, ions with opposite charge as well as nonelectrolytes remain quantitatively in the liquid phase, so that the conditions for the application of one-sided repetition are given.

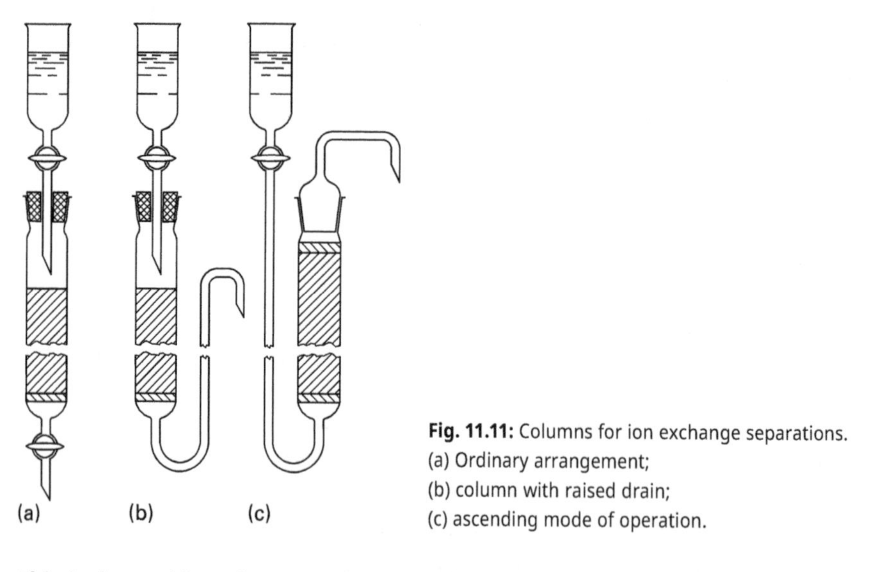

Fig. 11.11: Columns for ion exchange separations. (a) Ordinary arrangement; (b) column with raised drain; (c) ascending mode of operation.

This is done with exchanger columns through which the analysis solution is poured. While one type of ion is gradually retained in the various layers of the exchanger bed, the other flows unhindered through the column together with any nonelectrolytes present; residual solution is removed from the exchanger bed by rinsing with pure solvent.

The columns usually consist of a glass tube with the exchanger held in place by a frit or by a glass wool plug at the bottom. Above the column is usually a reservoir containing solvent, the supply of which can be controlled by a tap. To prevent the column from running empty, which would cause air bubbles that are difficult to remove to enter the exchanger bed, the outlet is often raised or the direction of flow reversed (see Fig. 11.11).

When making the columns, the tube must be about half filled with water (or organic solvent); then the exchanger is added with frequent shaking so that a uniform dense filling without air bubbles is obtained.

Synthetic resin exchangers must be pre-swollen before filling, otherwise the column may be burst by the swelling pressure.

For separations by one-sided repetition, a few separation steps are usually sufficient, so that the columns can be relatively short. The breakthrough capacity is of decisive importance: the column must contain enough exchanger to ensure that the quantity of ions to be separated is retained completely.

However, the breakthrough capacity of an ion exchange column depends not only on the amount of exchanger, but is also influenced by several other factors (flow rate of the solution, particle size of the exchanger, temperature, initial concentration of the solution, etc.). In particular, the exchange capacity for a given ion is reduced by the presence of other competing ions; this is especially true for H^+ ions (cf. Fig 11.12).

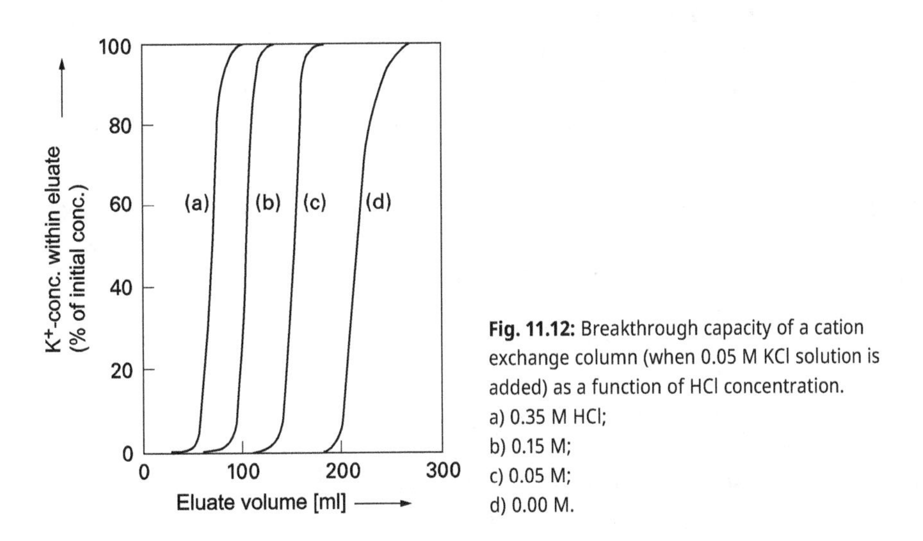

Fig. 11.12: Breakthrough capacity of a cation exchange column (when 0.05 M KCl solution is added) as a function of HCl concentration.
a) 0.35 M HCl;
b) 0.15 M;
c) 0.05 M;
d) 0.00 M.

The column dimensions must therefore be selected taking into account the circumstances in each case. For analytical separations on a normal scale, columns of about 20–30 cm in length and 1–2 cm in diameter are generally sufficient; for small weights, microcolumns (e.g., with a length of 5 cm and a diameter of 4–6 mm) have also been described.

Applications. By one-sided repetition, numerous important separations of anions and cations can be carried out, of which only the removal of the often interfering phosphate ion as well as cationic and anionic surfactants will be mentioned. The separation possibilities are increased by the fact that a large number of inorganic cations can be converted into anionic complexes.

Some specific applications are discussed in more detail below.

Complete desalination. If a solution is treated successively with a cation exchanger in the H^+ form and an anion exchanger in the OH^- form, all ions are removed from it and the solution is *desalinated*. The method is mainly used for the analytical examination of carbohydrate and protein solutions; it is also used on a technical scale for water treatment.

Conversion of salts into the equivalent amount of acid or base. If a salt solution is allowed to flow through a column with a cation exchanger in the H^+ form, the solution passed through contains an equivalent amount of acid which can be determined by titration.

In the same way, acid solutions with a specific content (normal solutions) can be prepared from weighed quantities of salts. Corresponding reactions can be carried out with anion exchangers in the OH^- form.

Separations by the sieve effect. The sieve effect can be used to separate small ions from large ones of the same charge sense; e.g., inorganic ions can be removed from solutions of high-molecular-weight organic dyes or S^{2-} – and CS_3^{2-}-ions from solutions of cellulose xanthate. The sieve effect is also used to purify colloidal solutions of low molecular weight electrolytes.

Precipitation exchange. The so-called *precipitation exchange*, in which exchange reactions are carried out on the surface of sparingly soluble inorganic precipitates, is also to be counted among the methods discussed here. However, since the precipitates in question behave somewhat differently from the actual ion exchangers (low capacity, lack of reversibility), this method will be dealt with later in the discussion of precipitation reactions (chap. 12).

11.4 Separations by systematic repetition: Column methods (ion exchange chromatography)

11.4.1 Principle

The above-mentioned differences in the exchange equilibria of equally charged ions allow separations to be carried out even of ions of the same charge sign. Since the separation effects are usually small, a larger number of separation stages is required than for the separation of unequally charged ions.

In *ion exchange chromatography*, the usual column arrangements are used. The substance mixture is applied to the column so that the ions to be separated are fixed in a narrow zone at the head of the separation bed, and eluted with a suitable liquid.

Acid, base or salt solutions are used for elution, by means of which the ions present on the exchanger are displaced to a greater or lesser extent.

11.4.2 Column material and dimensions

The separation columns used in ion exchange chromatography are usually made of glass; columns for eluting under pressure are usually made of stainless steel. The ratio of length to diameter should be at least 100:1; columns of about 80–200 cm length with 0.8–2 cm diameter have proved successful. For very small amounts of substance, a polyethylene column with only 0.2 mm^2 cross-section at 1.6 m length has been used. High-performance pressure columns have been described up to about 300 cm in length at diameters of 0.1–0.6 cm.

11.4.3 Stationary and mobile phase – loading

As with the other chromatographic methods already discussed, the particle size of the stationary phase should be as small as possible in ion exchange chromatography, since the efficiency of the column then increases.

Most inorganic exchangers accumulate as precipitates with extremely fine grain size. To prevent the flow resistance of columns with such fillings from becoming too great, these exchangers are usually mixed with an inert coarser material.

Resin exchangers are available in various grain sizes. It is favorable that they can be produced with very uniform spherical particles. For the usual separation columns, particle diameters of about 0.1–0.2 mm have proved successful; for micro and high-performance columns, the exchanger particles must be much smaller (about 20–50 μm diameter), and for separations with elution under pressure, particle sizes as small as 5–10 μm have even been used. The dependence of the separation level on the particle size is shown for a cation exchange column (Dowex 50) in Tab. 11.9.

Tab. 11.9: Separation plate height of a cation exchange column as a function of the particle size.

Grain size (mm)	Separation plate height (mm)
0,15–0,30	11
0,07–0,15	6
0,04–0,07	5
< 0,04	4

When eluting under higher pressures, the low mechanical stability of swollen resin exchangers becomes noticeable due to blockages of the columns; for this purpose, exchangers are used which are coated as a thin layer on a solid support, e.g. on glass beads.

An important parameter for resin exchange resins is the degree of crosslinking. On the one hand, swelling and pore size decrease with increasing crosslinking, and the granules become mechanically stronger. On the other hand, it becomes more and more difficult for the ions from the surrounding solution to penetrate into the interior of the resin particles, so that the separation efficiency decreases (cf. Tab. 11.10). As a compromise, exchangers with medium crosslinking are used.

Tab. 11.10: Separation level of an ion exchange column as a function of the degree of crosslinking.

Exchanger	Separation plate height (mm)
Dowex 50 × 2	1,5
× 4	3,4
× 8	4,8
× 12	11

Aqueous solutions of acids, bases or salts of various concentrations are predominantly used as mobile phases. Of great importance are additions of complexing agents, through which separation factors can often be substantially increased. Occasionally, additions of organic, water-miscible liquids (e.g. of lower alcohols, acetone, etc.) bring advantages (cf. Fig. 11.9); more rarely, separations are carried out by elution with purely organic solvents.

Only enough analyte should be loaded onto ion exchange columns to ensure that no more than approx. 5% of the separating bed is used; at higher loadings, the number of separation stages is reduced accordingly.

11.4.4 Influence of the flow rate of the mobile phase

The same considerations apply to ion exchange chromatography as to the other column methods: If the flow rate of the mobile phase is high, the equilibria are not fully established; if the flow velocity is very low, the separation effect suffers due to broadening of the bands as a result of diffusion in the liquid phase; in addition, the time required for a separation then becomes high.

The flow rate is usually expressed in ml eluate/min per cm^2 column cross-section; values of about 0.1–1 ml/min per cm^2 have been found to be favorable.

11.4.5 Influence of the temperature

By changing the temperature, both the separation plate height of the column and the separation factor of two ions can be affected. While one can usually expect the separation plate height to decrease with increasing temperature as a result of better equilibration, little can be predicted about the effect on the separation factor, and improvement of separations by increasing temperature cannot always be expected. On the other hand, elution can be faster at higher temperatures because of the decreasing viscosity of the moving phase.

Although an increase in temperature is usually likely to be favorable in ion exchange chromatography, for simplicity it is usually done at room temperature. This also reduces the attack of the elution solution on the exchanger.

11.4.6 Stepwise elution – gradient elution

To accelerate separations in ion exchange chromatography, elution is often carried out in steps with solutions of increasing efficiency. This usually involves increasing the acid or salt concentration of the liquid (see Fig. 11.13).

The same applies to gradient elution (cf. Fig. 11.15). Temperature gradients are used less frequently.

11.4.7 Displacement technique

Since any ion exchange reaction can be considered as a *displacement*, the difference between the elution and the displacement technique should be clarified again: In displacement, the mobile phase contains an ion that is bound much more tightly by the exchanger than any of the ions contained in the analytical sample, and is therefore the last to leave the column. In elution, there is one ion in the solution that is held less tightly by the exchanger than any of the ions contained in the analytical sample; it therefore migrates most rapidly through the column, but is still able to carry the other ions through the separation bed because it is always replenished by the inflowing elution solution.

The displacement technique is relatively seldom used in ion exchange, although it permits a stronger loading of the columns (up to about 50% of the capacity) and thus the throughput of larger substance quantities. A disadvantage is the overlapping of neighboring zones, which cannot be completely eliminated.

11.4.8 Isolation of components – detectors

The isolation of separated components is usually done with automatically operating fraction collectors.

More common is the automatic recording of elution curves; especially the measurement of light absorption in the visible or ultraviolet range (possibly after addition of color reagents, e.g. of ninhydrin) in flow-through cuvettes is suitable for this purpose. Furthermore, continuous measurement of electrical conductivity, refractive index, pH, radioactivity or continuous amperometric or voltammetric detection are used.

11.4.9 Efficacy and scope of the method

Since in ion exchange chromatography one usually works in the linear initial part of the distribution isotherms, the elution bands are usually symmetrical. The separation plate height is about 8–10 times the grain diameter of the stationary phase; for example, values of 0.1–0.4 (maximum 1.1) mm are given for grain sizes of about 10–50 µm, and separation plate heights of 2–3 mm can be expected for the frequently used grain sizes of 0.2–0.3 mm. The separation plate heights may increase sharply with very slowly exchanging resins (e.g. weakly basic anion exchangers) as a result of incomplete equilibration.

With the usual columns of 1–2 m length and 0.2–0.3 mm particle size of the exchanger, separation plate numbers of about 300–1000 can thus be achieved, with high-performance columns of 10 000 and more.

Of the numerous separations of inorganic as well as organic ions, only a few characteristic examples will be given.

A number of applications of inorganic exchangers are given in Tab. 11.11; these have proved particularly useful for the separation of alkali metal ions (cf. Tab. 11.12 and Fig. 11.13).

Tab. 11.11: Application of inorganic ion exchangers.

Exchanger	Examples of separations
Ammonium phosphomolybdate	$Na^+ - K^+ - Rb^+ - Cs^+$; $Cs^+ - Sr^{2+}$; $Ag^+ - Pd^{2+}$; $Tl^+ - Fr^+$
Ammonium phosphotungstate	$Na^+ - K^+ - Rb^+ - Cs^+$; $Cs^+ - Sr^{2+}$, Y^{3+}
Zirconium phosphate	$Rb^+ - Cs^+$; $Rb^+ - Sr^{2+}$; $Cs^+ - Ba^{2+}$; Amino acids
Zirconium molybdate	$Ca^{2+} - Sr^{2+} - Ba^{2+} - Ra^{2+}$; $Rb^+ - Cs^+$
Zirconium tungstate	$Li^+ - Na^+ - K^+ - Rb^+ - Cs^+$; $<Co^{2+} - Fe^{3+}$
Zirconia hydrate	Alkali metal ions – rare earth ions; $Na^+ - Cs^+$
Zirconium silicate	$F^- - OH^-$
Titanium(IV) phosphate	$Sr^{2+} - Y^{3+}$
Titanium dioxide hydrate	$PO_4^{3-} - CrO_4^{2-}$
Hydroxyapatite	$F^- - OH^-$
Glass	$F^- - OH^-$

Tab. 11.12: Partition coefficients α and separation factors β in the separation of alkali metal ions with different exchangers.

Exchanger	$^{\alpha}Cs$	$^{\alpha}Rb$	$^{\alpha}K$	$^{\alpha}Na$	$^{\beta}Cs/Rb$	$^{\beta}Rb/K$
Ammonium phosphomolybdate	5300	192	approx. 4	approx. 0	27,6	approx. 48
Ammonium phosphotungstate	3500	136	approx. 5	approx. 0	25,7	approx. 27
Organ. resin (Dowex 50, NH_4-form)	62	52	46	26	1,19	1,13

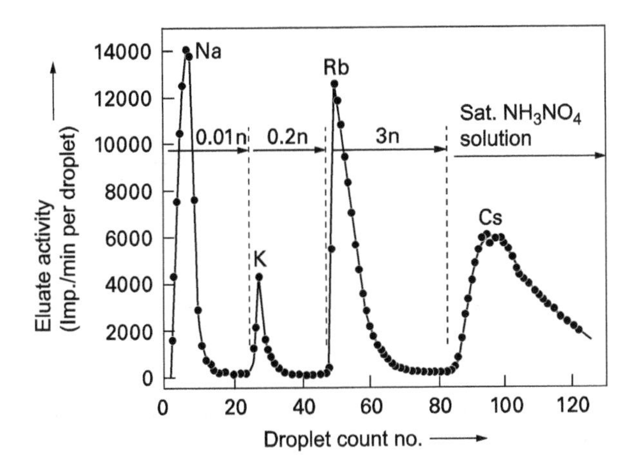

Fig. 11.13: Separation of alkali metal ions on an ammonium phosphomolybdate column; stepwise elution with ammonium nitrate solution.

An example of separations with the addition of complexing agents is given in Fig. 11.14. The alkaline earth metal ions cannot be readily separated on cation exchangers with strongly acidic functional groups; only by adding lactate to the mobile phase are the separation factors increased to such an extent that the separation Ca^{2+} – Sr^{2+} – Ba^{2+} is satisfactory.

Polyphosphates can be separated with columns of strongly basic anion exchangers; eluting with a buffer solution pH 8 and KCl gradient (Fig. 11.15).

Ion exchange chromatography has become extremely important in the field of biochemistry. Of the numerous applications, only the most important, the analysis of amino acid mixtures, should be mentioned; the process shows a remarkable separation efficiency, and it has also been possible to automate it (cf. Fig. 11.16). For the separation of other biological substances (e.g. enzymes, nucleotides, polysaccharides, etc.), the above-described modified cellulose exchangers continue to be used to a considerable extent.

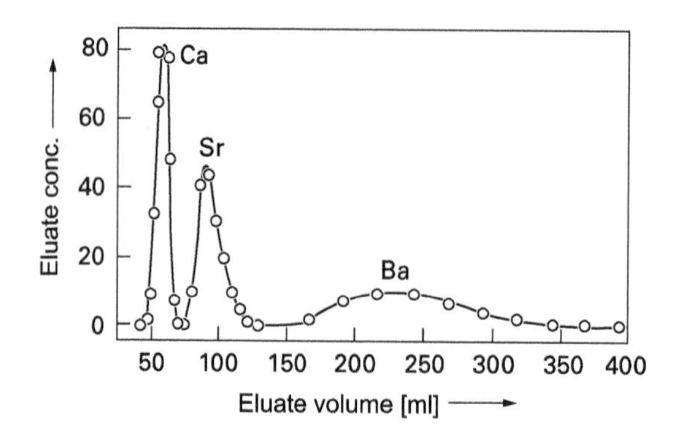

Fig. 11.14: Separation of alkaline earth metal ions on a cation exchange column; elution with lactate buffer solution.

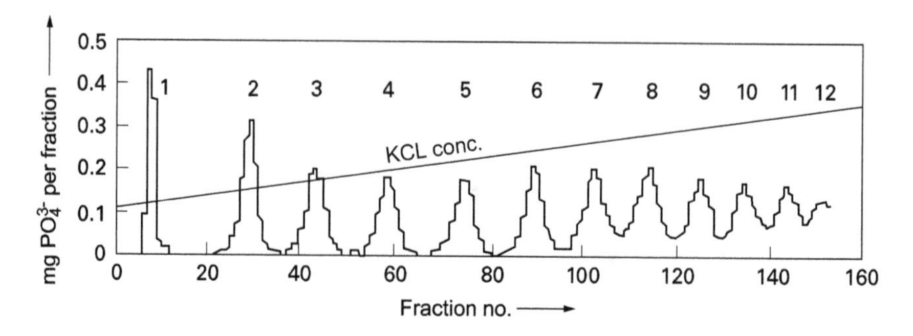

Fig. 11.15: Separation of polyphosphates on an anion exchange column; elution with KCI gradient. In the upper part, number of P atoms of the eluted compounds.

Finally, separations under ligand exchange reactions should be mentioned. One loads cation exchange columns with Cu^{2+}, Ni^{2+} or Ag^{+} and transfers the metal ions on the exchanger with NH_3 into the ammine complexes. Organic amines or amino acids can then be separated on such columns.

Separations by ligand exchange probably also include the processes known as *salt-exchange chromatography* and *solubilization chromatography*. In the former, alcohols, among others, are separated on ion exchangers by adding salts such as $(NH_4)_2SO_4$ to the aqueous starting solutions. The mechanism is considered to be the replacement of hydrate water in the exchanger with alcohol. Higher alcohols are kept very solid in this process; to elute them, *solubilization chromatography* is used by eluting with aqueous methanol solutions. Compounds of other substance classes (e.g. phenols, ketones, fatty acids, ethers, etc.) have also been separated by this method.

The advantages of ion exchange chromatography are the high separation efficiency and the extremely versatile applicability in both inorganic and organic analysis. A further advantage is the relatively high capacity of the resin exchangers in particular, which makes it possible to recover even somewhat larger quantities of substance.

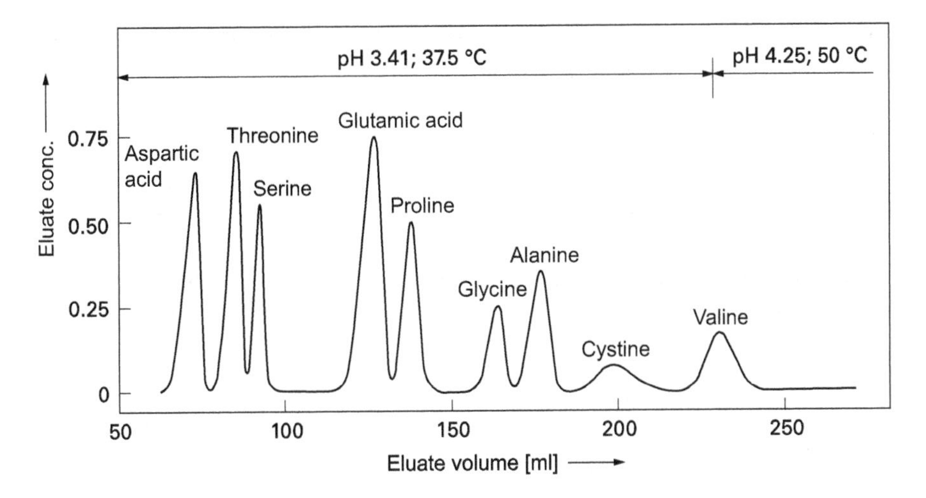

Fig. 11.16: Separation of amino acids with an ion exchange column; strongly acidic cation exchanger, Na^+ form.

However, the method is essentially limited to aqueous solutions and to the separation of ions. A disadvantage is the rather large time requirement due to the slow adjustment of the exchange equilibria. Furthermore, small amounts of the exchange material always go into solution, which can interfere with the further analysis of the eluates, and finally, certain proportions of the substances to be separated can be lost due to irreversible adsorption; however, this effect is only noticeable in trace analyses.

11.5 Separations by systematic repetition: Planar technique (thin-layer ion exchange chromatography)

Compared with the column technique, the thin-film method has played a relatively minor role in ion exchange to date. Thin-film plates can be made from granules of synthetic resin exchange resins fixed to the substrate by a binder, such as silica gel. Plates of zirconium phosphate, dioxide hydrate, hydroxyapatite, or of salts of heteropolyacids have also been recommended for special separations. Plates made of modified cellulose substitutes are easier to prepare.

More commonly used are papers impregnated with colloidal solutions of ion exchange resins in which precipitates with ion exchange properties have been produced or which consist entirely of modified celluloses.

With such exchange layers it is possible to develop chromatograms in ascending or descending order, as well as to work two-dimensionally or according to the method of radial chromatography.

The advantage of the method is that the separations are faster than when working with columns; however, the number of separation stages and the feed quantity are much lower.

11.6 Countercurrent method

For technical purposes, apparatuses have been developed in which the solution and solid exchanger are continuously passed against each other, but such arrangements have not attained analytical significance.

General literature

T. Cecchi, Ion-interaction chromatography:, in J. Cazes, Encyclopedia of Chromatography 2, 1276–1279 (2010).

J. Fritz, Early milestones in the development of ion-exchange chromatography: a personal account, Journal of Chromatography A *1039*, 3–12 (2004).

K. Gooding, Ion exchange: mechanism and factors affecting separation, in: J. Cazes, Encyclopedia of Chromatography *2*, 1258–1261 (2010).

K.L. Nash & M.P. Jensen, Analytical-scale separations of the lanthanides: a review of techniques and fundamentals, Separation Science and Technology *36*, 1257–1282 (2001).

A. Nordborg & E.F. Hilder, Recent advances in polymer monoliths for ion-exchange chromatography, Analytical and Bioanalytical Chemistry *394*, 71–84 (2009).

I.N. Papadoyannis & V.F. Samanidou, Ion chromatography: suppressed and non-suppressed, in J. Cazes, Encyclopedia of Chromatography *2*, 1247–1250 (2010).

J. Riviello, Water, electricity and ion exchange; how Hamish Small sustained the evolution of Ion Chromatography, Heliyon *7*, e07495 (2021).

Inorganic exchangers

H. Hikichi, K. Iyoki, Y. Yanaba, T. Okubo & T. Wakihara, Superior ion-exchange property of amorphous aluminosilicates prepared by a co-precipitation method, Chemistry – An Asian Journal 15(13), 2029–2034 (2020).

A. Mushtaq, Inorganic ion-exchangers: Their role in chromatographic radionuclide generators for the decade 1993–2002, Journal of Radioanalytical and Nuclear Chemistry *262*, 797–810 (2004).

S. Sarkar, P. Prakash & A.K. SenGupta, Polymeric-inorganic hybrid ion exchangers: preparation, characterization, and environmental applications, Ion Exchange and Solvent Extraction *20*, 293–342 (2011).

J. Siroka, P. Jac and M. Polasek, Use of inorganic, complex-forming ions for selectivity enhancement in capillary electrophoretic separation of organic compounds, Trends in Analytical Chemistry *30*, 142–152 (2011).

P. Vijayan, P, Chithra, S. Krishna, E. Ansar & J. Parameswaranpillai, Development and current trends on ion exchange materials, Separation and Purification Reviews, in press (2023).

Cellulose exchanger

N. Grubenhofer, Cellulose ion exchangers (Book Chapter), in: K. Dorfner, Ion Exchangers. De Gruyter, Berlin, 443–460 (2011).

S. Li, Y. Wang, L. Qiao & K. Du, Fabrication of self-reinforced polymorphic cellulose nanofiber composite microspheres for highly efficient adsorption of proteins, Cellulose *29*, 5191–520 (2022).

G. Salfate & J. Sánchez, Rare earth elements uptake by synthetic polymeric and cellulose-based materials: A Review, Polymers *14*, 4786 (2022).

F. Svec, Organic polymer support materials, Chromatographic Science Series *87* (HPLC of Biological Macromolecules), 17–48 (2002).

12 Solubility: Precipitation methods

12.1 General

12.1.1 Historical development

The formation of precipitates by adding reagents to solutions was already known in the Middle Ages, as was the phenomenon of cementation. Systematic studies of separations by precipitation were carried out primarily by Boyle, and later by Berzelius and C.R. Fresenius. The analytical application of electrolysis goes back to Gibbs (1864) and Luckow (1865).

The first organic precipitating reagent for inorganic ions, α-nitroso-β-naphthol, was recommended by Ilinski and Knorre (1885) for the precipitation of cobalt. The urea inclusion compounds were discovered by Bengen (1940).

We owe a detailed study of supersaturation phenomena primarily to v. Weimarn (1926). The precipitation from homogeneous solution was introduced by Chancel (1858), but only worked on in more detail by Moser (1922).

12.1.2 Definitions

A solution is *saturated* when it is in equilibrium with its soil body. It is *unsaturated* if it contains less, and *supersaturated* if it contains more solute than corresponds to equilibrium. (The often used expressions *dilute* and *concentrated* solution are not exactly defined).

The product of the concentrations of the individual ions of a dissociating substance (each in mol/l) in a saturated solution is the *solubility product.*

12.1.3 Auxiliary phases – precipitation reactions – precipitation pH values – speed of precipitation reactions

In precipitation, predominantly only one *auxiliary phase*, the solvent, is used. This results in deviations in the method of operation compared to the separation methods discussed so far: Although two phases are present after precipitation, one of them consists entirely of the deposited substance; a distribution in the sense mentioned above, in which the concentrations in both phases can be varied, is thus not present. Furthermore, the repetition of the separation in precipitation is more difficult, since the precipitates can usually only be dissolved again slowly.

https://doi.org/10.1515/9783111181417-012

Separations by precipitation in the presence of two auxiliary phases are possible, but are used only for special problems; peculiarities also occur in these procedures (see below).

Precipitation reactions. In many precipitations, the precipitate is produced by double reaction; here, the reactions of inorganic and organic ions belong above all. More rarely, precipitates are obtained by addition of two compounds (e.g. inclusion compounds).

Another group of precipitation processes proceeds under oxidation or reduction; such reactions can be brought about not only by the addition of reagents but also by electrolysis.

Finally, a component of a solution can be entrained with a precipitate of another compound generated in the solution; in this case, it is not the exceeding of the saturation concentration but essentially adsorption or solid solution formation that is decisive for the precipitation (co-precipitation).

Precipitation pH values. Many precipitations are significantly influenced by the pH value of the solution. This influence can be graphically represented in several ways.

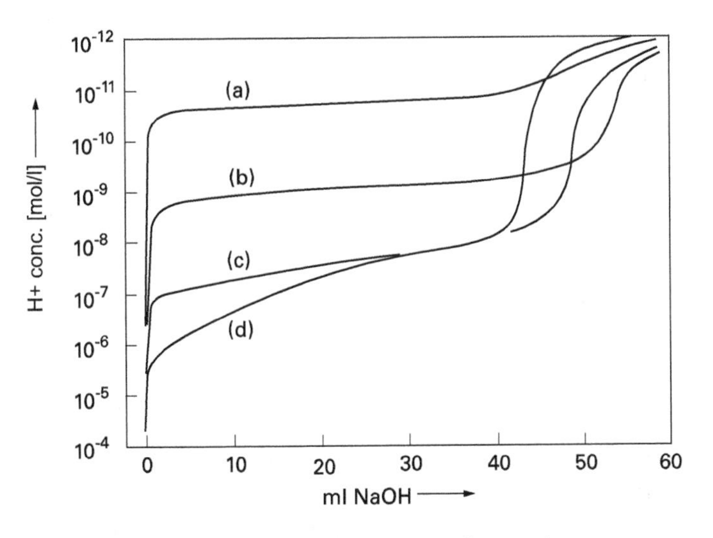

Fig. 12.1: Precipitation curves for the precipitation of magnesium, manganese, cobalt and iron(II) hydroxide with sodium hydroxide solution.
a) Mg^{2+}; b) Mn^{2+}; c) Co^{2+}; d) Fe^{2+}.

If, in the case of hydroxide precipitations, the amount of alkali added is plotted against the pH of the solution, a kink is obtained in the curve at the beginning and end of the precipitation, from which the pH interval of the precipitation can be read (Fig. 12.1). The initial kink, however, is dependent on the concentration of the solute, and furthermore an influence of the anion is often noticeable; the values nevertheless have a certain practical significance, since the precipitations are generally carried out under similar conditions (cf. also Tab. 12.1).

Tab. 12.1: Precipitation pH values of hydroxides for 0.02 M solutions.

Cation	Precipitation pH value	Cation	Precipitation pH value
Mg^{2+}	10,5	Pb^{2+}	6,0
Mn^{2+}	8,5–8,8	Fe^{2+}	5,5
La^{3+}	8,4	Cu^{2+}	5,3
Ag^{+}	7,5–8,0	Cr^{3+}	5,3
Cd^{2+} (chloride)	7,6	Al^{3+}	4,1
Hg^{2+} (chloride)	7,3	Th^{4+}	3,5
Zn^{2+}	6,8–7,1	Hg^{2+} (nitrate)	2
Co^{2+}	6,8	Sn^{4+}	2
Cd^{2+} (sulfate)	6,7	Fe^{3+}	2
Ni^{2+}	6,7	Ti^{4+}	2

Furthermore, the solubility of a compound can be plotted logarithmically against the pH value. For hydroxides, one obtains (because of $L = [Me^{n+}] \cdot [OH^-]^n$ and $[H^+] \cdot [OH^-] = K_W$). Straight line whose slope is given by the valence n of the cation and whose position in the diagram is given by the solubility product L (Fig. 12.2).

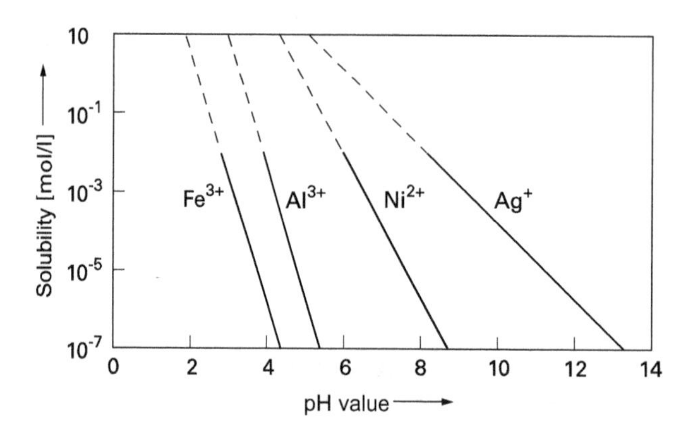

Fig. 12.2: Logarithmic solubility diagram for hydroxides.

Finally, the percentage of precipitate can be plotted as a function of the pH of the solution (Fig. 12.3). In this way, a clear overview of the conditions to be observed during precipitation is obtained, although the curves depend on the initial concentration of the compound present.

Rate of precipitation reactions. Inorganic and organic precipitations by ion reactions proceed very rapidly, and the formation of precipitates of inorganic ions with organic reagents also generally does not require any noticeable waiting times. In contrast, the formation of inner complex salts and of purely organic precipitates is likely to involve more or less considerable delays.

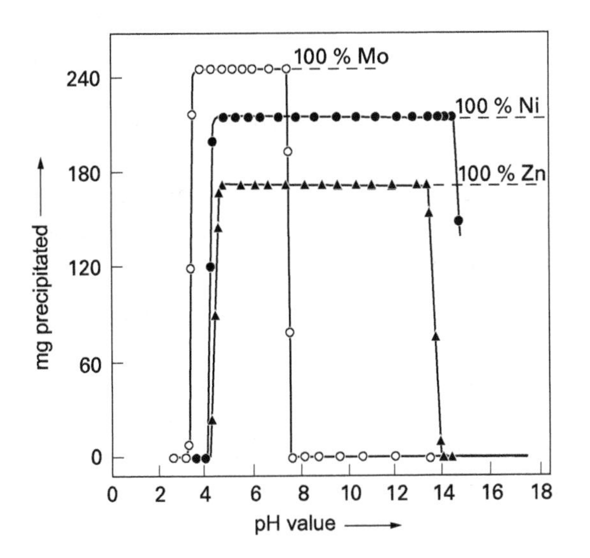

Fig. 12.3: pH ranges for the precipitation of some 8-hydroxyquinoline compounds.

In principle, it is possible to exploit differences in the rates of precipitation reactions for separations, but this effect has rarely been used in analytical chemistry because the separation effects tend to be small. As an example, consider the separation of zirconium and hafnium by precipitation of the o-phenylene-bis-dimethylarsine complexes of $ZrCl_4$ and $HfCl_4$ from tetrahydrofuran solution. In this process, the zirconium complex precipitates faster than that of hafnium, but only partial separation is achieved.

12.1.4 Solubility – influencing the solubility

Since every substance is at least slightly soluble in every solvent, no substance can be removed absolutely completely from a solution by a precipitation reaction. Nevertheless, precipitations can be used for analytical separations, since in many cases the fraction remaining in solution is negligible.

There are many precipitation reactions in which the precipitates are extraordinarily sparingly soluble (e.g. sparingly soluble sulfides, hydroxides, phosphates, compounds with organic reagents, etc.). In numerous other precipitations, however, noticeable losses occur which must be kept within tolerable limits by suitable measures (e.g. in the precipitation of AgCl, $BaSO_4$, $PbSO_4$, SiO_2 and others).

Influencing solubility. The solubility of a compound changes with changes in temperature, changes in pressure, addition of other substances to the solution and changes in the grain size of the solid remaining at the bottom of a solution.

Temperature dependence of solubility. As a rule, solubility increases with temperature to a greater or lesser extent, but negative temperature coefficients are also frequently observed.

The graphical representation of the dependence of solubility on temperature is described in Chap. 14 (Crystallization).

The solubility of solids or of liquids in liquid solvents is so little *dependent on pressure* that this effect is of no analytical significance. In contrast, the solubility of gases is strongly influenced by pressure (see section 8.1).

Change in solubility due to the addition of further substances to the solution. The change in solubility of a compound as a result of the addition of further substances to the solution can be due to various causes.

As a consequence of the validity of the mass action law, a decrease in solubility results for dissociating compounds due to equionic additions (cf. Fig. 12.4, left part of the curve). However, because of the simultaneous change in the activity coefficients, the solubility decrease is usually smaller than calculated from the simple form of the mass action law.

An increase in solubility can occur by complexation (cf. Fig. 12.4, right part of the curve; formation of $HAgCl_2$). In extreme cases, poorly soluble compounds can become easily soluble, e.g. $Fe(CN)_3$ by conversion to $K_3\,Fe(CN)_6$ or AgCl by formation of the $Ag(NH_3)_2$ Cl complex with NH_3.

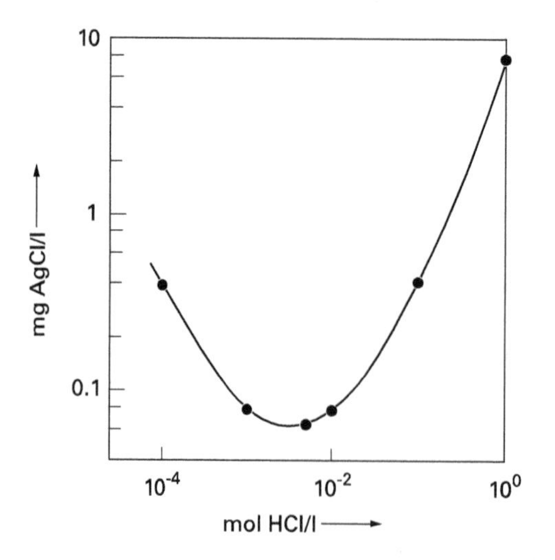

Fig. 12.4: Solubility of AgCl at 25 °C as a function of the HCl concentration of the solution.

The addition of non-equivalent ionic compounds to salt solutions often increases the solubility, since the activity of the salt dissolved first changes. An explanation for this effect is given by the Debye-Hückel theory.

On the other hand, the addition of salts in very high concentrations can reduce the solubility (salting-out effect). Finally, the solubility is often – as already mentioned – strongly dependent on the pH value.

The solubility of a compound can also be changed by adding a second solvent which is miscible with the original one. This is mainly used to reduce the solubility of precipitates.

In general, a strong decrease in solubility occurs with relatively small additions of the second solvent, while further additions have little effect. Finally, with very large additions, the decrease in solubility of the precipitate can be overcompensated by the increase in volume of the solution, which increases the absolute amount of solute.

Usually, about 30–40% of the original volume of acetone or a lower alcohol is added to aqueous solutions. The optimum amount of addition can be determined graphically if the solubility of the precipitate is known as a function of the mixing ratio of the two solvents. For this purpose, the tangent b is applied to the solubility curve a (Fig. 12.5) starting from the lower right corner of the diagram. The perpendicular c drawn from the point of contact to the abscissa gives the most favorable mixing ratio.

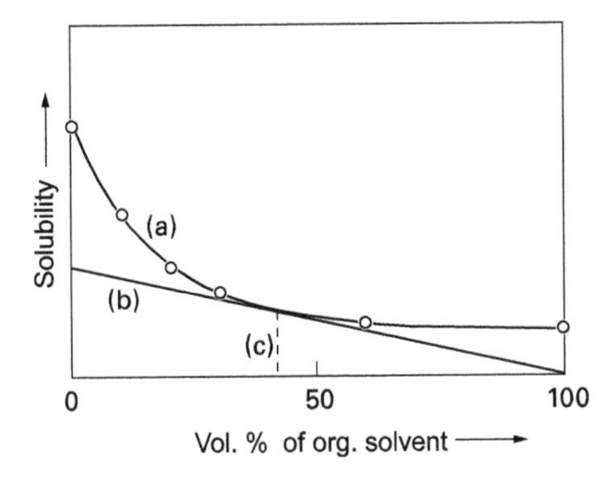

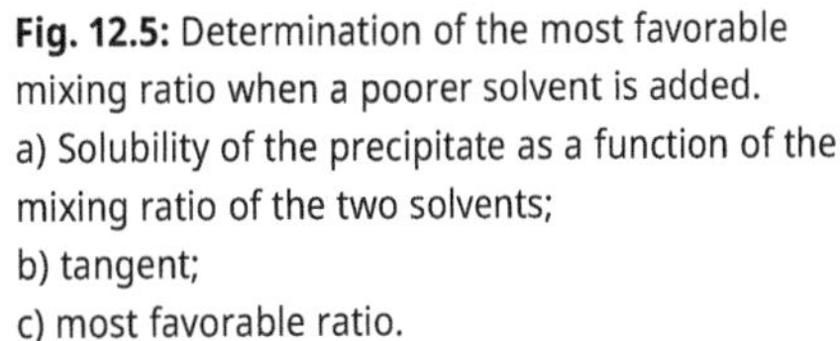

Fig. 12.5: Determination of the most favorable mixing ratio when a poorer solvent is added.
a) Solubility of the precipitate as a function of the mixing ratio of the two solvents;
b) tangent;
c) most favorable ratio.

Change in particle size of the remaining solid. The solubility increases with decreasing particle size of the remaining solid because the surface energy changes. However, this effect is only noticeable for particle diameters on the order of about 1 µm and below. For example, for $PbCrO_4$, solubility was found to be 70% higher than normal at an average particle size of 0.085 µm.

This solubility change plays a role in both precipitation of precipitates and grain enlargement due to longer contact time of solution and precipitate (see below).

12.1.5 Supersaturation – failure to precipitate at low concentrations

At the beginning of the formation of a precipitate, the precipitating particles are naturally very small. Since their solubility is greater than that of coarser precipitates, precipitation can only occur when the solution is supersaturated – with respect to coarser precipitate particles. The tendency to precipitate increases with the degree of supersaturation, i.e. the more supersaturated a solution is, the less durable it is (cf. Figure 12.6).

Supersaturation can be eliminated by *inoculating* the solution with precipitation particles. If supersaturated solutions stand for more or less a long time, crystal nuclei form even without inoculation, partly due to impurities in the solution, partly due to dust particles or roughness of the vessel walls.

Failure to precipitate at very low concentrations. In highly dilute solutions, precipitate formation may fail to occur even if the solubility product is exceeded; the reason for this may be, on the one hand, that the necessary degree of supersaturation is not reached, on the other hand, colloidal solutions may be formed, or the rate of formation of the compound in question may be greatly reduced. As a rule, therefore, even extremely sparingly soluble compounds can no longer be precipitated from highly dilute solutions.

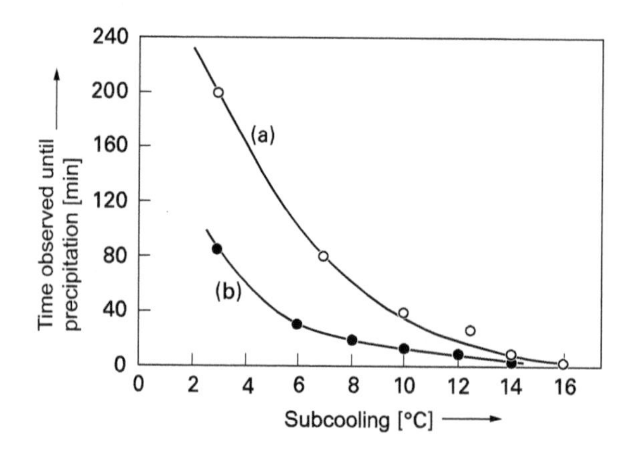

Fig. 12.6: Delay in precipitation of potassium dichromate as a function of subcooling.
a) For a solution saturated at + 30 °C;
b) for a solution saturated at + 60 °C.

Generally applicable numerical values for the lower limit of precipitability cannot be given, but concentrations of about 1 mg/l should normally be required. For example, the absence of precipitation of the potassium tetraphenylborate compound has been observed in the range of 0.5–2 mg K^+/l.

12.1.6 Entrainment effect – reducing the carry away

If precipitates are precipitated from solutions containing other dissolved substances, the precipitates are never completely pure, but always contain – usually small – proportions of the solution comrades. This effect is called *co-precipitation* or *entrainment effect*; it is caused by several factors which may interact to a varying extent from case to case:

- Mechanical confinement of solution residues in cavities of the precipitate;
- formation of compounds on the surface of the precipitate;
- formation of solid solutions; and
- adsorption on the surface of the precipitate.

Furthermore, it is occasionally observed that precipitates that remain in contact with the solution for a longer period of time subsequently absorb dissolved substances; this phenomenon is referred to as *post-precipitation*.

The extent of co-precipitation of foreign substances depends not only on the nature of the precipitate and that of the entrained material, but also in a complicated way on the precipitation conditions and on the presence of other solution comrades.

In amorphous precipitates, the influence of adsorption generally predominates; this is stronger the larger the surface area of the precipitate and the stronger the bond linking the co-precipitated ion to the oppositely charged component of the precipitate.

Furthermore, the smaller the concentration of the entrained component in the solution, the stronger is the co-precipitation by adsorption, a behavior which is understandable by the shape of the adsorption isotherm (according to Freundlich or Langmuir, cf. section 9.1). However, there are exceptions to this rule.

Crystalline precipitates have a particularly strong tendency to entrain if the co-precipitated material can be isomorphically incorporated into the lattice of the crystals or if the precipitate has a large surface area.

The formation of chemical compounds on the surface of precipitates has been perfectly demonstrated in only a few cases; examples include the compound $PbSO_4 \cdot K_2SO_4$, which forms in lead sulfate precipitations in the presence of potassium sulfate, and the compound $(PbCl)_2S$, which co-precipitates when lead sulfide precipitations are carried out in chloride-containing solutions.

The entrainment effect is one of the most important disturbances in separations by precipitation; therefore, numerous methods have been proposed to reduce or eliminate it.

Reduction of the entrainment effect by particle enlargement. Normally, small poorly formed crystals entrain foreign matter more than well-formed larger ones. Particle enlargement can most easily be achieved by slow precipitation in the heat, where the first precipitated particles can still grow during precipitation formation.

For example, in zinc sulfide precipitation, it was observed that the precipitate contained up to 20% cobalt when H_2S was introduced rapidly, but only 0.5% when precipitation was slow.

Another simple method of obtaining larger crystals is to leave the precipitate in contact with the solution for a long time, preferably in the heat. The small particles dissolve slowly as a result of their increased solubility, and the coarser ones grow at their expense. In the process, portions of foreign matter already entrained can be repelled again (cf. Tab. 12.2).

Tab. 12.2: Co-precipitation of NO_3^- with $PbSO_4$ as a function of the aging time of the precipitate; precipitation temperature 95 °C.

Aging time (h)	% NO_3^- in precipitation
0	0,64
1	0,12
6	0,04
24	0,01

However, the method is not generally applicable because when post-precipitates occur, the purity of the precipitate decreases; for example, the adsorption of Ni, Co and Zn on $Fe(OH)_3$ or that of Co on SnS_2 increases when the contact time of precipitate and solution is extended.

The so-called *precipitation from homogeneous solution* also achieves an effective grain enlargement of precipitates. In this case, the precipitating reagent is only slowly generated in the solution during precipitation, so that the strong supersaturation, which must inevitably occur at the drop-in point when precipitating with a reagent solution, does not occur and the crystals can grow undisturbed.

Precipitation reagents are usually produced in homogeneous solution by slow hydrolysis of esters, acid amides, etc., furthermore by oxidation or synthesis in the solution itself (e.g. the reagent cupron is produced from phenylhydroxylamine and $NaNO_2$); furthermore, volatile complexing agents can be slowly removed from the solution by evaporation or destroyed by boiling; examples of this are the homogeneous precipitation of AgCl, AgBr and Ag_3PO_4 by boiling the ammine complexes and the precipitation of tungstic acid by decomposing the H_2O_2 complex.

Probably the most generally applicable method is to raise the pH of the solution by boiling urea, which achieves the precipitation range of hydroxides and other poorly soluble compounds, e.g., the nickel-dimethylglyoxime complex (see Tab. 12.3).

Precipitation from homogeneous solution is usually carried out by prolonged boiling of the mixture after reagent addition, the precipitates separating in dense, easily filterable form and much purer than in the usual mode of operation.

Tab. 12.3: Precipitation from homogeneous solution (example).

Precipitating agent	Initiator	Precipitated compound
Urea	NH_3	
Hexamethylenetetramine	NH_3	Sparingly soluble hydroxides
Acetamide	CH_3COONH_4	
Thioacetamide	H_2S	
Thioformamide	H_2S	
Thiourea	H_2S	Heavy metal sulfides
Trithiocarbon	H_2S	
Thioformanilide	H_2	
Amidosulfonic acid	H_2SO_4	
Dimethyl sulfate	H_2SO_4	
Potassium methyl sulfate	H_2SO_4	Ca, Sr, Ba, Pb
$SO_2 + O_2$	H_2SO_4	
$K_2S_2O_8$	H_2SO_4	
Trimethyl phosphate	H_3PO_4	
Triethyl phosphate	H_3PO_4	Zr, Hf
Metaphosphoric acid	H_3PO_4	
Dimethyl oxalate	Oxalic acid	Ca, Mg, Zn, Th,
Diethyl oxalate	Oxalic acid	Rare earths,
Acetone dioxalic acid	Oxalic acid	Ac, U
Allyl chloride	HCl	Ag
8-Acetoxyquinoline	8-Hydroxyquinoline	Mg, Al, Zn, Th, Fe^{3+}, UO_2^{2-}, Co, Cu, Ni, Pb, Mn, Cd
Diacetyl + NH_2OH	Dimethylglyoxime	Ni, Pd
Phenylhydroxylamine + $NaNO_2$	Copperron	Fe^{3+}, Ti a. o.
$I_2 + ClO_3^-$	HIO_3	Th, Zr
$Cr^{3+} + BrO_3^-$	H_2CrO_4	Pb
$AsO_3^{3-} + HNO_3$	H_3AsO_4	Zr

A method that is similar in its objectives is the so-called *precipitation in a suitable medium*. In this method, the reagent solution and the sample solution are dropped simultaneously into a medium of optimum pH and temperature for precipitation, with intensive stirring. The method has mostly been used for the preparative preparation of various compounds, but has also proved to be efficient in the separation of scandium from ytter earths by hydroxide precipitation and in niobium-tantalum separation.

Co-precipitation can further be considerably reduced by *additions of surface-active reagents* that occupy the active centers of the precipitate. For example, SnS_2 precipitates practically completely free of cobalt if the solution contains some acrolein, and the purity of aluminum hydroxide precipitates is greatly improved by the presence of glycocoll.

The contamination of precipitates can also occasionally be reduced by *favorable selection* of the precipitation pH; for example, the higher the pH of the solution at the end of precipitation, the more copper is entrained in the precipitation of iron(III) hydroxide. One can now first precipitate the bulk of the iron from moderately strongly acidic solution in relatively pure form and then re-precipitate the remainder at higher pH. The

Tab. 12.4: Co-precipitation of Mn^{2+} (50 µg) with calcium phosphate (1.3–1.8 g) in the presence of complexing agents.

Added complexing agent	Quantity (ml 10% solution)	Manganese % co-precipitated
–	–	100
Ethylenediaminetetraacetic acid[x)]	10	20
	20	14
Diethylenetriaminepentaacetic acid[xx)]	10	20
	20	2
N-Hydroxyethylenediaminetriacetic acid[xxx)]	10	14
	20	0

x) $\begin{matrix}HOOC-CH_2\\HOOC-CH_2\end{matrix}>N-CH_2-CH_2-N<\begin{matrix}CH_2-COOH\\CH_2-COOH\end{matrix}$

xx) $\begin{matrix}HOOC-CH_2\\HOOC-CH_2\end{matrix}>N-CH_2-CH_2-N<\begin{matrix}CH_2-COOH\\CH_2-CH_2-N<\begin{matrix}CH_2COOH\\CH_2COOH\end{matrix}\end{matrix}$

xxx) $\begin{matrix}HO-CH_2-CH_2\\HOOC-CH_2\end{matrix}>N-CH_2-CH_2-N<\begin{matrix}CH_2COOH\\CH_2-COOH\end{matrix}$

total amount of copper entrained is then less than if the precipitate had been obtained all at once at higher pH.

Finally, co-precipitation can also be largely reduced or even practically completely prevented by *masking* (or complexation); e. g. trivalent iron remains quantitatively in solution when aluminum is precipitated as $AlCl_3 \cdot 6\ H_2O$, since it forms the complex $HFeCl_4 \cdot aq$ at high concentrations of hydrochloric acid; silver chloride precipitates free of palladium if the latter is converted to the ammine complex before precipitation, and the addition of complexing agents of the ethylenediaminetetraacetic acid type prevents the coprecipitation of manganese with calcium phosphate (cf. Tab. 12.4).

12.1.7 Separation of precipitate and solution: Filtration – centrifugation

Filter papers of various types, filter crucibles made of glass, porcelain or platinum, and filter rods made of glass are used for *filtration.*

Cellulose filters for quantitative analytical work are largely freed from non-volatile inorganic components by treatment with hydrochloric acid and hydrofluoric acid; the residual ash content can normally be neglected (see Tab. 12.5).

The ash (shown here, for example, for Whatman No. 42) consists essentially of Si (< 2 µg/g), Ca (13 µg/g), Mg (1.8 µg/g), S (< 5 µg/g), and Fe (5 µg/g). In addition, traces of

Tab. 12.5: Ash content of cellulose filters (according to Whatman; 2012).

Filter type	Ash content (%)
No. 1 qualitative application	0,06
No. 6 ditto	0,2
No. 40 quantitative application	0,007
No. 42 ditto	0,007
No. 540 hardened, quantitative application	0,006
No. 542 ditto	0,006

copper, aluminum, fluoride, and somewhat larger amounts of sodium and chloride are usually found; boron was also detected.

Furthermore, small amounts of ether-soluble and nitrogen-containing compounds are present, but their presence does not interfere – at least in inorganic analyses.

Very fine-grained precipitates can be filtered off with the addition of some loose filter floc, which increases the density of the filter.

Occasionally, disturbances occur due to precipitate particles creeping up the walls of the vessel or funnel. These can be eliminated by adding surface-active compounds.

Numerous types of filter paper are commercially available with different pore sizes (equivalent diameter, derived from the pressure drop) between about 1 µm and 12 µm. Of these, soft, fast-filtering grades with pore sizes of about 7–8 µm are most common for coarse-flocculated, amorphous precipitates and those with pore sizes of about 2 µm for very fine crystalline precipitates.

The usual filter papers are resistant to dilute acids and to alkaline solutions (up to about 8% NaOH). Swelling phenomena occur at higher alkali concentrations. For filtering strongly acidic solutions, *hardened* filters, i.e. those treated with concentrated nitric acid, or filters covered with plastics are used.

The adsorption on cellulose already mentioned in chap. 10 appears disturbingly when filtering highly diluted solutions ($< 10^{-5}$ M). Insofar as this is a reversible process, the retained fraction can be removed again by carefully washing out the filter. In contrast, components retained by ion exchange cannot be recovered with water, but only with solutions of electrolytes (e.g. acids). In addition, however, components of the solution can still penetrate into the interior of the cellulose fibers and then be difficult or impossible to wash out. This interference is not limited to inorganic ions or compounds, but has also been observed with organic substances.

Among a whole range of special filters, it is worth mentioning those made of polyvinyl chloride fibers (for filtration of strongly acidic or strongly alkaline solutions) and glass fiber filters resistant up to several hundred degrees. PTFE is also used.

Membrane filters. To filter extremely fine-grained precipitates or colloidal particles, so-called *membrane filters* are used, which can be made of cellulose, cellulose derivatives, polycarbonate, polyvinyl chloride, PTFE or glass fibers. These filters are manufactured with pore widths from 0.01 to about 10 μm and very high pore volume in the spatial unit, so that the filtration rate is relatively high. Nevertheless, the medium types have to be filtered under suction, the finer ones with overpressures up to about 30 bar.

In addition to narrow pores, the smooth surface of membrane filters is noteworthy; precipitates can often be quantitatively removed from the filter by hosing.

The membrane filters may contain traces of surfactants from their manufacture, which occasionally interfere with analyses. This contamination can be removed by washing.

Filter crucibles. Precipitates, which are attacked by the carbonizing organic matter during the ashing of paper filters, are filtered off by suction through filter crucibles. The first such crucibles were made by Gooch from porcelain or platinum crucibles, the perforated bottoms of which were covered with a layer of asbestos. Later, a layer of finely divided platinum was used in place of asbestos, enabling the filtration of hydrofluoric acid solutions[1] . These crucibles, which were somewhat cumbersome to manufacture, have now been largely replaced by glass and porcelain crucibles with bottoms made of porous frit.

Glass filter crucibles are available with pore sizes of 5–100 μm, of which those with 5–10 μm are mainly used for fine precipitates and those with 20–30 μm for coarser precipitates in analytical chemistry.

The resistance of glass filter crucibles to water and the common acids is very good, they are somewhat more affected by alkaline solutions (with the exception of dilute NH_3 solutions). Glass filter crucibles can be heated up to about 300 °C, but usually only lower temperatures are used (about 110–150 °C).

Filter cups with melted frit are suitable for filtering small quantities of liquid (Fig. 12.7). The precipitate is precipitated in the vessel and the liquid is aspirated by tilting the vessel through the lateral attachment.

Porcelain filter crucibles (less frequently quartz filter crucibles) are used for filtering precipitates that must be heated to high temperatures; these are stable up to about 1000 °C. For analytical work, pore sizes of 6–8 μm are most favorable. Resistance to acids is very good; strongly alkaline solutions attack the porcelain noticeably.

Microfiltrations can be carried out using glass or porcelain filter rods according to Emich (Fig. 12.8). The solution is aspirated through the rod, with the precipitate remaining partly in the precipitation vessel and partly on the frit of the rod.

Centrifugation. Small amounts of precipitate can be made to settle by centrifugation in simple electrically or manually operated centrifuges. The supernatant liquid is

1 The term *Neubauer crucible* is commonly used in European literature.

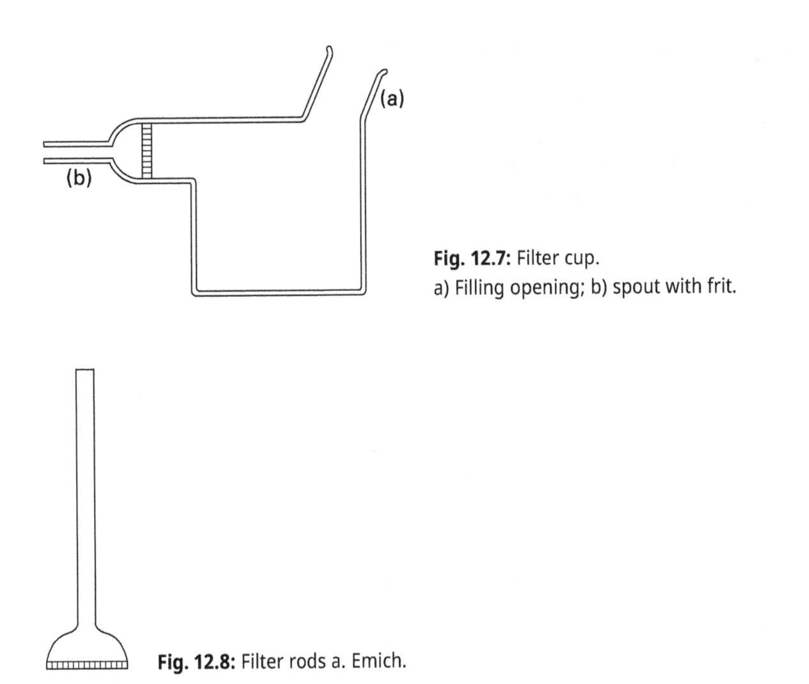

Fig. 12.7: Filter cup.
a) Filling opening; b) spout with frit.

Fig. 12.8: Filter rods a. Emich.

removed, the precipitate is cleaned by washing with fresh solvent and centrifuged again. Since the usual centrifuge tubes have a capacity of only 15 ml (larger centrifuges are probably only used for preparative purposes), this method is mainly used for microanalyses.

12.2 Separations by one-time equilibration: Working with an auxiliary phase

12.2.1 Inorganic precipitation by double conversion

The numerous precipitates used in gravimetric determinations of inorganic substances are mainly produced by slow addition of the reagent solution to the hot analysis solution, occasionally also by precipitation from homogeneous solution. The properties of the precipitates are often improved by allowing the solution with precipitate to stand in the heat for a longer period (see above).

Where possible, precipitation reactions are used which give stoichiometrically compounded compounds either directly or after annealing. Some of the most important precipitations are listed in Tab. 12.6.

Tab. 12.6: Examples of precipitation reactions of inorganic compounds.

Reagent	Precipitated
NH_3, NaOH	Ag^+, Ba^{2+}, Al, In, Tl^{3+}, Rare Earths, Sn^4, Ti, Zr, Hf, Th, Nb, Ta, Bi, Cr^{3+}, U^{6+}, Mn^{4+}, Fe^{3+}, Rh^{3+}
SO_4^{2-}	Sr, Ba, Pb^{2+}
Cl^-	Ag
PO_4^{3-}	Ag, Mg, Zn, Cd, Mn^{2+}, Co^{2+}
ClO_4^-	K
SCN^-	Cu^+
S^{2-}	As^{3+}, Sb^{3+}, Hg^{2+}
Mo^{6+}	Pb^2, PO_4^{3-}
Ag^+	Cl^-, Br^-, I^-, PO_4^{3-}, SCN^-
Ba^{2+}	SO_4^{2-}
Cd^{2+}	S^{2-}
Pb^{2+}	F^- (as PbClF)
Ca^{2+}	F^-
Mg^{2+}	PO_4^{3-}, AsO_4^{3-}

12.2.2 Organic precipitation reagents for inorganic ions

Organic reagents are widely used to precipitate inorganic substances. The precipitation reactions are partly acid-base reactions, partly inner complex compounds are formed, and partly precipitation occurs through adsorption effects (e.g. precipitation with tannin).

In many cases, organic reagents can be used to detect entire groups of inorganic ions, but sometimes they can also be used for relatively selective separations (see Tab. 12.7). Selectivity can often be increased by favorable pH adjustment and masking reactions (Tabs. 12.8 and 12.9).

The precipitates obtained with organic reagents are often characterized by remarkable low solubility; furthermore, they usually have a high molecular weight, which is advantageous in gravimetric determinations. Since the precipitates usually no longer have a salt-like character, the entrainment effects are unusually low in many cases.

Tab. 12.7: Organic precipitation reagents for inorganic ions (examples).

Reagent	Formula	Precipitate
Oxalic acid	$COOH-COOH$	Ca, Rare earths etc.
Dimethylglyoxime	$H_3C-C=NOH$ / $H_3C-C=NOH$	Ni, Pd, Fe^{2+}, Pt^{2+}, Bi
Copperron	$C_6H_5-N(ONH_4)-N=O$	Fe^{3+}, Ti, Zr, Hf, V^{5+}, Mo^{6+}, U^{4+}, Nb, Ta, Pd, Ga, Sn^{4+}, Bi, Sb^{3+}, Cu, Th, Tl, Al, Hg^{+}
Anthranilic acid	$C_6H_4(COOH)(NH_2)$	Cu, Cd, Co, Ni, Mn, Hg, Pb, Zn
Microlonic acid	$O_2N-CH-C(-CH_3)=N-N(NO_2)-C(=O)$ (ring)	Ca, Pb, Th (Cu, Fe, Ni, Co, Mn, Ba)
Mercaptobenzthiazole	$C_6H_4<(N=C(-SH)-S)$	Cu, Tl, Cd, Pb, Bi, Ag, Au, Hg
α-Benzoinoxime	$C_6H_5-CH(OH)-C(=NOH)-C_6H_5$	Cu, Mo^{6+}, W^{6+} (Cr^{6+}, Pd, V, Ta)
8-Hydroxyquinoline	$C_9H_6N(OH)$	Ag, Al, Bi, Ca, Co, Cu, Cd, Fe^{3+}, Hg^{2+}, Mg, Mn, Mo^{6+}, Ni, Pb, Ti, U^{6+}, V^{5+}, W^{6+}, Zn
Sodium tetraphenyloborate	$(C_6H_5)_4BNa$	K, NH_4^+, Rb, Cs, Ag, Tl^{+}
Nitron	C_6H_5-N, C_6H_5N, HC, N, C, $N-C_6H_5$ (ring)	NO_3^-, ReO_4^-, ClO_4^-, Br^-, I^-, NO_2^-, CrO_4^{2-}, ClO_3^-
Tetraphenylarsonium chloride	$(C_6H_5)_4AsCl$	MnO_4^-, ClO_4^-, $AuCl_4^-$, $PtCl_6^{2-}$, $HgCl_4^{2-}$ a. o.

Tab. 12.8: pH ranges of the precipitation of 8-hydroxyquinoline compounds.

Fallen ion	Complete precipitation	Fallen ion	Complete precipitation
Cu^{2+}	pH 5.3–14.6	Bi^{3+}	pH 5.0–8.3
Ag^{+}	6,1–11,6	V^{5+}	2,0–5,3
Mg+	9,5–12,7	Mo^{6+}	3,3–7,6
Ca^{2+}	9,2–12,7	W^{6+}	5,0–5,7

Tab. 12.8 (continued)

Fallen ion	Complete precipitation	Fallen ion	Complete precipitation
Zn^{2+}	4,7–13,5	U^{6+}	5,7–9,8
Cd^{2+}	5,7–14,5	Mn^{2+}	5,9–9,5
Hg^{2+}	4,8–7,4	Fe^{3+}	2,8–11,2
Al^{3+}	4,2–9,8	Co^{2+}	4,2–11,6
Pb^{2+}	8,4–12,3	Ni^{2+}	4,6–10,0
Ti^{4+}	4,8–8,6		

Tab. 12.9: Improvement of the selectivity of the precipitation of metals with thionalide by masking.

Initial solution	Isolation of
Mineral acid solution	Ag, Cu, Au, Hg, Sn, As, Bi, Pt, Pd, Ru
Tartrate – KCN – solution	Au, Tl, Sn, Pb, Sb, Bi
Tartrate – NaOH solution	Cu, Au, Hg, Cd, Tl
Tartrate – KCN – NaOH solution	Tl

12.2.3 Precipitation reactions for organic substances: Double reaction – addition – condensation – inclusion compounds – antigen-antibody reaction – protein precipitation for structure elucidation

Most precipitation reagents for organic compounds act on specific functional groups; therefore, they can generally only be used to separate entire classes of substances, more rarely individual compounds. Apart from double reactions (often between acids and bases), addition and condensation reactions are used (cf. Tab. 12.10).

A special type of precipitation consists in the formation of *inclusion compounds*. These are compounds formed by the incorporation of atoms or molecules into lattice cavities of solid substances. The most important prerequisite for their formation is a suitable size ratio between the cavity and the shape of the incorporated foreign molecule, while the chemical properties of the latter are generally not decisive.

Analytically important inclusion compounds are formed mainly by urea and thiourea. Deoxycholic acid, hydroquinone, 4,4′-dinitrodiphenyl, cyclodextrins as well as water, chabasite, montmorillonite, graphite and others play a minor role as host molecules.

Solid urea usually forms a tetragonal lattice; however, in the presence of certain elongated foreign molecules, a hexagonal lattice is energetically favored, and the foreign molecules in question are incorporated into the central cavities of this lattice (cf. Fig. 12.9).

Whether an inclusion compound is formed depends not only on the diameter but also on the chain length of the foreign molecule; if the molecules are too short, the

Tab. 12.10: Precipitation reactions of organic compounds (examples).

Precipitation reagent	**Fallen compounds**
Picric acid	Amines, alkaloids, protein, naphthalene, anthracene, etc.
Microlonic acid	Amines, alkaloids
Sulfonic acids	Amino acids
Oxalic acid	Alkaloids, urea
Perchloric acid	Protein
Trichloroacetic acid	Protein
Sulfosalicylic acid	Protein
Phosphotungstic acid	Alkaloids, betaines
Silicotungstic acid	Alkaloids
$KBiI_4$	Alkaloids, betaines
Reinecke salt	Amines
$KI \cdot I_2$	Alkaloids
$HAuCl_4$, H_2PtCl_6, $HFeCl_4$	Amines, alkaloids
$K_4Fe(CN)_6$	Amines, alkaloids
$NaB(C_6H_5)_4$	Amines, alkaloids, betaines
Benzidine	Sulfonic acids
$AgNO_3$	Fatty acids, sulfonic acids
$Pb(CH_3COO)_2$	Fatty acids, mercaptans
$Hg(CN)_2$	Mercaptans
$HgSO_4$	Olefins
Benzoyl chloride	Alcohols
Dimedon	Aldehydes
$NaHSO_3$	Aldehydes
Hydroxylamine	Aldehydes, ketones
Semicarbazide	Aldehydes, ketones
2,4-Dinitrophenylhydrazine	Aldehydes, ketones
Phenylhydrazine	Aldehydes, ketones, carbohydrates

Tab. 12.10 (continued)

Precipitation reagent	Fallen compounds
Br_2	Phenols
Benzoquinone	Phenols
Digitonin	Steroids

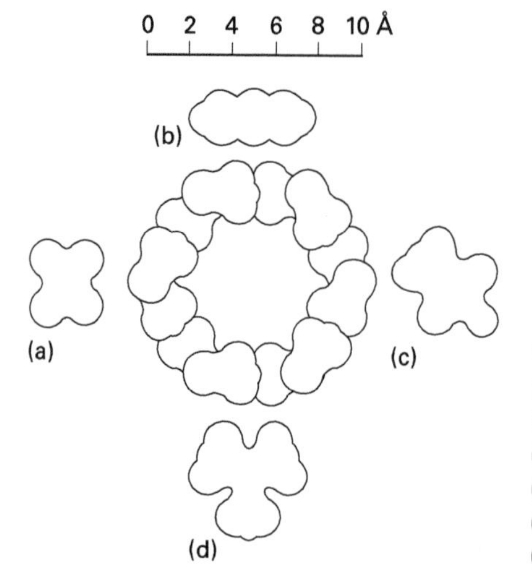

Fig. 12.9: Sectional view through a channel of the hexagonal urea lattice and through some hydrocarbon molecules.
(a) n-Octane (to be incorporated);
(b) benzene (no inclusion compound);
(c) 3-methylheptane (no inclusion compound);
(d) 2,2,4-trimethylpentane (no inclusion compound).

recoverable energy is not sufficient to form the hexagonal urea lattice, and on the other hand, molecules that are too long (e.g., high polymers) often do not give inclusion compounds either.

The inclusion compounds are obtained by precipitation with saturated aqueous or methanolic urea solutions, by reacting the liquid organic mixture with solid urea (finely powdered) or by crystallization of hot urea-saturated solutions. The precipitations are not quite complete; lowering the temperature favors the formation of the adducts.

In addition to straight-chain saturated hydrocarbons from C_6, urea inclusion compounds are obtained with straight-chain alcohols from C_6, straight-chain monocarboxylic acids from C_4, straight-chain ketones from C_3, as well as with various olefins, primary alkyl halides, esters, secondary alcohols, ethers, dicarboxylic acids, aldehydes, mercaptans, thioethers, amines, nitriles, and others. etc. Weakly branched kerosenes can also be incorporated, especially if the branch is on a longer straight chain.

Strongly branched aliphatic hydrocarbons and their derivatives, cycloparaffins and aromatics and their derivatives, do not form urea inclusion compounds.

The urea inclusion compounds are only stable in the solid state; if a solvent for urea is added, they decompose. Highly volatile incorporated components can be distilled off by raising the temperature. This class of substances has occasionally been used to separate straight-chain kerosenes and n-fatty acids from mixtures. Since branched molecules, even if they do not form such compounds by themselves, can be incorporated to a considerable extent, the value of the method is limited.

Antigen-antibody reaction. Finally, the precipitation reaction, the *antigen-antibody reaction*, which is important for the field of bioanalysis, should be mentioned. If a solution containing the specific antigen is added to a serum containing antibodies, a precipitate is formed. The special feature of this reaction is that precipitation is reduced or fails to occur in the case of a larger excess of both the one and the other reactant (cf. Fig. 12.10).

Protein precipitation (protein crystallization) for structure elucidation. One of the most important methods for the structure elucidation of proteins is the application of X-ray structure analysis. For this, however, protein single crystals are required if possible, but these are difficult to produce due to their highly complex composition. The poor mechanical properties are due to a large number of cavity structures, which are also the reason for the incorporation of water of crystallization or similar solvents. The production of protein crystals already requires highly pure protein solutions, which are then mixed with precipitants (highly concentrated salt solutions, alcohols, polyethylene glycol) in very small volumes (nl or µl). The protein/precipitant mixture in the phase diagram must already be in the supersaturated range (nucleation range). The precipitation process can be initiated slowly and over long periods of time (up to months) by diffusive mixing processes. In the meantime, the spatial elucidation of proteins and protein complex structures (e.g. antigen-antibody complex) has often been achieved with protein single crystals generated in this way.

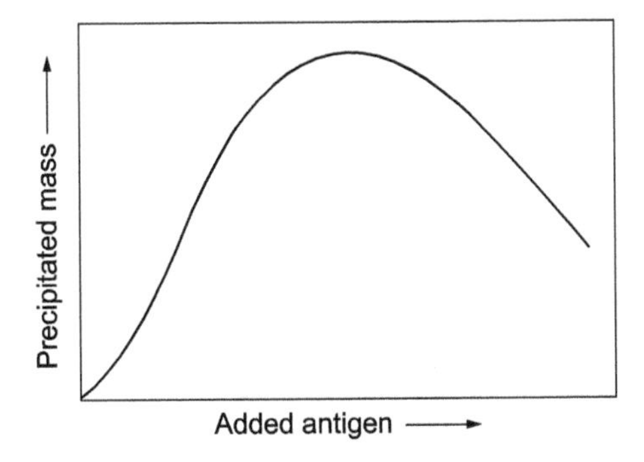

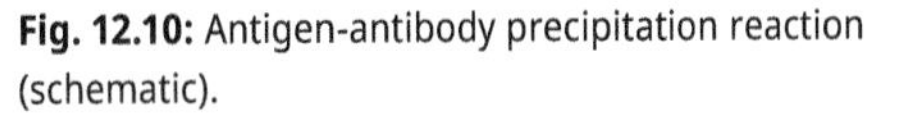
Fig. 12.10: Antigen-antibody precipitation reaction (schematic).

Sumner succeeded in protein crystallization for the first time with the enzyme urease and with the proteins concanavalin A and B. Together with Northrop (pepsin crystallization), he received the Nobel Prize in Chemistry in 1946.

The formation of protein crystals is also of considerable importance for formulation as an active ingredient (control of pharmacokinetics in dissolution processes) and for increasing protein storage life.

12.2.4 Precipitation by reduction: Gaseous and dissolved reducing agents

Precipitation using reducing reagents is mainly used in the analysis of noble metals, but some less noble elements can also be precipitated from solutions in this way (see Tab. 12.11). The precipitations usually take place quite slowly, the solutions should therefore be heated for a longer time after reagent addition.

12.2.5 Precipitation by reduction or oxidation: Electrolysis

Electrolytic precipitation is a special type of precipitation method, in which no reagent but electric current is used for precipitation.

Understanding these methods requires knowledge of the terms *decomposition voltage, redox buffer, electrode potential,* and *overvoltage.*

Tab. 12.11: Precipitation by gaseous and dissolved reducing agents (examples).

Fallen	Reducing agent
Cu	$NaBH_4$, H_3PO_2, $VOSO_4$, $Na_2S_2O_4$
Ag	HCHO, H_3PO_2, $SnCl_2$, ascorbic acid, Hg^+, $NaBH_4$, $Na_2S_2O_4$, hydrazine, glycerol, and others.
Au	SO_2, $FeSO_4$, hydrazine, hydroxylamine, oxalic acid, ascorbic acid, hydroquinone, $NaBH_4$, $VOSO_4$, HCHO, HCOOH, $TiCl_3$
Cd	$NaBH_4$
Hg	H_3PO_2, H_3PO_3, $SnCl_2$, As^{3+}, HCHO, HCOOH, hydrazine, $Na_2S_2O_4$, $NaBH_4$, SO_2, V^{2+} sulfate
Pb	$NaBH_4$, $CrCl_2$, $Na_2S_2O_4$
As	H_3PO_2, $SnCl_2$, $TiCl_3$, $CrSO_4$, Hg^+, $Na_2S_2O_4$
Sb	$Na_2S_2O_4$, H_3PO_2, V^{2+} sulfate
Bi	H_3PO_2, $Na_2S_2O_4$, $CrCl_2$, Na_2SnO_3, HCHO, $TiCl_3$, hydrazine, V^{2+} sulfate
Se	SO_2, hydrazine, hydroxylamine, $SnCl_2$, ascorbic acid, $TiCl_3$, H_3PO_3, glucose, H_3PO_2, thiourea, V^{2+} sulfate, $Na_2S_2O_4$
Te	SO_2, hydrazine, hydroxylamine, $TiCl_3$, $SnCl_2$, V^{4+} sulfate, H_3PO_3, semicarbazide, V^{2+} sulfate.
Ni	$NaBH_4$
Rh	Hydrazine sulfate, HCOOH, $TiCl_3$, VCl_2, $CrCl_2$
Pd	Hydrazine sulfate, HCOOH, H_2, $TiCl_3$, SO_2, CO, H_3PO_2, V^{2+} sulfate, ethylene.
Ir	HCOOH
Pt	Hydrazine hydrochloride, HCOOH, H_2, $TiCl_3$, V^{2+} sulfate

Decomposition potential. If an increasing DC voltage is applied to a metallic conductor, the current strength increases proportionally to the voltage (Ohm's law); if, on the other hand, two chemically inert electrodes are immersed in the solution of an electrolyte and the applied voltage is increased, only a minimal, so-called *residual current* initially flows. Only after a certain voltage has been exceeded does the current increase more steeply (see Fig. 12.11). If the rising part is extended backwards to the abscissa, the so-called *decomposition potential* of the electrolyte is obtained. This is mainly dependent on the type of dissolved ions, but also on other variables.

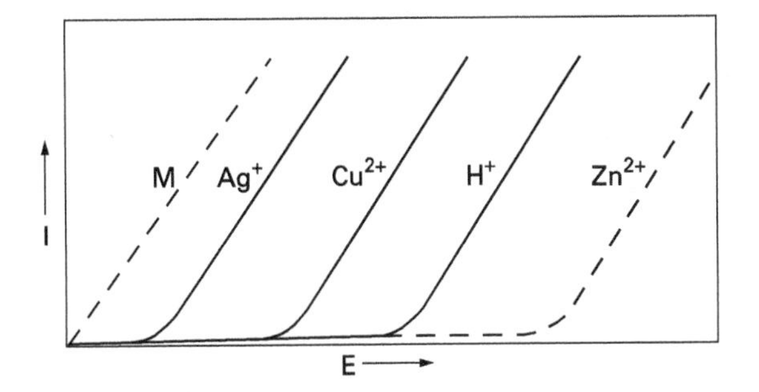

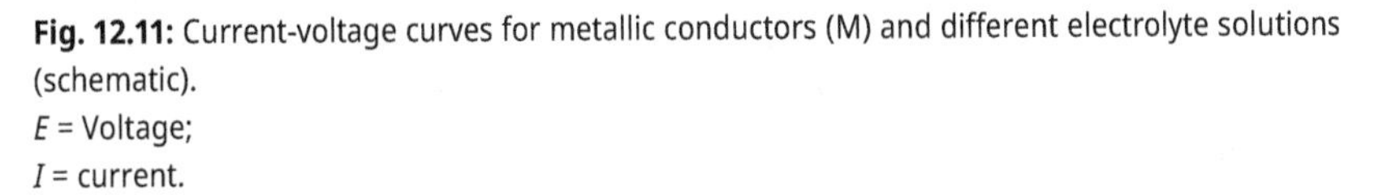
Fig. 12.11: Current-voltage curves for metallic conductors (M) and different electrolyte solutions (schematic).
E = Voltage;
I = current.

In contrast to the conditions when current passes through metallic conductors, chemical changes of the solution, so-called *electrode reactions*, occur at the electrodes in electrolyte solutions. In principle, these are very simple:

At the *cathode,* electrons are added to the solution; ions that accept them are *reduced.*

At the *anode,* electrons are removed from the solution; ions that release the electrons are *oxidized* in the process.

Since there can never be a significant surplus or a noticeable deficit of electrons in the electrolyte, both processes must always occur simultaneously and in a stoichiometric ratio.

As a result of the large number of possible oxidation and reduction reactions, numerous electrode reactions are known, especially since both cations and anions can be reduced at the cathode and cations and anions can be oxidized at the anode. Furthermore, the conditions can become quite complicated due to the formation of unstable reaction products which continue to react secondarily.

Of particular analytical importance are electrode reactions in which an element or a compound can be quantitatively separated from the solution. The most favorable conditions for this have been determined empirically for a whole number of elements.

An important disorder of electrolytic depositions is that one type of ion in the solution is reversibly oxidized and reduced by the electric current without any deposition occurring. In the course of electrolysis, the higher valent form is reduced at the cathode, the soluble reaction product reaches the anode by diffusion or by stirring, where it is oxidized again, and so on. In such cases, the electrolyte contains a so-called *redox buffer system*, and the current flowing through the solution only causes useless heating in the end.

An example of this is an iron(III) sulfate solution. After switching on the current, reduction to iron(II) sulfate occurs at the cathode, which is then oxidized again at the anode.

Electrode potential. The decomposition potential could be used to characterize the electrochemical behavior of an electrolyte solution, but this is not very suitable for this purpose because it depends in a complicated way on various factors (e.g. on the type of anion). Instead, the term *electrode potential* is used.

If a metallic electrode is immersed in a solution of one of its salts, an electric potential is formed on the surface, the magnitude of which is characteristic of the metal in question under certain conditions.

However, the absolute potentials of such single electrodes cannot be determined experimentally, but only the difference of the potentials of two electrodes can be measured. In order to be able to compare different electrodes with each other, all potentials are measured against the so-called *normal (standard) hydrogen electrode* or at least related to it. The potential of this reference electrode is arbitrarily set equal to 0.0000 V.

The measuring arrangement is shown in Fig. 12.12. Vessel a contains the metal electrode to be measured in a solution of one of its salts (e.g. a copper rod in a copper sulfate solution). Vessel b contains the normal hydrogen electrode (a platinized platinum sheet surrounded by gaseous hydrogen of 1 bar pressure; electrolyte: 2 N H_2SO_4 solution). Both vessels are electrically conductively connected by bridge c, which is filled with a saturated salt solution (e.g. KCl solution) to largely eliminate the formation of an additional potential jump at the interface of the two electrolytes. The potential difference between the two electrodes is measured with the high-impedance voltmeter d.

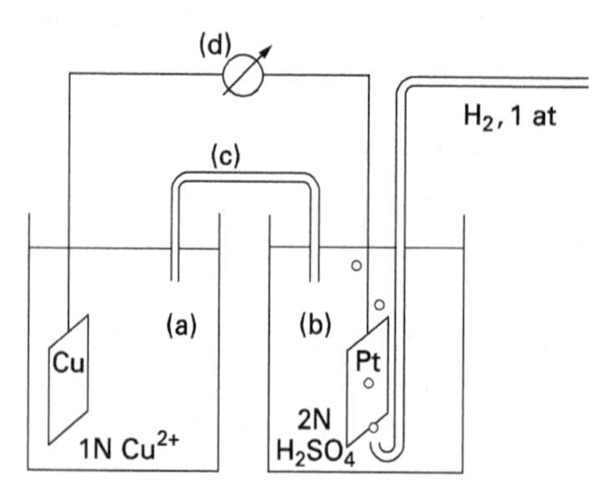

Fig. 12.12: Arrangement for potential measurement (principle).
a) Vessel with the metal electrode to be measured;
b) Normal Hydrogen Electrode;
c) salt bridge;
d) voltmeter.

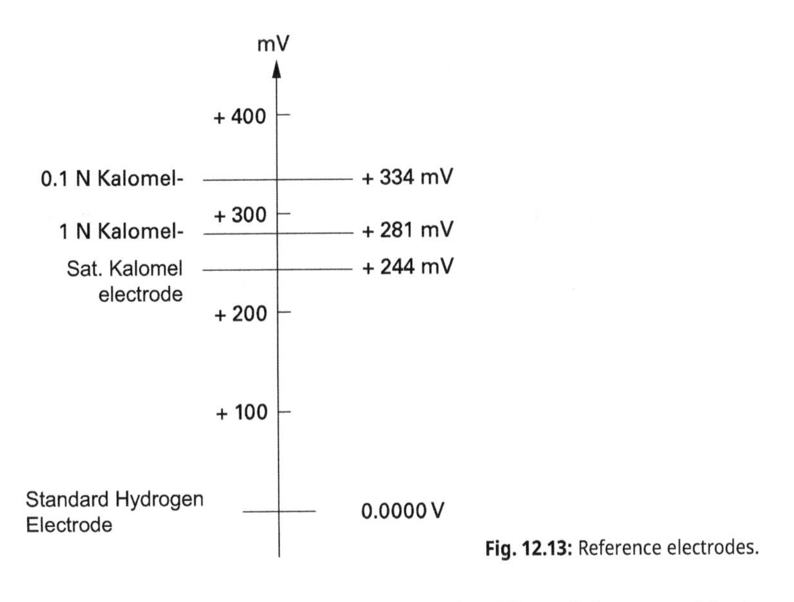

Fig. 12.13: Reference electrodes.

Because of the somewhat awkward handling of the normal hydrogen electrode, other reference electrodes are often used whose potentials are accurately measured against the normal hydrogen electrode. Three of the most important are shown with the position of their potentials in Fig. 12.13. The potential against the standard hydrogen electrode is obtained by adding the eigenvalue of the reference electrode to the measured value.

The influence of temperature and concentration of the solution on the potential measured in this way is represented in a first approximation[2] by *Nernst's potential equation*:

$$E = E_0 + \frac{RT}{nF} \ln c, \tag{1}$$

where E is the potential (measured against the standard hydrogen electrode), E_0 is a constant characteristic of the electrode in question, R is the gas constant, T is *the* absolute temperature, n is *the* change in charge as the metal passes into the ion, F is Faraday's constant = 96,500 coulombs, and c is the concentration of the metal ions in the solution (in molarities m = mole / 1000 g water).

Substituting the numerical values for R and F, converting the natural logarithm to the decadic logarithm, and choosing a temperature of 25 °C, we get:

$$E = E_0 + \frac{0,059}{n} \log c \tag{2}$$

2 In the exact equation, the concentration c must be replaced by the activity a.

E_0 can be determined by measuring the potential difference E at a known concentration c and a known n. E_0 is obtained directly from the potential difference at $c = 1$ (mol / 1000 g water), since the second element of equation (2) is then omitted due to $\log 1 = 0$.

The constant E_0 is called the *normal potential of* a metal; it means, as a first approximation, the voltage of the metal electrode in a 1 molar solution of one of its completely dissociated salts[3] versus the normal hydrogen electrode.

If the metals are arranged in the order of their normal potentials, the so-called *potential series* is obtained (Tab. 12.12).

Tab. 12.12: Potential series.

Metal	Normal potential	Metal	Normal potential
Ca	−2,84	Sn	−0,14
Na	−2,71	Pb	−0,13
Al	−1,66	H_2	0,0000
Mn	−1,05	Bi	+0,3
Cr	−0,86	Cu	+0,34
Zn	−0,76	Hg	+0,80
Fe	−0,44	Ag	+0,80
Cd	−0,40	Pt	+1,2
Co	−0,27	Au	+1,4
Ni	−0,23		

The more negative the normal potential of a metal is, the more difficult it is to reduce; the more positive it is, the easier it is to reduce. If several metal ions are present in a solution at the same time, the *noblest* one, i.e. the ion with the most positive potential, is deposited first.

The normal potentials are idealized quantities, for the determination of which the activities rather than the concentrations are used. In the practical implementation of electrolysis, however, the potentials that actually occur in the solutions are of greater importance. They are referred to as *real potentials*. These can deviate considerably from the normal potentials.

Overvoltage. The decomposition potential of an electrolyte is ideally equal to the difference between the anode and cathode potentials.

Often, however, a higher decomposition voltage is observed than is calculated from the two potentials. This anomaly is attributed to inhibition phenomena during deposition and is referred to as *overvoltage*. Such overvoltages can occur in both cathodic and anodic reactions. In the deposition of metals, the overvoltage is generally

3 More precisely: solution of activity 1.

small (a few tenths of a volt); depending on the cathode material, it can assume larger amounts in the deposition of hydrogen (Tab. 12.13).

Tab. 12.13: Overvoltage of hydrogen deposition on various metals at a current density of 0.01 A/cm^{2}.

Metal	Overvoltage (V)
Pd	0,04
Pt (blank)	0,16
Ni	0,3
Cu	0,4
Pb	0,4
Sn	0,5
Zn	0,7
Hg	1,2

The hydrogen overvoltage is of great importance because hydrogen can always be deposited during the electrolysis of aqueous solutions. As a result of the overvoltage, it is possible to deposit metals from aqueous solutions which are less noble than hydrogen; in particular, it is possible to take advantage of the high H_2 overvoltage on mercury (see below).

Analytical applications of electrolysis. In the application of electrolysis in analytical chemistry, a distinction is made between two different working methods:
- deposition at controlled current and
- deposition at controlled voltage or potential.

Deposition under controlled current intensity. For depositions under controlled amperage, circuits according to Fig. 12.14 are used.

The current delivered by the direct current source a flows through the electrolysis vessel c via the ammeter f after the switch b is closed. The voltage can be controlled by the slide resistor d and measured with the voltmeter e.

In practical work, it is not normally necessary to keep the current exactly constant; it is set to the desired value at the start of electrolysis and then readjusted from time to time. Besides pure direct current, pulsating direct current from a rectifier with or without smoothing can also be used.

The electrodes are usually made of a platinum-iridium alloy with about 10% iridium, which is chemically particularly resistant. Of numerous other materials given in the literature, only metallic mercury plays some role.

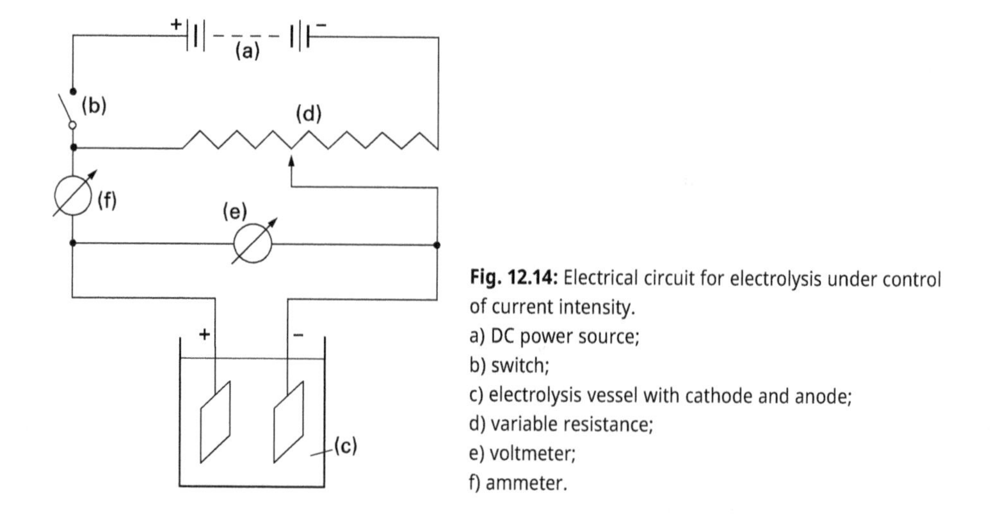

Fig. 12.14: Electrical circuit for electrolysis under control of current intensity.
a) DC power source;
b) switch;
c) electrolysis vessel with cathode and anode;
d) variable resistance;
e) voltmeter;
f) ammeter.

The platinum cathodes are mainly used in the form of cylindrical nets with fixed stems and surfaces of about 100–150 cm^2; the associated anodes consist either of a thick, spirally wound platinum wire or of a second net standing inside the first. Such electrodes have the advantage that the solution can easily circulate through the openings of the net and the precipitates adhere firmly to the wires that surround them tubularly.

For electrolysis on a smaller scale, the dimensions of the nets are reduced; depositions from a few milliliters of solution are carried out with simple wire electrodes.

For electrolytic deposition, solutions containing sulfuric acid, nitric acid, sulfuric acid + nitric acid or perchloric acid are usually chosen. Hydrochloric acid and chlorides interfere, since chlorine is formed anodically, which attacks the electrodes.

Practically the most important precipitation is that of copper, but in addition to this, numerous other elements are almost completely precipitated. Furthermore, many hydroxides can be precipitated by electrolysis by evolving hydrogen by increasing the voltage, thereby raising the pH at the surface of the cathode. Some other hydroxides, e.g., $U(OH)_4$, are precipitated by reducing higher valence states of the element in question in weakly acidic solutions. Finally, precipitates can also be formed at the anode if hydrolyzing higher valence ions are formed by oxidation (cf. Tab. 12.14).

Various metals (Bi, Cd, Ga, Hg, In, Sn, Zn) alloy superficially with the platinum of the cathode during electrolysis; therefore, when depositing these elements, it is expedient to first coat the electrode with a layer of copper or silver.

The deposition of hydroxides and hydrated oxides plays a role especially in radiochemistry, since in this way it is possible to deposit very small amounts of radioactive substances uniformly on flat electrodes. This provides a geometrically favorable arrangement for measuring activity. Such depositions can be carried out in beaker-shaped vessels (cf. Fig. 12.15).

Tab. 12.14: Electrolytic depositions on Pt electrodes at controlled current (rarely used depositions bracketed).

Cathodically depositable as elements: Ag, Bi, Cd, Cu, Co, Hg, Ni (As, Au, Ga, In, Pb, Pd, Pt, Rh, Ru, Sb, Se, Sn, Te, Tl, Zn).
Cathodically depositable as compounds: (Al, Am, Cd, Cm, Cr, Fe, La, Mg, Mn, Mo, Np, Pu, Sn, Tc, Th, Ti, U, Zr and others as hydroxide or oxide hydrates; As, Ge, Sb as hydrides).
Anodically depositable as compounds: Pb^{4+} (Ag^{2+}, Co^{3+}, Mn^{4+}, Tl^{3+}) as oxides and oxide hydrates, respectively.

Even extremely small, imponderable amounts of radioactive substances can be deposited electrolytically, but considerable difficulties arise from the formation of radiocolloids, adsorption on the vessel walls, and the introduction of impurities.

The *deposition rate* decreases more and more during electrolysis, so that the last substance residues are removed from the solution only very slowly. Stirring greatly accelerates the deposition, and furthermore the surface area of the electrode should be as large as possible in relation to the solution volume. Increasing the current intensity is also generally favorable, although hydrogen evolution increases at high current intensities. Moderate H_2 development accelerates metal deposition because the diffusion layer on the electrode is repeatedly ruptured, allowing fresh solution components to reach the surface. However, if hydrogen evolution is too strong, metal deposits become porous and tend to fall off the cathode, and moreover, part of the current is increasingly consumed uselessly.

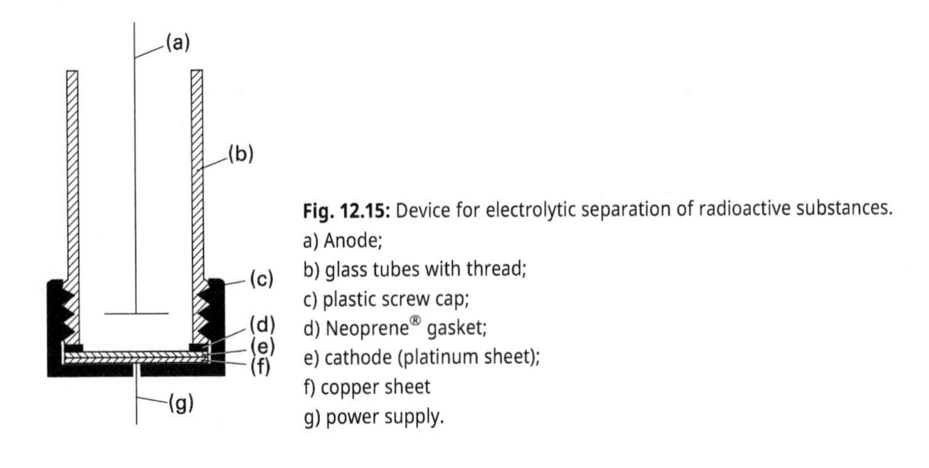

Fig. 12.15: Device for electrolytic separation of radioactive substances.
a) Anode;
b) glass tubes with thread;
c) plastic screw cap;
d) Neoprene® gasket;
e) cathode (platinum sheet);
f) copper sheet
g) power supply.

Electrolysis on mercury cathodes. The mercury cathodes introduced by Gibbs have several advantages over the platinum cathodes: First, the deposition of various metals is facilitated because dilute solutions in mercury or intermetallic compounds are formed under energy recovery; furthermore, because of the high hydrogen overpotential on Hg (cf. Tab. 12.13), several base metals can be deposited that cannot be precipitated on platinum cathodes; and finally, large amounts of depositable metals can be removed from solutions because the precipitates cannot fall off the electrode. An overview of the performance of the method is given in Tab. 12.15.

Tab. 12.15: Metal deposits from sulfuric acid solutions on mercury cathodes.

Separable, quantitative in mercury: Ag, Au, Bi, Cd, Co, Cr, Cu, Fe, Ga, Hg, In, Ir, Mo, Ni, Pd, Po, Pt, Rh, Sn, Tc, Tl, Zn
Quantitatively removed from solution, but not completely in mercury: As, Os, Pb, Se
Incompletely isolated: Ge, Mn, Re, Sb, Te.

Mercury cathodes are used primarily when large quantities of a precipitable element are to be separated from traces of other elements which remain quantitatively in the solution. Electrolysis is carried out almost exclusively in sulfuric acid or perchloric acid solutions.

The devices used include simple beakers with a mercury cathode on the bottom; a platinum wire is immersed in the solution as the anode. Somewhat more elaborate are vessels with a cooling jacket, drain cock and fused-in platinum pin as current supply to the cathode (cf. Fig. 12.16).

Silver and cadmium anodes. Mention should also be made of silver anodes, which have been proposed for the deposition of chlorine, bromine and iodine from the corresponding halide solutions. Precipitates of the silver halides are obtained, which adhere firmly to the anode if the composition of the electrolytes is favorable (e.g. alkaline tartrate solutions). Some other anions can also be removed from solutions by electrolysis, e.g., sulfide on cadmium anodes. However, all these processes have not gained any major importance.

Electrolysis at controlled potential. If a solution containing several compounds with different decomposition potentials (cf. Fig. 12.11) is electrolyzed, the electrolysis potential can be controlled so that the most easily deposited element alone is deposited.

The simplest voltage control is to use a current source with a certain maximum voltage, which is applied directly to the two electrodes without the interposition of a resistor (*short-circuit electrolysis*). For example, if a single cell of a lead collector is connected,

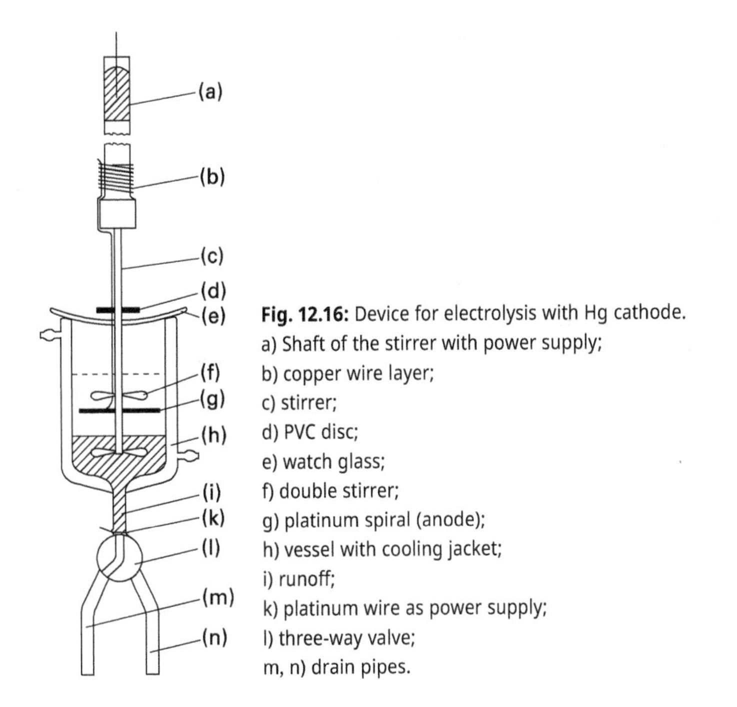

Fig. 12.16: Device for electrolysis with Hg cathode.
a) Shaft of the stirrer with power supply;
b) copper wire layer;
c) stirrer;
d) PVC disc;
e) watch glass;
f) double stirrer;
g) platinum spiral (anode);
h) vessel with cooling jacket;
i) runoff;
k) platinum wire as power supply;
l) three-way valve;
m, n) drain pipes.

the voltage cannot increase above 2 V. In this way, various more noble metals can be separated from less noble ones.

More versatile is the circuit given above for electrolysis at controlled current (Fig. 12.14). It can be used to set a certain voltage at the beginning of the electrolysis (instead of the current) and to readjust this voltage during the course of the separation. Some examples of separations which can be achieved with this are given in Tab. 12.16.

Tab. 12.16: Electrolysis at controlled potential (application examples).

Deposited	Not deposited	Electrolyte	Voltage (V)
Ag	Bi, Cd, Co, Cu, Ni, Zn	dil. H_2SO_4	1,2
Au	Cd, Cu, Ni, Zn	dil. HCl	1,3
Bi	Cd, Ni, Zn	dil. H_2SO_4	2,0
Cd	As, Co, Cu, Zn	KCN solution	2,7
Cu	Cd	dil. $HClO_4$	2,2
Cu	Sn	NH_3 + tartrate	1,8
Hg	Bi, Cu	dil. HNO_3	1,3

Electrolysis at constant potential takes somewhat longer than electrolysis at constant current because of the lower current intensity. The process is particularly important for the separation of nobler metals such as Ag, Au, Bi, Cu and Hg. Separations of anions (Cl^-, Br^- and I^- on silver anodes) have also been described.

Potential control. Electrolysis at constant potential cannot separate elements with small differences in decomposition potentials because as electrolysis proceeds, the current drops, changing the deposition overvoltages. One can eliminate this interference by keeping the cathode potential constant during deposition rather than the voltage.

The simplest method of potential control uses the principle of the galvanic element: a more noble metal is deposited by dissolving a less noble one. For example, if a zinc and a platinum rod are dipped into the solution to be electrolyzed and both are short-circuited, the zinc assumes a certain potential, which is determined by the zinc ion concentration in the solution (and the temperature). At the platinum electrode, the same potential is established as a result of the conductive connection. Ions of metals more noble than zinc are reduced and deposited metallically on the platinum electrode, with the equivalent amount of zinc going into solution (see Fig. 12.17). The process is referred to as *internal electrolysis*.

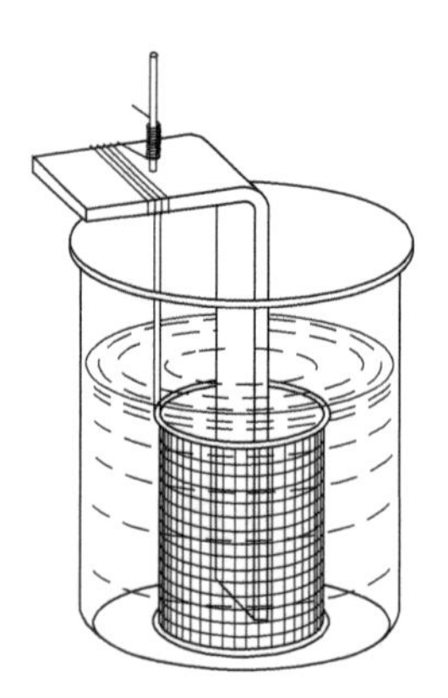

Fig. 12.17: Internal electrolysis without diaphragm.

In itself, the potential of the zinc bar does not remain constant, since the zinc ion concentration in the solution is continuously increased. However, if the solution already contains a relatively high zinc concentration at the beginning of electrolysis and the amount of metal deposited is small, the concentration of dissolved zinc hardly changes and the potential of the two electrodes can be regarded as practically constant.

In this mode of operation, the deposited element should be deposited not only on the platinum electrode, but also on the zinc electrode. However, if only small amounts of metal are present in the solution, this is not the case. The explanation for this is probably to be found in higher overvoltage of the metal deposition on the zinc compared to the platinum.

For the separation of somewhat larger amounts of metal by internal electrolysis, vessels are used in which the two electrodes are separated by a porous septum, a so-called *diaphragm*. The base electrode is placed in a clay cylinder or dialysis sleeve or covered with a collodion layer. The counter electrode can be arranged concentrically around it.

Instead of zinc, other metals can also be used as base electrodes, which makes the potential of the system selectable within certain limits. Furthermore, a base potential can be achieved by placing an inert platinum electrode in a reducing dissolved redox system, and finally, anodic deposition of higher oxides or oxide hydrates can also be carried out by selecting a strongly oxidizing system as the counter electrode (cf. Table 12.17).

The main advantage of internal electrolysis is the simplicity of the equipment. It also permits the separation of metals from chloride-containing solutions, since no free chlorine is formed. A disadvantage is the slowness of the depositions, so that even devices with diaphragm and agitation only allow the separation of small amounts of metal (maximum about 10 mg).

Better monitoring and control of electrolysis is achieved by continuous measurement and control of the cathode potential. For this purpose, the salt bridge of a calomel reference electrode is brought close to the cathode and its potential is measured with a high-impedance voltmeter (Fig. 12.18).

Tab. 12.17: Internal electrolysis (application examples).

Deposited	Diaphragm	Anode	Not deposited
Ag	Parchment	$Cu/Cu(NO_3)_2$	As, Bi, Cu, Fe, Ni, Pb
Ag, Bi, Cu	–	$Pt/V^{3+} + V^{2+}$	–
Bi, Cu	Al_2O_3	$Pb/Pb(NO_3)_2$	As, Cd, Fe, Pb, Sb, Sn, Zn
Cd	–	$Zn/ZnSO_4$	Zn
Cu	Parchment	$Fe/FeSO_4$	As, Cd, Co, Fe, Ni, Pb, Zn
Hg	Parchment	$Cu/CuSO_4$	Cu, Zn
Ni	Parchment	$Zn/Zn(NH_3)_6Cl_2$	Zn
Pb	Clay	$Zn/Zn(NO_3)_2$	Zn
PbO_2	Collodion	Coal/$K_2S_2O_8$ solution	–
Zn	Collodion	Na amalgam/Na_2SO_4	–

The circuit largely corresponds to that given in Fig. 12.14 for electrolysis at controlled current. In addition, the calomel comparison electrode g is present. The potential measuring device e must have a high internal resistance so that the measurement is practically currentless; otherwise the reference electrode would be damaged.

In order to avoid the inconvenience of continuous supervision of the electrolysis, self-regulating devices, so-called *potentiostats*, have been developed. With these, an optionally specified potential can be kept constant to within a few millivolts.

In the case of quantitative deposition, the aim will be that at most about 0.1% of the amount of metal originally present remains in the solution. This corresponds to a decrease in concentration by three powers of ten; since, according to Eq. (2), the potential of an electrode changes by 58 mV for a change in concentration in the electrolyte by one power of ten (assuming monovalent ions), it must become less noble by $3 \cdot 58 = 174$ mV for practically quantitative deposition. Consequently, if two monovalent ions are to be separated, their deposition potentials must differ by at least this amount at the start

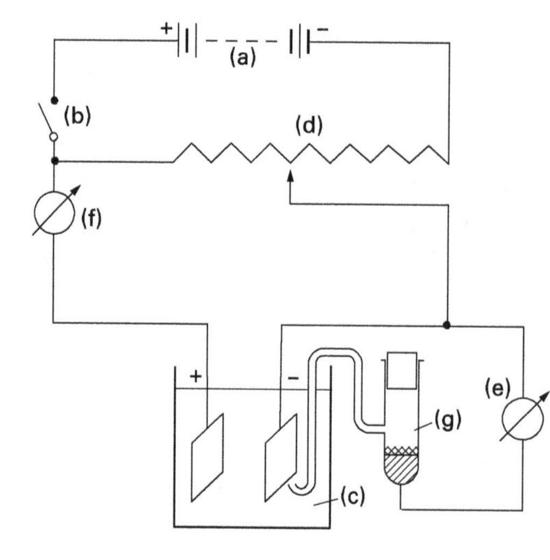

Fig. 12.18: Circuit for electrolysis at controlled potential (principle).
a) Power source;
b) switch;
c) electrolysis vessel;
d) sliding resistance;
e) high impedance voltmeter;
f) ammeter;
g) calomel electrode.

of electrolysis. However, because of unavoidable errors in potential control and because of the change in cathodic overvoltage with current, a potential difference of at least 0.2 V is normally required for separation. For the separation of divalent ions, this difference only needs to be about 0.1 V, and for the separation of trivalent ions, 0.07 V.

With the aid of potential control, the selectivity of electrolytic depositions can be significantly improved. Since it is possible to shift the deposition potentials by adding complexing agents, further improvements can be achieved by such additives.

The composition of the solution to be selected for a particular separation and the cathode potential to be set must be determined empirically, since the real potentials can deviate greatly from the normal potentials. (When mercury cathodes are used, the potentials can be taken from the polarographic half-step potentials, which are often known.) Some examples of electrolytic separations under potential control are given in Tab. 12.18.

Performance of electrolytic separation methods. The main advantage of electrolytic separations is that separations are possible without the addition of chemicals and without filtrations, whereby the concentrations in the starting solutions can vary within wide limits. Furthermore, the entrainment effects are generally very small, so that exceptionally high separation factors can be achieved. In many separations, even the main component can be separated without trace elements in solution being noticeably incorporated into the metal precipitate. Mercury cathodes are particularly suitable for such separations, since large quantities of metal can be separated on them.

This is countered by some disadvantages and difficulties. The complete removal of an ion from a solution is not possible; detectable amounts always remain undepleted, the amounts of which vary depending on the element, the composition of the solution, and the electrolysis conditions. Other unavoidable losses occur at the termination of electrolysis: in this process, without interrupting the current, one removes

Tab. 12.18: Electrolysis at controlled potential (examples).

Deposited	Electrolyte	Potential vs. sat. calomel el. (V)	Not deposited
Cu	$HCl + HNO_3$ + hydrazine + urea + tartrate + succinate, pH 5.3	−0,30	Bi, Cd, Ni, Pb
Bi	$HCl + HNO_3$ + hydrazine + urea + tartrate + succinate, pH 5.3	−0,40	Cd, Ni, Pb
Pb	$HCl + HNO_3$ + hydrazine + urea + tartrate + succinate, pH 4–5	−0,60	Cd, Ni, Sb, Sn, Zn
Sn	$HCl + HNO_3$ + hydrazine + urea + tartrate + succinate, pH 2	−0,60	Cd, Fe, Mn, Zn etc.
Cu	$HClO_4$ + tartrate + hydrazine + NH_3	−0,40	Cd, Pb, Zn
Pb	$HClO_4$ + tartrate + hydrazine + NH_3	−0,60	Cd, Zn
Ag	approx. 1 m NH_3	−0,24	Cd, Cu
Cu	approx. 1 m NH_3	−0,73	Cd
Rh	3.5 M NH_4Cl	−0,4	Ir
Bi	H_2SO_4 + hydrazine + Na citrate, pH 3	−0,2	Sb, Sn
Sb	H_2SO_4 + hydrazine + Na citrate + HCl, pH 3	−0,3	Sn

the electrode from the solution and rinses it rapidly with water; even when working very carefully and rapidly, a noticeable portion of the precipitate, which is easily attacked in its freshly deposited state, is redissolved. For nobler elements, such as copper, this amount is about 0.03–0.1 mg, but it can increase to more than 1 mg for less noble metals such as tin or lead.

If, in addition to metal deposition on the electrodes, gas evolution occurs, which is usually the case in the controlled current mode of operation, some solution sprays through the surface of the liquid when the fine gas bubbles break through, and the droplets formed may be carried out of the electrolysis vessel. This disturbance is remedied by covering the vessel with a watch glass broken in two, the underside of which is sprayed off from time to time.

Furthermore, some deposits tend to fall off the electrode, residues of solution or of oxides may be trapped in the precipitate, and finally, small amounts of the anode material (and of the cathode material if mercury cathodes are used) enter the solution during electrolysis.

Interference with deposition can occur in the presence of complexing agents as well as redox buffer systems in high concentrations. The interference by chloride has already been mentioned.

12.3 Separations by one-time equilibration: Working with two auxiliary phases

12.3.1 Precipitation

The entrainment effect described earlier can be exploited for analytical separations if the entrained component can be removed virtually completely from a solution. The co-precipitation is often quantitative when the entrained compound is present in low concentration. The method is therefore particularly suitable for the enrichment of trace constituents, and the entrained precipitate is accordingly referred to as a *trace catcher*.

Since two auxiliary phases, the solvent and the precipitate, are present in this method, *entrainment isotherms* can be determined, which formally correspond to adsorption or ion exchange isotherms, but in many cases there is no equilibrium setting. To determine these isotherms, a certain amount of precipitate is generated under specified conditions and the distribution of the entrained matter between precipitate and solution is determined as a function of concentration. Curve shapes corresponding to Freundlich or Langmuir isotherms are often obtained[4] (cf. Fig. 12.19a). Nernst distributions have also been observed in crystalline precipitates.

The entrainment effect can be explained – as already mentioned – by adsorption as well as compound or solid solution formation, but the interpretation in the concrete case is usually not possible, and also the isotherm shape does not usually allow clear decisions.

Another mechanism of co-precipitation occurs when a poorly soluble precipitate is entrained that is present in such small quantity that it is not filterable (and possibly not visible). In such cases, an additional precipitate produced in larger quantity acts as a *collector*; the isotherm then consists of a vertical line whose distance from the ordinate corresponds to the solubility of the entrained compound (Fig. 12.19b).

Various sparingly soluble inorganic precipitates are used as trace traps, which are usually precipitated in amounts of about 10–100 mg. As can be seen from Tab. 12.19, mainly coarse-flaked non-crystalline precipitates with a large inner surface are suitable for entrainment.

Another type of trace catchers consists of precipitates of pure organic compounds; such collectors are obtained in various ways.

For example, ethanol or acetone solutions of water-insoluble organic compounds can be poured into the aqueous solution of the trace element. Suitable are e.g. phenolphthalein, β-naphthol, m-nitrobenzoic acid, β-hydroxynaphthoic acid etc. Many complex compounds of metals with organic reagents are entrained, e.g. 8-hydroxyquinoline, anthranilic acid or dimethylglyoxime compounds.

4 In another experimental method, the entrained precipitate is fractionally precipitated and the distribution of the entrained material among the individual fractions is examined.

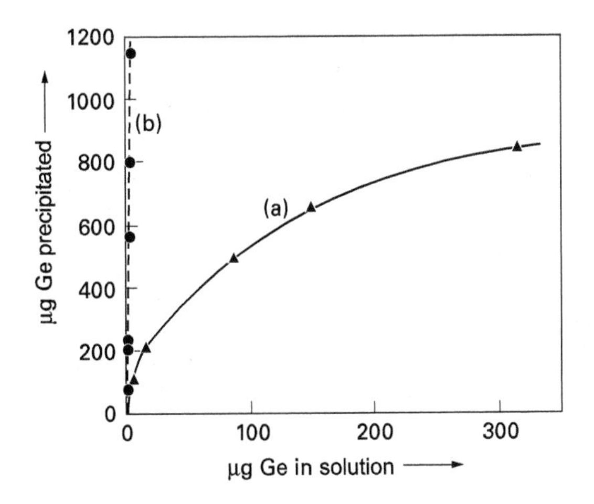

Fig. 12.19: Co-precipitation of germanium(IV) with iron(III) hydroxide and tin(IV) sulfide.
(a) Precipitate 20 mg Fe from 50 ml solution as $Fe(OH)_3$;
(b) 10 mg Sn precipitated from 20 ml solution as SnS_2.

Tab. 12.19: Co-precipitation of trace elements with inorganic precipitates (examples).

Precipitation	**Co-precipitated**
$Al(OH)_3$	Be, Bi, Co, Cr^{3+}, Ga, In, PO_4^{3-}, Sn^{4+}, Ti, V^{5+}, W^{6+}, U^{6+}
$Fe(OH)_3$	Al, Ag, AsO_4^{3-}, Be, Bi, Cd, Co. Cr^{3+}, Mn, Mo^{6+}, Nb, Ni, Pb, Pd, PO_4^{3-}, Rh, Sb, Se^{4+}, Te^{6+}, Ti, Tl^{3+}, Rare Earths, U^{6+}, W^{6+}, Zr
MnO_2	AsO_4^{3-}, Bi, Co, Cr, Mo^{6+}, Sb, Fe, Nb, Sn^{4+}, Pa, Tl^{3+}
CuS	Ag, Bi, Cd, Ge, Hg^{2+}, Pb, Pd, Pt, Sb, Tc, Zn
HgS	Ag, Bi, Cd, Cu, Pb, Tl^{I}, Zn
Te	Ag, Au, Hg, Pd, Pt, Se
As	Se, Te
Hg_2Cl_2	Ag, Au
LaF_3	Th, transuranics
$Zn[Hg(SCN)_4]$	Co, Cu, Ni
$Hg(IO_4)_2$	Th
K-Rhodizonate	Sr, Ba, Ra, Pu
NH_4 – Dipicrylaminate	K, Rb, Cs
$CaCO_3$	Fe, Pb, Sr.

A second group of precipitates is obtained by adding chloride, bromide, iodide or thiocyanate to aqueous solutions of various organic bases. As such, dyes such as methyl violet, crystal violet, malachite green, fuchsin, methylene blue, rhodamine B, safranin, various polymethine dyes and others are used.

These precipitates mainly entrain chloro, bromo and thiocyanato complexes of metals, e.g. the chloro complexes of Tl^{3+}, Sb^{5+} and Au^{3+} as well as the iodo and thiocyanato complexes of the same metals and of Bi, Cd, Hg, In, Zn and others.

The organic bases mentioned above can also be precipitated by adding water-soluble organic acids, e.g. sulfonic acids.

Finally, mention should be made of precipitates for which a colloidal chemical precipitation mechanism can be assumed; this is especially true for tannin, which can be precipitated by the addition of methyl violet.

In most cases, several hundred milligrams of the organic precipitate are produced; in many cases, extraordinarily small amounts of trace elements can still be co-precipitated (cf. Tab. 12.20). The selectivity can be improved by a favorable choice of the pH value and by masking.

Tab. 12.20: Co-precipitation of inorganic elements with organic precipitates (examples).

Trace catcher	Medium	Entrainment (90–100%) & smallest concentration
Methyl violet + SCN^-	Dil. HCl	0.1 µg U/l
Methyl violet + I^-	Dil. H_2SO_4	2 µg Bi/l
Methyl violet + I^-	0.5 N H_2SO_4	0.05 µg In/l
Crystal violet + SCN^-	0.05 N HCl	10 µg Zn/l
p-Dimethylamino-azobenzene + methyl orange	0.2 N HCl	0.1 µg Tl^{3+}/l
8-Hydroxyquinoline + β-naphthol	Neutral	2 µg Ni/l
Eriochrome blue black T + methyl violet	pH 4	40 µg Cr^{3+}/l
Methyl violet + tannin	0.6 N H_2SO_4	0.5 µg Sn^{4+}/l
Methyl violet + tannin	0.2 N HCl	1 µg Zr/l
Methyl violet + tannin	0.2 N HCl	0.0001 µg Nb/l
Methyl violet + tannin	0.2 N HCl	0.7 µg Ta/l
Methyl violet + tannin	0.2 N HCl	0.3 µg Mo^{6+}/l
Methyl violet + tannin	0.2 N HCl	0.1 µg W^{6+}/l

The extent of co-precipitation of trace elements is influenced by numerous variables. Figure 12.20 shows the co-precipitation of vanadate with a number of precipitates as a function of the amount precipitated. Considerable differences occur between the various trace collectors, and it is also evident that in these examples the entraining precipitate must be produced in relatively large quantities.

Furthermore, an important variable is the pH of the solution after completion of precipitation; as an example of its influence, consider the co-precipitation of molybdate with MnO_2 (Fig. 12.21).

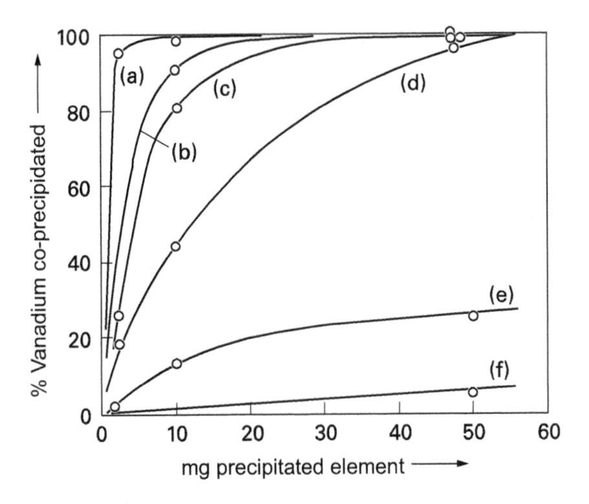

Fig. 12.20: Co-precipitation of 43 µg vanadium(V) from 50 ml solution with different collectors. (a) $Cr(OH)_3$; (b) $Fe(OH)_3$; (c) $Al(OH)_3$; (d) $SiO_2 \cdot aq$; (e) calcium phosphate; (f) $BaSO_4$.

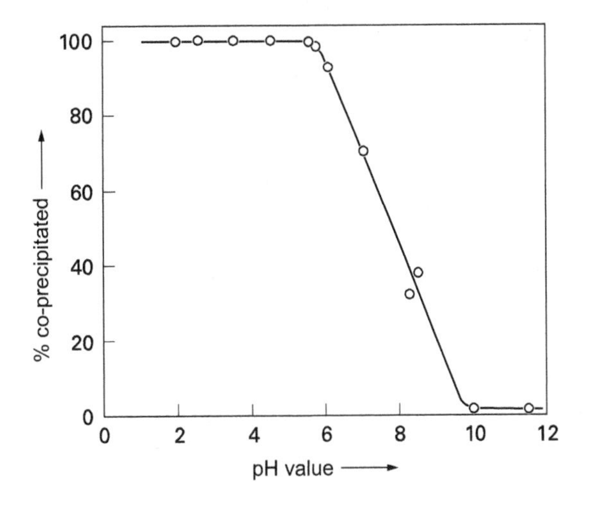

Fig. 12.21: Co-precipitation of 10 µg molybdenum(VI) with 5 mg MnO_2 as a function of solution pH.

Furthermore, the importance of the surface charge of the precipitate should be pointed out; this can be changed by the precipitation conditions. For example, if silver chloride precipitation is produced in an $AgNO_3$ solution with increasing amounts of NaCl, Ag^+ ions are first adsorbed on the surface of the solid particles and a positive charge is imparted to them. Positive ions present in the solution, e.g. Pb^{2+}, are not adsorbed, while the negatively charged Pb-ethylenediamine-tetraacetic acid complex is virtually completely entrained (Fig. 12.22).

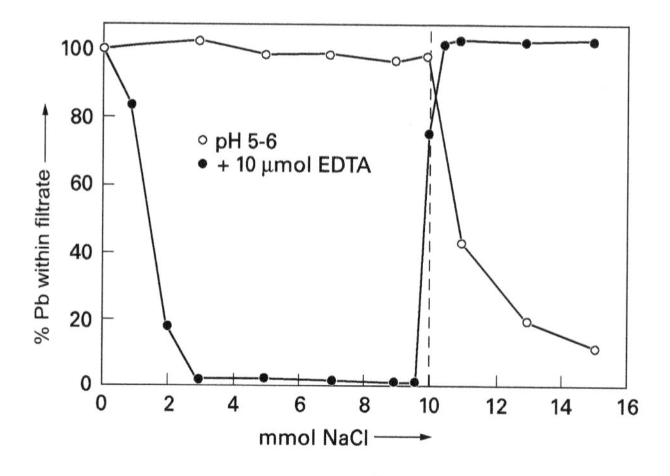

Fig. 12.22: Co-precipitation of lead (as Pb^{2+} and as Pb-EDTA complex) with AgCl as a function of the amount of NaCl added; 10 mMol $AgNO_3$ in solution.

Conversely, when the amount of NaCl used for precipitation exceeds the equivalence point, the AgCl precipitate acquires a negative surface charge due to adsorption of Cl^- ions, and now the positive Pb^{2+} ions are adsorbed while the negative Pb-EDTA complex is repelled.

Occasionally, a pronounced time effect is observed, in that the co-precipitated compound is only completely adsorbed on the precipitate (*post-precipitation*) after prolonged stirring of the preparation; the cause is likely to be slow compound formation at the surface of the solid phase.

The importance of the co-precipitation method lies in the possibility of effectively enriching trace components even from extremely dilute solutions. The advantage here is the simple and fast execution.

A major disadvantage of the method is its low selectivity, which often limits its applicability. Furthermore, there have been hardly any studies to date on the extent to which other components of a solution can block the adsorbent centers of the collector and thus hinder co-precipitation.

Another difficulty is that the entrained trace components usually have to be separated again from the collector used for entrainment. Only in the case of the organic collectors mentioned above (and in the case of mercury precipitates) can the trace entrainment be eliminated by simple incineration.

12.3.2 Precipitation with solid reagents – precipitation exchange

Closely related to entrainment methods is a method in which trace components are removed from solutions by precipitation on the surface of solid reagents. For example, silver can still be practically completely precipitated in amounts of 0.05–50 µg from 2 liters of solution on a few milligrams of solid p-dimethylamino-benzylidene rhodanine, and a corresponding reaction can also be carried out with solid dithizone. The added reagents should be in as voluminous a form as possible; they may be produced by precipitation immediately before use or by crystallization in the solution itself.

Precipitation reactions on the surface of sparingly soluble inorganic precipitates are called *precipitation exchange* because they involve ion exchange, i.e., a double conversion. However, this process usually uses the method of one-way repetition, so it will be discussed in more detail later.

12.3.3 Two-phase precipitation

If the precipitation reagent does not dissolve in the solvent in which the analyte is present, *two-phase precipitation* can be used. The sample is dissolved in one solvent (e.g. water) and the reagent in another (e.g. benzene). If the two solutions are brought into intimate contact by stirring or shaking, the precipitate forms at the phase interface if it is sparingly soluble in both liquids. The method has so far only gained minor importance.

12.3.4 Cementation

In precipitation by *cementation*, ions of more noble elements are reduced by less noble solid metals and precipitated from solutions. The method was already used in ancient times for the precipitation of copper ions with metallic iron, but was not investigated in detail until Bergman. The decisive factor for the reaction is the position of the two elements involved in the voltage series. The method is therefore closely related to internal electrolysis.

When carrying out separations by cementation, the base metal is added to the analytical solution in the form of rods, chips or powder and the whole is stirred for a while. Some examples of use are given in Tab. 12.21.

In the absence of side reactions (e.g., H_2 evolution), the deposited amount of the more noble metal is equivalent to the amount of the less noble metal in solution. One can use this relationship for quantitative analyses, e.g., in the determination of elemental silicon in addition to SiO_2:

$$Si + 4\,AgF \rightarrow 4\,Ag + SiF_4.$$

The precipitated silver is determined.

The advantage of the cementation process is that it is simple and quick to carry out. Selectivity can be improved to a certain extent by choosing the less noble element.

Tab. 12.21: Deposits by cementation (examples).

Deposited element	Metallic reducing agent
Cu	Al, Cd, Fe, In, Zn
Ag	Al, Cu, Fe, Zn
Au	Ag, Zn
Cd	Fe, Zn
Hg	Cu, Fe, Zn
In	Zn
Tl	Mg, Zn
Ge	Fe, Mg, Zn
Sn	Zn
Pb	Al, Fe, Zn
As	Zn
Sb	Al, Cd, Fe, Sn, Zn
Bi	Al, Cu, Cd, Fe, Mg, Pb, Sn, Zn
Te	Fe, Zn
Fe	Zn
Ni	Fe, Zn
Ru	Mg
Rh	Cu, Mg, Sb, Zn
Pd	Ag
Os	Mg, Zn
Ir	Mg, Zn
Pt	Ag, Al, Cu, Hg, Mg, Zn

A disadvantage is that the analysis solution is contaminated by the added element. So far, hardly any investigations are available on the completeness of precipitation by cementation.

12.4 Separations by one-sided repetition

12.4.1 Re-precipitation

The purity of a precipitate can be increased by re-dissolving and re-precipitation. This process, known as *re-precipitation*, largely removes mechanically trapped solution residues and also reduces the content of adsorbed impurities held by solid solution formation; however, the purification effect usually decreases as the concentration of the impurities decreases, especially in the presence of steep adsorption isotherms.

The losses of precipitated compound increase during precipitation, so that only precipitates with very low solubility are suitable for this purpose. For this reason, as well as because of the above-mentioned poor efficiency at very low impurity concentrations and the somewhat time-consuming procedure, precipitation is usually repeated only once, rarely more often.

12.4.2 Fractionated precipitation

Fractional precipitation is a method of operation in which precipitates are produced in step-by-step by adding several small portions of precipitating reagent or by slowly increasing the pH of the solution, and the individual fractions are filtered off or centrifuged off separately.

Ideally, the components of the solution should be completely separated from each other; in practice, however, only shifts in concentration within the fractions compared to the initial mixture are usually achieved.

The method is mainly used to study complicated mixtures of components similar to each other, e.g. partial separation of protein compounds by increasing salt additions; fractions are precipitated from solutions of synthetic high-molecular polymers by adding several portions of a liquid of poorer solubility. By comparing the average molecular weights of the starting material and the fractions, one can approximate the molecular weight distribution of the original polymer.

12.4.3 Precipitation exchange – precipitation with ion exchangers

A number of freshly precipitated precipitates are able to take up ions of other compounds from solutions and bind them to their surface, whereby an equivalent amount of the equally charged ion of the precipitate goes into solution. Some sparingly soluble sulfides have proved particularly suitable for this purpose, but such exchange reactions can also be carried out with other precipitates (Tab. 12.22). To increase the available surface area, the precipitates are often finely distributed on inert supports (e.g. silica gel, cellulose, asbestos).

These exchange reactions proceed very rapidly; the solutions containing the ions to be separated are sucked through thin layers or short columns of the precipitates. However, the capacity of the solid phase is low, so that the process is limited to the separation of small quantities.

Tab. 12.22: Separations by *precipitation exchange* (examples).

Precipitation	Ions separated from solutions and lowest concentration
CdS	Bi^{3+} (approx. 0.4 mg/l); Ag^{+} (5 mg/l); Hg^{2+} (0.1 mg/l)
CuS	As^{3+}
Ag_2S	Au^{3+} (30 µg/l)
AgCl	Br^{-} (0.8 mg/l)
AgBr	I^{-} (approx. 2 µg/l)
Ag_2CrO_4	Cl^{-} (2 mg/l)
Pb oxalate	S^{2-} (32 mg/l)
$BaSO_4$	Sr^{2+}
CaF_2	Rare earths; Cm

A variant of the method involves exchanging radioactive ions from solutions for inactive ions of the same type in the precipitate. For example, radioactive $^{110}Ag^{+}$ (carrier-free) can be removed from extremely dilute solutions by sucking the liquid through a layer of freshly precipitated AgI.

Both reactions, ion and isotope exchange, can be used to isolate short-lived radioactive nuclides.

Precipitation with ion exchangers should also be mentioned at this point. These can be carried out, for example, by allowing the analysis solution to flow through an anion exchanger column in the OH form. Ions that form sparingly soluble hydroxides are precipitated on the exchanger. Thus, Y^{3+} was separated from Cs^{+}, La^{3+} from Ba^{2+} or Ce^{3+} from Na^{+}. Accordingly, sparingly soluble sulfates can be precipitated with anion exchangers in the sulfate form.

The importance of precipitation exchange lies mainly in the fact that radioactive elements can be removed from solutions in extremely low concentrations.

12.5 Separations by systematic repetition: Precipitation chromatography

Precipitation chromatography corresponds to the other column chromatographic techniques described above. For example, various metal ions can be separated on 8-hydroxyquinoline or violuric acid columns; the metal compounds precipitate in the order of their solubilities, and the zones can be eluted with dilute acids.

Various other attempts of this kind have so far been mostly unsatisfactory; for example, an acid solution of heavy metal salts can be applied to an agar-agar column containing some sodium sulfide and a buffer solution; as the solution diffuses into the column packing, a pH gradient is formed, and the heavy metals are precipitated successively in separate zones as sulfides. Attempts to separate metals by cementation or electrolysis in columns have also been described.

Precipitation also occurs – unintentionally – during the chromatography of inorganic ions on Al_2O_3, MgO and others. Precipitates of hydroxides or basic salts can be precipitated by alkali residues on the column in individual zones.

12.6 Separations by systematic repetition: Thin-layer technique (precipitation paper chromatography)

Chromatographic separations can also be carried out with paper supports in which precipitation reagents are incorporated. A distinction is made between papers containing soluble reagents and those containing sparingly soluble precipitates.

The former are prepared by impregnation with reagent solutions and drying, while in the latter precipitation reactions are carried out within the paper, here the separations proceed as precipitation exchange reactions.

When working with such papers, one applies one or a few drops of analytical solution to the paper strip in the usual way and develops the chromatogram with suitable solutions; the radial chromatographic technique has also been used.

With this method, mainly inorganic ions have been separated from each other, more rarely organic ones. However, the method is not very powerful and has not yet gained any major importance.

General literature

Supramolecular compounds

E. Chernykh & S. Brichkin, Supramolecular complexes based on cyclodextrins, High Energy Chemistry *44*, 83–100 (2010).

S. Fakayode, M. Lowry, K. Fletcher, X. Huang, A. Powe & I. Warner, Cyclodextrins host-guest chemistry in analytical and environmental chemistry, Current Analytical Chemistry *3*, 171–181 (2007).

K. Harris, Fundamental and applied aspects of urea and thiourea inclusion compounds, Supramolecular Chemistry *19*, 47–53 (2007).

G. Hembury, V. Borovkov & Y. Inoue, Chirality-sensing supramolecular systems, Chemical Reviews *108*, 1–73 (2008).

E. Krieg, M. Bastings, P. Besenius & B. Rybtchinski, Supramolecular polymers in aqueous media, Chemical Reviews *116*, 2414–2477 (2016).

J. Lagona, P. Mukhopadhyay, S. Chakrabarti & L. Isaacs, The cucurbit[n]uril family, Angewandte Chemie – International Edition *44*, 4844–4870 (2005).

W. Lustig, S. Mukherjee, N. Rudd, (. . .), J. Li & S. Ghosh, Metal-organic frameworks: Functional luminescent and photonic materials for sensing applications, Chemical Society Reviews *46*, 3242–3285 (2017).

J. Mosinger, V. Tomankova, I. Nemcova, & J. Zyka, Cyclodextrins in analytical chemistry, Analytical Letters *34*, 1979–2004 (2001).

J. Steed & J. Atwood, Supramolecular Chemistry: Second Edition, J. Wiley & Sons, New York (2009).

I. Terekhova & O. Kulikov, Cyclodextrins: physical-chemical aspects of formation of complexes *host-guest* and molecular selectivity in relation to biologically active compounds, Chemistry of Polysaccharides 38–76 (2005).

L. You, D. Zha & E. Anslyn, Recent advances in supramolecular analytical chemistry using optical sensing, Chemical Reviews *115*, 7840–7892 (2015).

Electrolysis

A. Bard & C. Zoski, Electroanalytical Chemistry, CRC Press, Boca Raton (2012).

D. Crow, Principles and Applications of Electrochemistry, Taylor & Francis, London (2017).

F. Scholz, Electroanalytical methods: Guide to experiments and applications, Springer, Heidelberg (2010).
J. Wang, Analytical Electrochemistry, J. Wiley & Sons, New York (2005).

Examples

Detailed information on the technique of precipitation can be found in any textbook of quantitative analysis.

Precipitation from homogeneous solution

T. Mori, K. Takao, K., Sasaki (. . .), T. Arai & Y. Ikeda, Homogeneous liquid–liquid extraction of U(VI) from HNO3 aqueous solution to betainium bis(trifluoromethylsulfonyl)imide ionic liquid and recovery of extracted U(VI), Separation and Purification Technology *155*, 133–138 (2015).
S. Tovstun, A. Ivanchikhina, M. Spirin, E. Martyanova & V. Razumov, Studying the size-selective precipitation of colloidal quantum dots by decomposing the excitation-emission matrix, Journal of Chemical Physics *153*, 084108 (2020).
C. Tzschucke, C. Markert, W. Bannwarth, (. . .), A. Hebel & R. Haag, Modern separation techniques for the efficient workup in organic synthesis Angewandte Chemie – International Edition *41*, 3964–4000 (2002).
N. Wang, Q. Wang, Y. Geng, D. Wu & Y. Yang, Recovery of Au(III) from Acidic Chloride Media by Homogenous Liquid-Liquid Extraction with UCST-Type Ionic Liquids ACS Sustainable Chemistry and Engineering *7*, 19975–19983 (2019).

Filtration

A. Gasper, D. Öchsle & D. Pongratz, Handbuch der Industriellen Fest/Flüssig-Filtration, Wiley-VCH, New York (2000).
J. Han, J. Fu & R. Schoch, Molecular sieving using nanofilters: Past, present and future, Lab on a Chip *8*, 23–33 (2007).
B. Kowalczyk, I. Lagzi & B. Grzybowski, Nanoseparations: Strategies for size and/or shape-selective purification of nanoparticles Current Opinion in Colloid and Interface Science *16*, 135–148 (2011).
L. Moskvin & T. Nikitina, Membrane methods of substance separation in analytical chemistry, Journal of Analytical Chemistry 59, 2–16 (2004).
A. Rushton, A. Ward & R. Holdich, Solid-Liquid Filtration and Separation Technology, J. Wiley & Sons, New York (2008).
D. Thomas, A. Charvet, N. Bardin-Monnier & J.-C. Appert-Collin, Aerosol Filtration, Elsevier, Amsterdam (2016).

Co-precipitation with organic precipitation

Y. Aktaş & H. Ibar, Preconcentration of some trace elements on bentonite modified with trioctyl phosphine Oxide, Fresenius Environmental Bulletin *13*, 156–158 (2004).

K. Čundeva, T. Stafilov, G. Pavlovska, I. Karadjova & S. Arpadjan, Preconcentration procedures for trace cadmium determination in natural aqueous systems prior to Zeeman ETAAS, International Journal of Environmental Analytical Chemistry 83, 1009–1019 (2003).

S. Hoeffner, J. Conner, and R. Spence, Stabilization/solidification additives, in R. Spence and C. Shi, Stabilization and Solidification of Hazardous, Radioactive, and Mixed Wastes, CRC Press, Boca Raton, 177–198 (2005).

Organic precipitation reagents for inorganic ions

S. Gürmen, S. Timur, C. Arslan & I. Duman, Production of pure tungsten oxide from scheelite concentrates Scandinavian Journal of Metallurgy *31*, 221–228 (2002).

D. Kara & M. Alkan Preconcentration and separation of copper (II) with solvent extraction using N,N′- bis (2-hydroxy-5-bromo-benzyl)1,2 diaminopropane, Microchemical Journal *71*, 29–39 (2002).

S. Zarco-Fernández, M. Mancheño, R. Muñoz-Olivas & C. Cámara, A new specific polymeric material for mercury speciation: Application to environmental and food samples Chimica Acta *897*, 109–115 (2015).

Literature to the text

Precipitation with ion exchange resins

S. Dulanská, B. Remenec, J. Bilohuščin, F. Mátel & M. Bujdoš, Development of ^{126}Sn separation method by means of anion exchange resin and gamma spectroscopy, Applied Radiation and Isotopes *123*, 128–132 (2017).

S. Fernandes, I. Romão, C. Abreu, M. Quina & L. Gando-Ferreira, Selective separation of Cr(III) and Fe(III) from liquid effluents using a chelating resin Water Science and Technology *66*, 1968–1976 (2012).

T.-Y. Kim, S.-Y. Jeung, S.-Y. Cho, Y. Kang & S.-Y. Kim, Removal and regeneration of heavy metal ions using cation exchange column, Journal of Industrial and Engineering Chemistry *10*, 695–700 (2004).

Precipitation exchange

T. Neumann, Fundamentals of aquatic chemistry relevant to radionuclide behavior in the environment, Woodhead Publishing Series in Energy *42* (Radionuclide Behaviour in the Natural Environment), 13–43 (2012).

Precipitation chromatography

E. Breynaert & A. Maes, Column precipitation chromatography: An approach to quantitative analysis of eigencolloids Analytical Chemistry *77*, 5048–5054 (2005).

Y. Ito & L. Qi, Centrifugal precipitation chromatography, Journal of Chromatography B: Analytical Technologies in the Biomedical and Life Sciences *878*, 154–164 (2010).

Y. Ito, Centrifugal precipitation chromatography: novel fractionation method for biopolymers, based on their solubility, Journal of Liquid Chromatography & Related Technologies *25*, 2039–2064 (2002).

Filtration

M. Al-Aseeri & Q. Bu-Ali, Filtration systems: classification & selection, TCE *778*, 44–46 (2006).
M. Dosmar u. S. Pinto, Crossflow filtration, Drugs and the pharmaceutical sciences *174* (Filtration and Purification in the Biopharmaceutical Industry), 495–542 (2008).
G. Ghugare, V. Nimkande & D. Khairnar, Isolation and Enrichment of Bacteriophages by Membrane Filtration Immobilization Technique Current Protocols in Cell Biology *79*, e41 (2018).
J. Jönsson & L. Mathiasson, Membrane extraction in analytical chemistry, Journal of Separation Science *24*, 495–507 (2001).
T. Meltzer, R. Livingstone, and M. Jornitz, Filters and experts in water system design, Ultrapure Water *20*, 26–28, 30 (2003).
Y. Zhang, S. Zhang & T.-S. Chung, Nanometric Graphene Oxide Framework Membranes with Enhanced Heavy Metal Removal via Nanofiltration, Environmental Science and Technology *49*, 10235–10242 (2015).

Fractionated precipitation of (bio)polymers

C. A. Meyer, X. Zhang, L. Yu, H. Bi, (. . .), Y. Zhou & G. Tai, Total fractionation and characterization of the water-soluble polysaccharides isolated from Panax ginseng Carbohydrate Polymers *77*, 544–552 (2009).
A. Torres, L. Foster, and G. Leland, Fractionation of serum proteins: in the search for disease biomarkers, American Biotechnology Laboratory *28*, 8–10, 13 (2010).
F. Yang & Y. Ito, Method for the fractionation of dextran by centrifugal precipitation chromatography Analytical Chemistry *74*, 440–445 (2002).

Kinetics of precipitation reactions

B. Fritz u. C. Noguera, Mineral precipitation kinetics, Reviews in Mineralogy & Geochemistry *70* (Thermodynamics and Kinetics of Water-Rock Interaction), 371–410 (2009).
J. Schoot, O. Pokrovsky, & E. Oelkers, The link between mineral dissolution/precipitation kinetics and solution chemistry, Reviews in Mineralogy & Geochemistry *70* (Thermodynamics and Kinetics of Water- Rock Interaction), 207–258 (2009).
J. Xue, Z. Jia, X. Jiang, (. . .), X. Zhu & D. Yan, Kinetic separation of polymers with different terminals through inclusion complexation with cyclodextrin Macromolecules *39*, 8905–8907 (2006).

Protein precipitation for structural analysis

B. Rupp, Biomolecular Crystallography: Principles, Practice, and Application to Structural Biology, Garland Science, Taylor & Francis Group, New York (2010).

13 Solubility: Extraction and phase analysis

13.1 General – definitions – auxiliary phases

Extraction is the process of removing individual substances from a mixture of solids using a suitable solvent.[1] The process can be regarded as the reverse of a precipitation operation. Accordingly, only one auxiliary phase, the solvent, is used in the extraction process.

The extraction process may involve a chemical reaction (e.g., extracting $CaCO_3$ from silicates or fluorides with dil. HCl or CH_3COOH). If the extraction proceeds without chemical reaction, the method allows the isolation of individual undecomposed compounds, which are either recovered from the extract, or which remain unaffected in the residue. In the field of inorganic analysis, the method is also referred to as *phase analysis*.

Since separations by extraction can be performed effectively only when the solvent is in contact with every particle of the soluble component, the starting material must be finely ground or shaped into the form of a thin film before or during extraction.

In the presence of solid solutions, effective separation by extraction is not possible.

13.2 Separations by one-time equilibrium adjustment

In the case of separations by one-time equilibration, the sample is triturated intensively with the solvent or – especially when organic substances are present – finely comminuted in the presence of the solvent in a *mixer* with rotating knives.

The one-time extraction has the disadvantage that some solution always remains attached to the residue, which worsens the separation effect.

13.3 Separations by one-sided repetition

13.3.1 Repeated extraction

The aforementioned interference due to adhesion of solution to the insoluble components can be eliminated by extracting the sample several times; this method of working is therefore used very frequently.

1 Shaking out of solutions (*liquid-liquid extraction*), which is often referred to as extraction in the literature, does not fall under this definition.

https://doi.org/10.1515/9783111181417-013

Ring-oven method. Instead of discontinuous extraction with several portions of solvent, the liquid can also be allowed to flow continuously through the sample. One application of this principle is the *ring-oven method*: A drop of the solution to be analyzed is placed in the center of a small circular filter, which is placed over the central bore of an aluminum heating block that is about 105 °C hot (Fig. 13.1). One then precipitates some of the solutes on the filter (e.g., by passing H_2S gas) and washes the unprecipitated compounds radially outward with a solvent dripping from a pipette. The washing liquid evaporates as it approaches the inner edge of the heating block bore; the solutes remain in the paper in the form of a sharp ring and can be sensitively detected with color reagents.

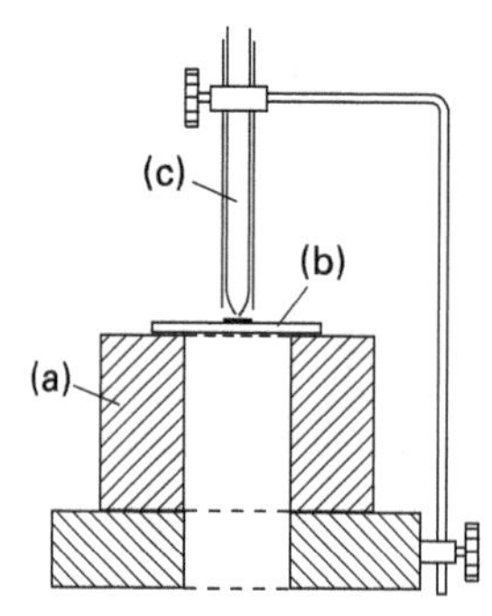

Fig. 13.1: Ring-oven method.
a) Heating block with hole;
b) filter paper with substance mixture;
c) wash pipette with holding tube.

Supercritical fluid extraction (SFE). The extraction of soluble components from porous solids using *supercritical fluids* (mostly CO_2 has become a very efficient method.

The basis is the change in solvent power properties when the so-called critical point P_c is exceeded in a p,T phase diagram (see Fig. 13.2 for the example of CO_2).

Carbon dioxide takes on properties that are comparable to the gas phase in terms of viscosity, but similar to those of a liquid in terms of dissolving properties. The latter depends on the density, which can be varied. The high flowability makes a supercritical medium an ideal solvent, since on the one hand the high diffusivity allows penetration into porous solids, and on the other hand the removal of the supercritical medium after extraction is easily accomplished by pressure relief. Caffeine extraction using SFE has been known on a large scale for years.

A system for SFE (see Fig. 13.3) essentially consists of a pump which, in the case of CO_2, transports it as a liquid at 5 °C and >74 bar through an extractor column containing the material to be extracted. After the extractor column, there is a restriction capillary which serves to maintain the pressure. From this, the extraction mixture enters the so-called separator, a recipient at lower pressure, in which the expansion of CO_2 results in continuous separation from the extracted material.

In combination with a stationary phase, *supercritical fluid chromatography* can be set up, which is of great importance for the separation of polymer mixtures.

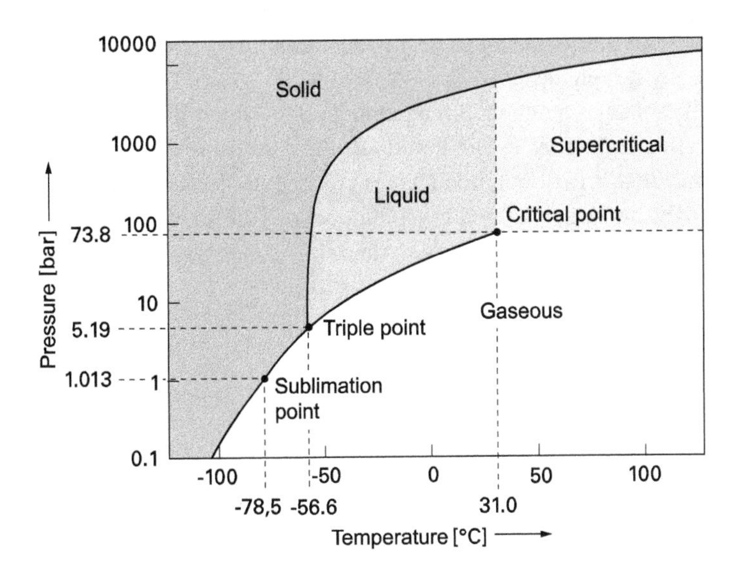

Fig. 13.2: p,T phase diagram of carbon dioxide.

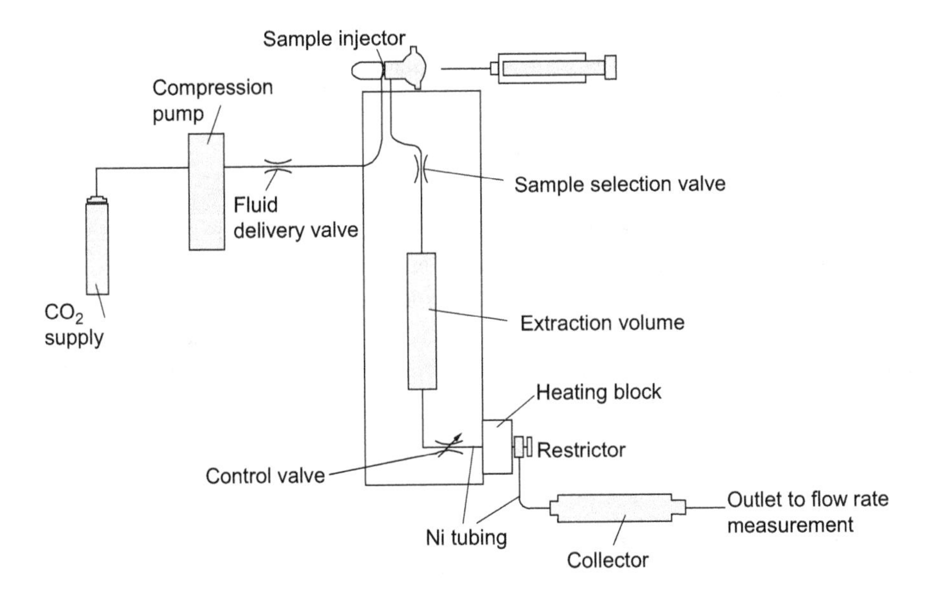

Fig. 13.3: Schematic structure of an SFE apparatus.

13.3.2 Circulation method

Extraction methods are often used in which the moving phase is circulated; the extractant is continuously purified by distillation and allowed to flow through the sample again.

A device with which the individual steps of this operation are carried out automatically paragraph by paragraph has been given by Soxhlet (Fig. 13.4). The substance mixture is contained in a sleeve made of filter paper b, into which the extraction agent drips from the reflux condenser c. The liquid level in the sleeve is determined by the reflux condenser. When a certain liquid level is reached in the sleeve, the solution is drawn off through the siphon tube d by siphon action; the solvent is evaporated in the flask a and passes through the tube e to the reflux condenser.

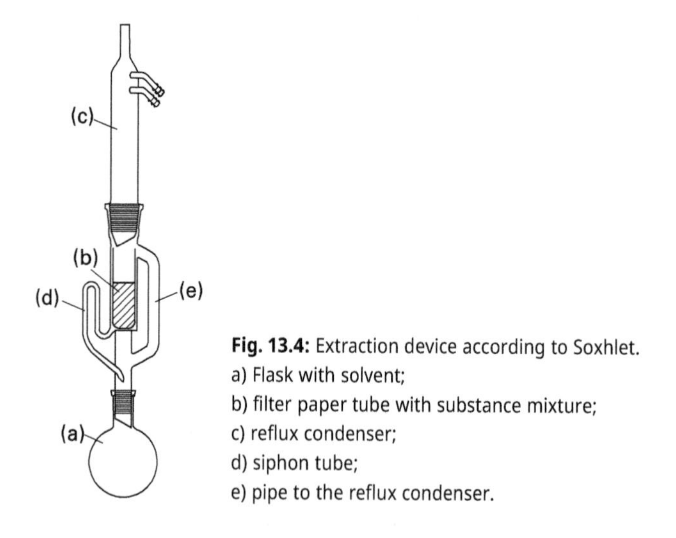

Fig. 13.4: Extraction device according to Soxhlet.
a) Flask with solvent;
b) filter paper tube with substance mixture;
c) reflux condenser;
d) siphon tube;
e) pipe to the reflux condenser.

Instead of the batch-wise extraction according to Soxhlet, a continuous mode of operation can also be used; of the numerous arrangements described for this purpose in the literature, a simple device for the extraction of small amounts of substance is shown (Fig. 13.5). The sample is contained in a glass filter crucible placed under a reflux condenser so that it is continuously flushed by the condensate. A device which is basically the same, in which extraction can be carried out either in an inert gas stream or under vacuum, is shown in Fig. 13.6.

13.3.3 Fractionated extraction – gradient extraction

Fractionated extraction is a variant of repeated extraction. In this process, substance mixtures are extracted one after the other with different solvents in order to obtain several components separately.

The method plays a role mainly in the chemistry of high polymers; it is possible to obtain more uniform fractions from mixtures of different degrees of polymerization. The main difficulty here is to comminute the polymer as finely as possible, which is the prerequisite for the effectiveness of the method. Either the starting mixture is used as a thin

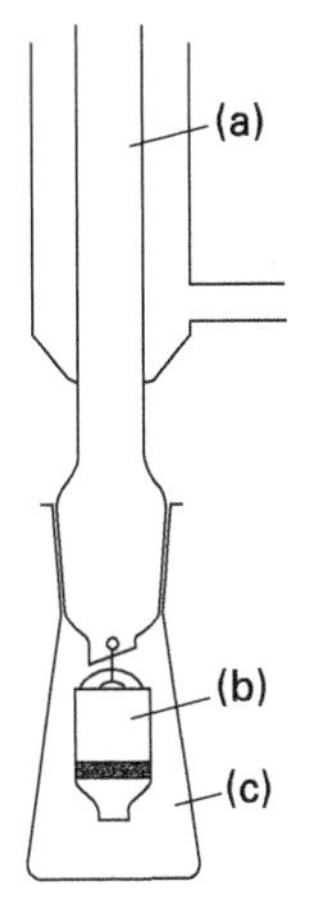

Fig. 13.5: Continuous microextraction.
a) Reflux condenser;
b) filter crucible with sample;
c) flask with extraction agent.

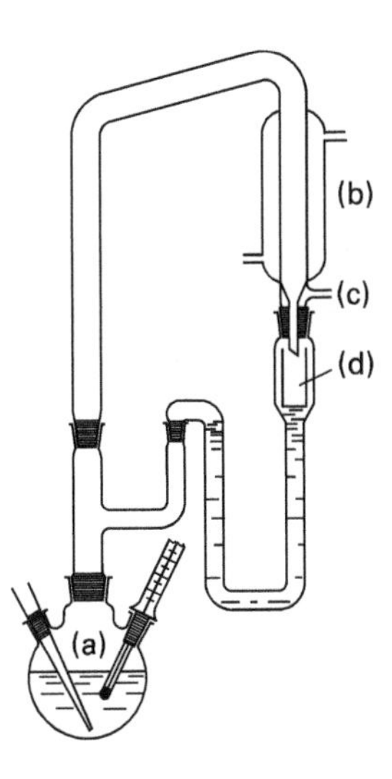

Fig. 13.6: Device for extraction in an inert gas stream or under vacuum.
a) Flask with gas inlet and thermometer;
b) cooler;
c) vacuum connection;
d) extraction sleeve or filter crucible with substance.

film on a metal support or as a coating on small granules of an inert support (e.g. diatomaceous earth). The various solvents are prepared by mixing a liquid that dissolves the polymer with a non-solvent liquid in different mixing ratios; for example, mixtures of methyl acetate and petroleum ether are used to fractionate polyvinyl acetate.

Gradient extraction is also possible here. The high polymer is precipitated onto diatomaceous earth, the whole is filled into a column and extracted with a solvent mixture, the solvent properties of which are continuously improved by a mixing chamber system.

In another type of gradient extraction, the composition of the solvent is kept constant but the temperature is continuously increased.

13.3.4 Scope of application and effectiveness of the method

In inorganic analysis, extraction methods are used relatively rarely, since solid solutions often occur whose components cannot be separated in this way. Some examples of applications are given in Tab. 13.1.

Tab. 13.1: Separations of inorganic mixtures by extraction (examples).

Mixture	Solvent	Dissolved
Alkali metal chlorides	Lower alcohols, pyridine, etc.	LiCl
$NaClO_4 + KClO_4$	Ethanol	$NaClO_4$
$Ca(NO_3)_2 + Sr(NO_3)_2 + Ba(NO_3)_2$	Ethanol	$Ca(NO_3)_2$
Cement	5% Boric acid	Plaster
Ores, rocks	CCl_4	S (elementary)
$CaSO_4$ + Silicate	HI	$CaSO_4$
Au-Ag alloys	HNO_3	Ag
Diverse minerals	Dioxane	H_2O

Selective extraction for the investigation of ores and metallurgical products has gained greater importance. In many cases, the procedure known as *phase analysis* is the only method by which the inorganic compounds (not the elements) present in mixtures can be determined (cf. Tab. 13.2).

Tab. 13.2: Phase analysis (examples).

Mixture	Solvent	Dissolved
Potassium salts (chlorides)	CH_3OH, C_2H_5OH	$MgCl_2 \cdot aq$
Cement, mortar, quicklime	Ethylene glycol, glycerol	CaO
$Pb + PbO + PbSO_4$	Ricinoleic acid	PbO
$Pb + PbO + PbSO_4$	Ammonium acetate lsg.	$PbO + PbSO_4$
$Cu + Cu_2O + CuO$	NH_3 solution	Cu_2O
Cu + CuO	Dil. H_2SO_4	CuO
SeO_2 + Selenite	CH_3OH	SeO_2
Se + Selenide	Na_2SO_3 solution	Se
Fe + Iron oxides	$CH_3OH + Br_2$	Fe
Quartz + silicates	HBF_4 solution	Silicate
Quartz + clay	H_3PO_4 solution	Clay

Extraction methods are much more important in the field of organic trace analysis than in inorganic analysis, since many organic substances, e.g. biological material, plastics, dyes, etc., are amorphous, so that solid solutions cannot be formed; often certain compounds can also be dissolved from the interior of coarser particles into which the solvent can penetrate with swelling phenomena.

When extracting biological material, solvents that are miscible with water, e.g., acetone or lower alcohols, are often advantageous because they diffuse into the interior of cells better than water-insoluble liquids.

Examples for the extraction of biological substances can be found in Tab. 13.3.

Tab. 13.3: Separation of organic mixtures by extraction (examples).

Material	Solvent	Dissolved
Food	Ether, petroleum ether, trichloroethylene, etc.	Fat
Bones	Gasoline, benzene	Fat
Palm kernels	Gasoline, benzene	Fat
Sugar beets	Water (hot)	Sugar
Wood, barks	Water	Tannins
Food	Hydrochloric acid (10%)	Vitamin B_1
Food	Water; acetic acid (0.2%)	Vitamin C
Coffee, tea	Benzene	Caffeine
Flowers, fruits	Ether, petroleum ether	Ether. oils
Plant material	HCl diverse concentr.	Alkaloids
Plant material	Acetone, $CHCl_3$ et al.	Insecticides

Fractional extraction of high polymer plastics is performed to obtain fractions that are more uniform in degree of polymerization than the starting material; moreover, by determining the average molecular weight of each fraction, as in fractional precipitation, it is possible to calculate the polymer distribution of the original sample.

Special extraction solutions are used for the analysis of fertilizers and soils to determine the so-called *plant-available* nutrient content, since it is not the total content that is important here, but only the proportion that can be taken up by the roots of the plants (Tab. 13.4).

The effectiveness of separations by extraction is conditioned by the differences in the solubilities of the respective substances. Useful results can only be obtained if the poorly soluble portion of the analysis sample has an extremely low solubility. Furthermore, the extraction agent must be able to come into contact with all particles of the easily soluble substance.

Accordingly, when a number of extraction instruments were examined, errors of about ±2% were observed fairly consistently. Another source of error occurs in instruments with recirculation of the solvent: During evaporation of the extract, a small portion of the solution sprays out, and the fine droplets may be partially carried by the vapor stream to the reflux condenser, thus re-entering the substance to be extracted. The extent of this disturbance depends mainly on the concentration of the solution in the flask and on the distillation speed.

Tab. 13.4: Extraction solutions for the determination of plant-available nutrients (examples).

Sample material	Extraction solution	Extracted nutrient
Fertilizer	Ammonium citrate; pH 7	Phosphate ion
Peat	1 M NH_4Cl solution	Na^+, K^+, Mg^{2+}, Ca^{2+}
Soilr	Sodium citrate (35 g/l)	Phosphate ion
Soil	1 N CH_3COONH_4, pH 7	K^+, Fe^{2+}, Mn^{2+}
Soil	1.7 M NaCl + HCl (pH 2.5)	NH_4^+
Soil	0.5 N H_2SO_4	Cu^{2+}, Co^{2+}, Zn^{2+}, Mn^{2+}
Soil	H_2O (hot)	Borate ion
Soil	0.1 N Ammonium lactate + 0.04 N CH_3COOH (pH 3.7)	Phosphate ion

13.4 Countercurrent method

Extractions can be carried out by the countercurrent method, in which the extraction material is guided against the flow of the solvent with the aid of a screw. Such separations have found wide application in technology, but are of no significance for analytical chemistry.

General literature

N. Aoki & K. Mae, Extraction, Micro Process Engineering *1*, 325–345 (2009).

A. Chaintreau, Simultaneous distillation-extraction. From birth to maturity. Review, Flavour and Fragrance Journal *16*, 136–148 (2001).

T. Cunha & R. Aires-Banos, Large-scale extraction of proteins, Methods in Biotechnology *11* (Aqueous Two-Phase Systems), 391–409 (2000).

J. Dean & R. Ma, Pressurized fluid extraction, Handbook of Sample Preparation 163–179 (2010).

L. Garcia-Ayuso & M. Luque de Castro, Employing focused microwaves to counteract conventional Soxhlet extraction drawbacks, TrAC, Trends in Analytical Chemistry *20*, 28–34 (2001).

H. Kataoka, A. Ishizaki, Y. Nonaka, and K. Seito, Developments and applications of capillary microextraction techniques: A review, Analytica Chimica Acta *655*, 8–29 (2009).

J. Luque-Garcia & M. Luque de Castro, Where is microwave-based analytical equipment for solid sample pre-treatment going?, TrAC – Trends in Analytical Chemistry *22*, 90–98 (2003).

M. Mikutta, M. Kleber, K. Kaiser, and R. Jahn, Review: Organic matter removal from soils using hydrogen peroxide, sodium hypochlorite, and disodium peroxodisulfate, Soil Science Society of America Journal *69*, 120–135 (2005).

A. Pfennig, D. Delinski, W. Johannisbauer & H. Josten, Extraction technology, Industrial Scale Natural Products Extraction 181–220 (2011).

Q. Pham & F. Lucien, Extraction process design, in: J. Ahmed u. M. Rahman, Handbook of Food Process Design, Wiley-Blackwell, Chichester, UK *2*, 871–918 (2012).

Literature to the text

I. Seabra, M. Braga, M. Batista & H. De Sousa, Effect of solvent (CO_2/ethanol/H_2O) on the fractionated enhanced solvent extraction of anthocyanins from elderberry pomace Journal of Supercritical Fluids *54*, 145–152 (2010).

Phase analysis

C. van Alphen, Automated mineralogical analysis of coal and ash products – Challenges and requirements Minerals Engineering *20*, 496–505 (2007).

O. Omotoso, D. McCarty, S. Hillier, & R. Kleeberg, Some successful approaches to quantitative mineral analysis as revealed by the 3rd Reynolds Cup contest, Clays and Clay Minerals *54*, 748–760 (2006).

K. Rao, The role of solids characterization techniques in the evaluation of ammonia leaching behavior of complex sulfides, Mineral Processing and Extractive Metallurgy Review *20*, 409–445 (2000).

S. Vassilev & C. Vassileva, Methods for characterization of composition of fly ashes from coal-fired power stations: A critical overview, Energy & Fuels *19*, 1084–1098 (2005).

Ring-oven method

W. Chu, Y. Chen, W. Liu, L. Zhang & X. Guo, Three-dimensional ring-oven washing technique for a paper-based immunodevice, Luminescence *35*, 503–511 (2020).

Y. Hou, Y. Guo, X. Ma, (. . .), B. Li & W. Liu, Ring-Oven-Assisted in situ synthesis of Metal-Organic Frameworks on the Lab-On-Paper Device for chemiluminescence detection of nitrite in whole blood, Analytical Chemistry *95*, 4362–4370 (2023).

Soxhlet device

M. Luque de Castro & F. Priego-Capote, Soxhlet extraction: Past and present panacea, Journal of Chromatography A *1217*, 2383–2389 (2010).

M. Luque de Castro & F. Priego-Capote, Focused microwave-assisted Soxhlet extraction, in: Marchetti, Microwaves: Theoretical Aspects and Practical Applications in Chemistry, Transworld Research Network, TC, Kerala, India, 227–247 (2011).

A. Zygler, M. Słomińska & J. Namieśnik, Soxhlet extraction and new developments such as soxtec (Book Chapter), in: Comprehensive Sampling and Sample Preparation *2*, 65–82 (2012).

Supercritical fluid extraction

X. Ding, Q. Liu, X. Hou & T. Fang, Supercritical Fluid Extraction of Metal Chelate: A Review Critical Reviews in Analytical Chemistry *47*, 99–118 (2017).

M. Henry & C. Yonker, Supercritical fluid chromatography, pressurized liquid extraction, and supercritical fluid extraction, Analytical Chemistry *78*, 3909–3915 (2006).

J. Kroon & D. Raynie, Supercritical Fluid Extraction, in: J. Pawlyszin & H. Lord, Handbook of Sample Preparation, Wiley – Blackwell, Hoboken, 191–196 (2010).

J. Lee, E. Fukusaki & T. Bamba, Application of supercritical fluid carbon dioxide to the extraction and analysis of lipids Bioanalysis *4*, 2413–2422 (2012).

Y. Marcus, Supercritical Carbon Dioxide, Nova Science Publishers, Hauppauge (2019).

F. Pena-Pereira & M. Tobiszewski, The Application of Green Solvents in Separation Processes, Elsevier, Amsterdam (2017).

A. Sánchez-Camargo, F. Parada-Alonso, E. Ibáñez & A. Cifuentes, Recent applications of on-line supercritical fluid extraction coupled to advanced analytical techniques for compounds extraction and identification, Journal of Separation Science 42, 243–257 (2019).

14 Solubility: Crystallization

14.1 General (definitions – auxiliary phases – melting and solubility diagrams)

Precipitation by changing the temperature of a melt or solution without the addition of reagents is called *crystallization*[1]. Crystallizations are predominantly carried out by cooling, rarely by heating the systems concerned.

Crystallizations from melts proceed without auxiliary phases; in crystallizations from solutions, an auxiliary phase, the solvent, is present.

Melting diagrams show the conditions during the transition from the liquid to the solid state in multicomponent systems. Of the numerous types, only two-substance systems with a gapless series of solid solutions and eutectic systems will be mentioned here (Fig. 14.1).

If a melt of composition X is cooled in a eutectic system, part of component B is precipitated pure when the melting curve is reached. The remaining melt accumulates A in the course of deposition, and at the eutectic point A and B finally crystallize together. The same applies to the cooling of a melt with composition Y, from which pure A is precipitated until the eutectic point is reached.

Thus, by cooling down melts of eutectic systems, only a part of one component can be obtained pure; which one depends on the composition of the starting mixture and the position of the eutectic point. Such systems are therefore not suitable for quantitative separations.

If a system with a complete series of solid solutions is present (Fig. 14.1b), precipitates are obtained during cooling whose composition differs from that of the original melt. Here, however, no pure component at all can be obtained by one-time crystallization; furthermore, the separation effects are usually only small.

For the graphical representation of the behavior of solutions[2] two different methods of application are used, which are obtained by swapping the coordinates; either the abscissa or the ordinate is chosen as the temperature axis (cf. Fig. 14.2). The latter application corresponds to that shown in Fig. 14.1 for melts; it is usually used when the entire solvent – solute system is to be represented. Often, however, only a part of the diagram is shown (usually between about 0 °C and the boiling point of the solution); in this case, the temperature is usually plotted on the abscissa.

1 This definition also includes non-crystalline precipitates due to temperature change. (Crystallizations due to evaporation of the solvent are not dealt with here).

2 A generally valid distinction between solutions and melts is hardly possible. Solutions are usually understood to be systems in which one component (the solvent) is liquid at ordinary temperature, while melts are systems whose components all solidify at ordinary temperature.

https://doi.org/10.1515/9783111181417-014

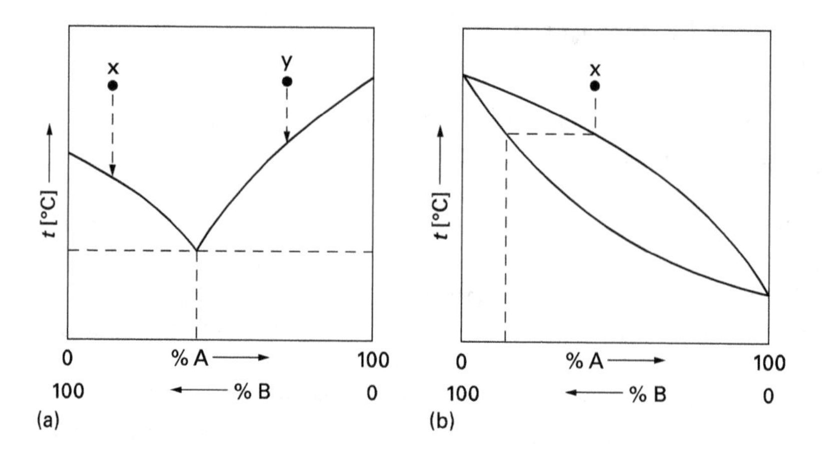

Fig. 14.1: Melting diagrams.
a) Eutectic system;
b) system with complete order of solid solutions.

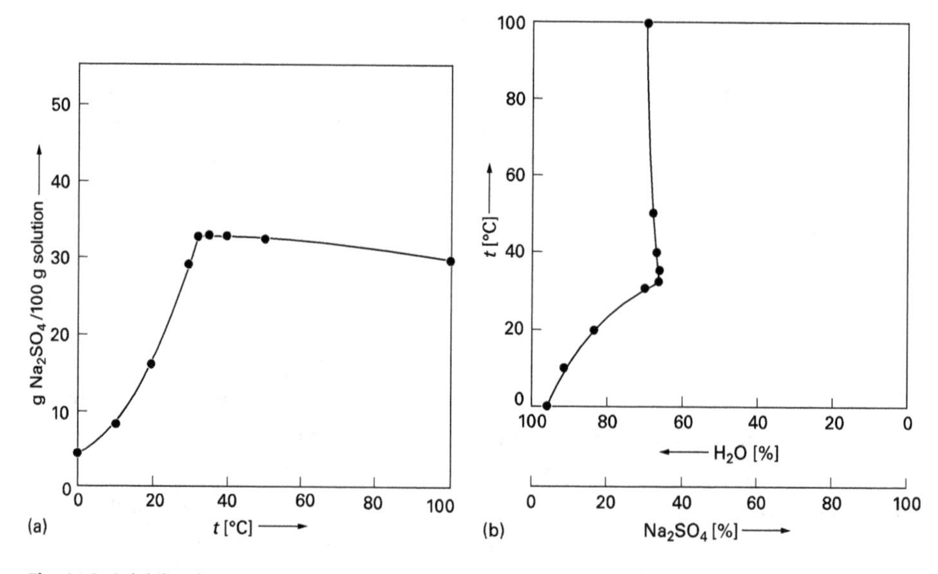

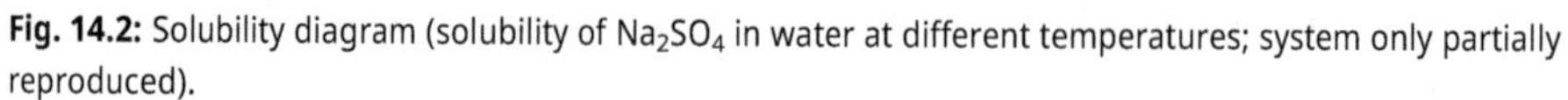

Fig. 14.2: Solubility diagram (solubility of Na_2SO_4 in water at different temperatures; system only partially reproduced).
a) Display with the abscissa as the temperature axis;
b) display with the ordinate as the temperature axis.

14.2 Separations by one-time crystallization

One-time crystallization is used to a considerable extent in preparative and technical chemistry, but it hardly plays a role for analytical separation problems because, as a rule, considerable amounts of the crystallized compounds still remain in the solution and the yields are thus too low.

Directional solidification. In contrast, the somewhat reversed method is occasionally used, in which not the dissolved substance but the solvent is largely removed by crystallization (*solidification* by *directional freezing out*). Since low temperatures usually have to be applied for this purpose, the method is referred to as *directional solidification*; it is only applicable if no solid solutions occur in the system concerned.

The experimental procedure is to lower a vertical tube, closed at the bottom, containing the solution slowly into a cold bath until only a small amount of liquid phase is present in which the compounds to be determined are located (Fig. 14.3). It is crucially important that the solution is stirred continuously during the crystallization process, otherwise noticeable portions of the solution, which is relatively concentrated at the interface with the solid phase, can become mechanically trapped. For example, the production of a 99.8% hydrogen peroxide solution is possible in this way.

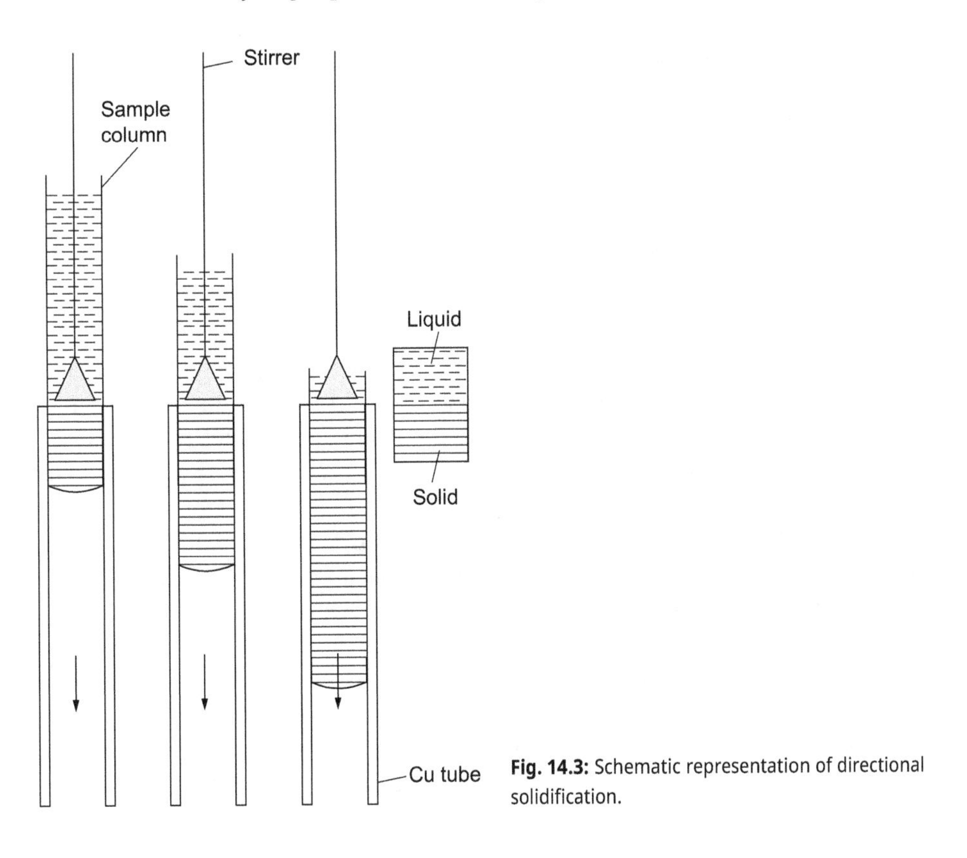

Fig. 14.3: Schematic representation of directional solidification.

The advantage of the process lies in the possibility of gently enriching easily decomposable trace components of solutions in the mother liquor. Losses of volatile substances have been observed, but this defect can be largely eliminated by covering the surface during freezing.

The method has been used for preparative recovery of extremely pure solvents, but can also be used for gentle enrichment of microorganisms or isotope-labeled substances

14.3 Separations by one-sided repetition

14.3.1 Repeated crystallization from melts (zone melts)

The difficulties which stand in the way of the application of crystallization from melts in analytical chemistry have been removed to some extent by the *zone melting method* introduced by Pfann (1964). In this method, the mixture of substances to be separated is in solid form in an elongated boat, and an annular furnace is drawn from one end of the boat to the other so that a narrow zone of melt migrates through the substance (Fig. 14.4).

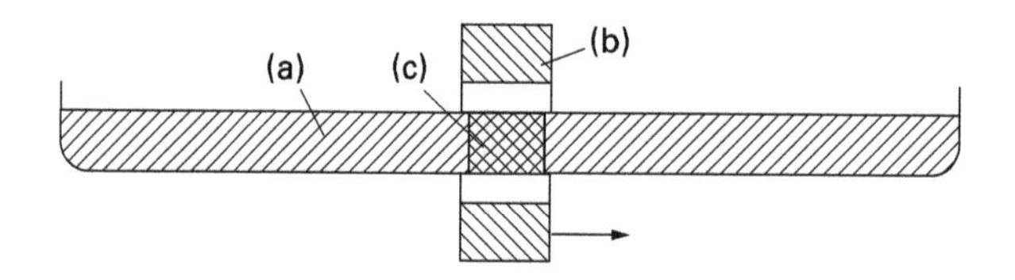

Fig. 14.4: Zone melting apparatus (schematic).
a) Boat with substance; b) heating device; c) melting zone.

The substance may contain an impurity which forms solid solutions with the main component and which remains preferentially in the melt when the solid phase crystallizes out. Therefore, the portions crystallizing out at the beginning of the boat at the start of the experiment contain less, and the melt zone contains more impurity than the starting material. As the furnace moves on, so much impurity is eventually accumulated in the melt that the crystallizate acquires the same composition as the initial mixture ahead of the melting zone. From then on, as much impurity is dissolved on the front side as is precipitated on the back side (see Fig. 14.5a). Thus, a state of equilibrium is formed which changes only when the melt zone reaches the end of the boat. Now, no new portions of substance can be melted, and the amount of melted material decreases. On the one hand, the impurity continues to accumulate in the residual melt, and on the other hand, its concentration also increases in the crystals; the highest content is found in the portions that crystallize last (Fig. 14.5b). The hatched areas at the beginning and end of the curve in Fig. 14.5b are the same.

If further melting zones are successively drawn through the substance in the same way, the impurity is pushed more and more to the end of the boat (Fig. 14.5c), until finally, after many passes, a limit state is reached which no longer changes and in which the initial and final slopes of the concentration curve merge (Fig. 14.5d).

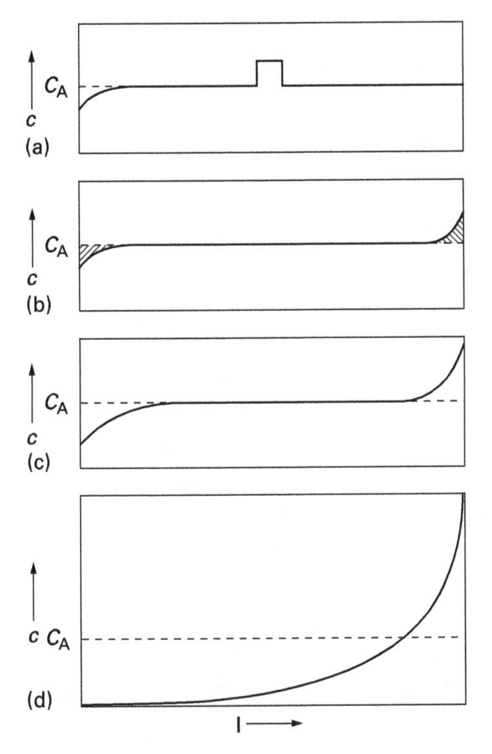

Fig. 14.5: Concentration c of impurities in the crystallizate along the boat during zone melting; C_A = initial concentration.
a) Conditions during the passage of the first melt zone;
b) after the passage of the first melting zone;
c) after repeated passage of a melting zone;
d) after passing through very many melt zones.

The efficiency of the enrichment (or depletion) of the impurity depends on the one hand on its distribution coefficient between crystals and melt, on the other hand it is approximately proportional to the quotient of the total length of the boat and the length of the melting zone. The apparatus will therefore be set up so that the material is present in a very elongated form and the melt zone is as narrow as possible.

Zone melting apparatus must be adapted to the melting point of the material to be cleaned. For high-melting metals (e.g. Si), horizontal graphite or quartz boats are often used and the melting zones are generated by inductive heating. Lower melting, especially organic materials, are better cleaned in vertical, not too thin-walled glass tubes of 1–2 m length and 0.5–3 cm diameter. In these, a melting zone of about 5 cm in length is drawn through the sample at a rate of 2–3 cm per hour. To speed up the process, several heating and cooling devices are usually arranged alternately one after the other. Some results obtained in the purification of organic compounds are given in Tab. 14.1.

The main application of the zone melting method is in the preparative preparation of extremely pure substances. The method does not permit quantitative separations, but it is suitable for the enrichment of impurity traces, which can then be more easily detected and determined, and has thus also acquired a certain importance for analytical chemistry.

Tab. 14.1: Purification of organic compounds by zone melting (examples).

Substance	Contamination	Conc. C_A (%)	Passages	Final conc. at start of sample C_n (%)	C_A/C_n
Benzene	Thiophene	10^{-1}	15	10^{-2}	10
Benzene	CH_3COOH	2	10	10^{-1}	20
H_2O	CH_3CH_2COOH	18	13	2	9
p-Bromotoluene	o-Bromotoluene	2	9	$2 \cdot 10^{-2}$	100
p-Xylene	o-Xylene	10	6	$5 \cdot 10^{-1}$	20

14.3.2 Repeated crystallization from solutions

Repeating a crystallization one or more times is referred to as *recrystallization*. The process is widely used for the preparative purification of numerous compounds, but is hardly useful for analytical chemistry because the yields decrease further with each crystallization. An example of application is the separation of n-paraffins from wax by repeated crystallization of the urea inclusion compounds.

The situation is more favorable when solvents are removed by freezing. If the crystallizate still contains a portion of the dissolved substances, the first entrained portion can be recovered by heating and renewed partial crystallization; however, the overall degree of enrichment in the solution is reduced.

The method of one-way repetition has also been carried out in column setups. If, for example, the solution of a fatty acid mixture in i-octane + 1% methanol is allowed to flow through a urea column, the straight-chain stearic acid is retained in the upper part of the column and cannot be eluted with this solvent mixture either; the branched tuberculostearic acid, on the other hand, flows through the column free of stearic acid at about 80%.

14.4 Systematic repetition (crystallization in the triangular scheme – separation series – column method)

Crystallization in the triangular scheme was originally developed for the preparative separation of rare earths; however, because of its cumbersome operation, the method is hardly used today.

The separation series derived from the triangular scheme (cf. section 1) has recently been used for the separation of fatty acids with the aid of urea inclusion compounds (cf. Fig. 14.6).

In a series of 24 flasks, 10 g of solid urea was added to each, and the fatty acid mixture (15 g) and 250 ml of solvent (methanol – ethyl acetate 70:30, cold saturated with urea) were also added to the first. Then it was heated until a clear solution was formed, slowly cooled to 20 °C and the mother liquor was decanted from the crystals into the next vessel. In this one, the crystallization was repeated and so on, until the liquid phase had passed through all the vessels. Finally, the fatty acid content was determined in each flask; the acids used had been practically completely separated, and impurities had accumulated in the intermediate fractions.

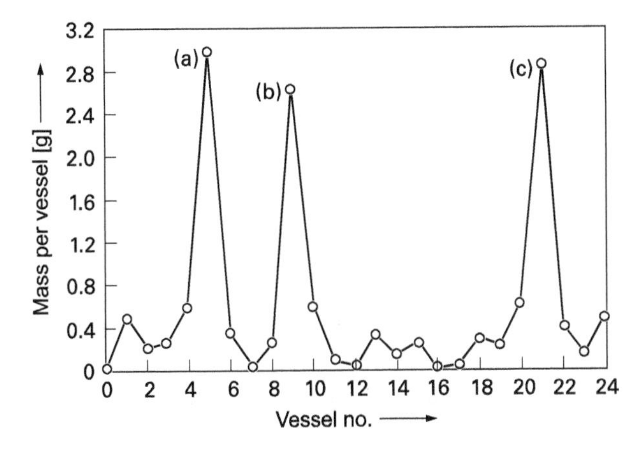

Fig. 14.6: Separation of a fatty acid mixture by crystallization of the urea inclusion compounds. a) Stearic acid; b) palmitic acid; c) oleic acid.

Of greater importance for analytical chemistry is a column method given by Baker & Williams, which consists of a combination of repeated dissolution and repeated crystallization in a temperature field[3] (cf. Fig. 14.7).

The substance mixture is placed on the top of a column filled with an inert carrier material (e.g. glass beads), which is heated at the top and cooled at the bottom.

Separation is achieved by eluting with a solvent that is poor for the compounds in question, to which a good one is continuously mixed in increasing concentration. The compound sample is initially dissolved, but precipitates again in the next colder zone of the column, is redissolved by the improvement of the solvent, and so on. Mixtures of

3 The process is also referred to as *precipitation chromatography*, but is to be included here among the crystallization methods.

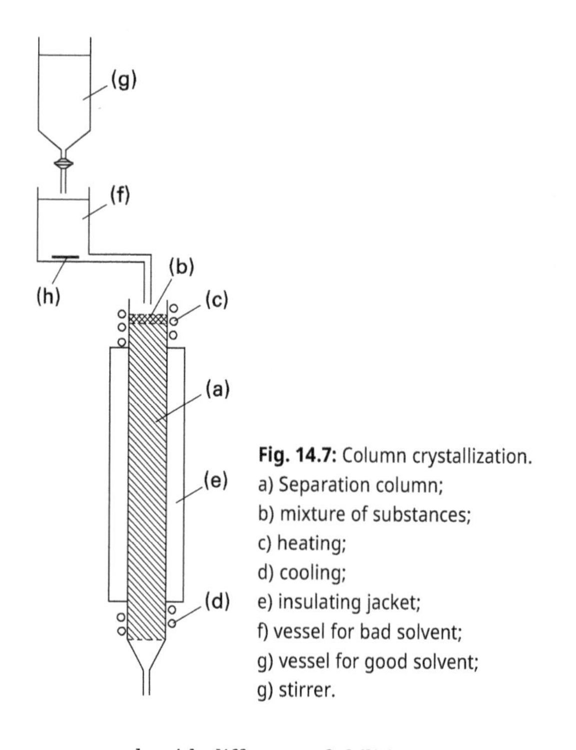

Fig. 14.7: Column crystallization.
a) Separation column;
b) mixture of substances;
c) heating;
d) cooling;
e) insulating jacket;
f) vessel for bad solvent;
g) vessel for good solvent;
g) stirrer.

compounds with different solubilities are pulled apart as they pass through the column and, in favorable cases, elute separately.

In one variation of the method, a periodic back-and-forth temperature is applied in place of the temperature gradient along the column.

The method has proved particularly useful in the fractionation of polymer-uniform high polymers, but it has also occasionally been used to separate low-molecular-weight compounds, e.g. polyphenylenes or steroids.

Columns containing substances capable of forming inclusion compounds as a stationary phase were indicated on various occasions, e.g. columns with nickel complexes for the separation of nitro compounds or columns with urea for the separation of high-molecular-weight esters and – with temperature programming – of n-paraffins. Finally, it is worth mentioning a proposal to carry out the zone melting method with an auxiliary phase. A column is filled with a suitable solvent and allowed to solidify. The substance mixture is then applied to the top of the column and narrow melt zones are drawn through the solid phase. Compounds with different partition coefficients should migrate rapidly between the melt and the crystallizate and be pulled apart into zones.

14.5 Countercurrent method

For the sake of completeness, crystallization in countercurrent should also be mentioned. This method of operation can be realized, among other things, by placing a saturated solution in an annular gap between two concentric tubes. A temperature gradient along the tube causes partial crystallization of the solute, with a spiral moving the crystals against the solution and achieving repeated dissolution and precipitation. However, the method has not yet become important for analytical chemistry.

General literature

Y. Delannoy, Purification of silicon for photovoltaic applications, Journal of Crystal Growth *360*, 61–67 (2012).

D. Erdemir, A. Lee, and A. Myerson, Nucleation of crystals from solution: classical and two-step models, Accounts of Chemical Research *42*, 621–629 (2009).

W. Genck, A clearer view of crystallizers, Chemical Engineering (Rockville, MD, United States) *118*, 28–32 (2011).

M. Hursthouse, L. Huth & T. Threlfall, Why do organic compounds crystallize well or badly or ever so slowly? Why is crystallization nevertheless such a good purification technique?, Organic Process Research & Development *13*, 1231–1240 (2009).

J. Putman & D. Armstrong, Recent advances in the field of chiral crystallization, Chirality *34*, 1338–1354 (2022).

Y. Wang & A. Chen, Enantioenrichment by crystallization organic process research and development *12*, 282–290 (2008).

Literature to the text

Freeze out

R. Agrawal, D. Herron, H. Rowles, & G. Kinard, Cryogenic technology, Kirk-Othmer Encyclopedia of Chemical Technology (5th Edition) *8*, 40–65 (2004).

S. Morgalev, A. Lim, T. Morgaleva, S. Loiko & O. Pokrovsky, Fractionation of organic C, nutrients, metals and bacteria in peat porewater and ice after freezing and thawing, Environmental Science and Pollution Research *30*, 823–836 2023).

M. Rahman, M. Ahmed & X. Chen, Freezing-melting process and desalination: review of present status and future prospects, International Journal of Nuclear Desalination *2*, 253–264 (2007).

H. Xiao, J. Wang & B. Wang, Application and progress of freezing technology in purification and remediation of heavy metal pollution in water and soil, Journal of Glaciology and Geocryology *44*, 340–351 (2022).

Directional solidification

F. Fazaieli, M. Afshar Mogaddam, M. Farajzadeh, B. Feriduni & A. Mohebbi, Development of organic solvents-free mode of solidification of floating organic droplet-based dispersive liquid–liquid microextraction for the extraction of polycyclic aromatic hydrocarbons from honey samples before their determination by gas chromatography–mass spectrometry, Journal of Separation Science *43*, 2393–2400 (2020).

X. Hou, X. Zheng, C. Zhang, (. . .), Q. Ling & L. Zhao, Ultrasound-assisted dispersive liquid-liquid microextraction based on the solidification of a floating organic droplet followed by gas chromatography for the determination of eight pyrethroid pesticides in tea samples, Journal of Chromatography B: Analytical Technologies in the Biomedical and Life Sciences *969*, 123–127 (2014).

F. Huang, L. Zhao, L. Liu, (. . .), R. Chen & Z. Dong, Separation and purification of Si from Sn-30Si alloy by electromagnetic semi-continuous directional solidification Materials Science in Semiconductor Processing *99*, (2019).

W. Yu, Y. Xue, J. Mei, (. . .) M. Xiong & S. Zhang, Segregation and removal of transition metal impurities during the directional solidification refining of silicon with Al-Si solvent, Journal of Alloys and Compounds *805*, 198–204 (2019).

Recrystallization of urea inclusion compounds

K. Harris, Fundamental and applied aspects of urea and thiourea inclusion compounds, Supramolecular Chemistry *19*, 47–53 (2007).

Y. Yang & L. Lu, Novel urea/thiourea-betaine inclusion compounds consolidated by host-guest hydrogen bond, Molecular Crystals and Liquid Crystals *722*, 47–57 (2021).

Zone melting

M. Duan, J. Zhao, B. Xu, **(. . .)**, H. Wan & C. Fu, Study on the behavior of impurities in zone melting of aluminum, Journal of Materials Research and Technology *21*, 3885–3895 (2022).

G. Lalev, J.-W. Lim, N.-R. Munirathnam, (. . .), K. Mimura & M. Isshiki, Concentration behavior of non-metallic impurities in Cu rods refined by argon and hydrogen plasma-arc zone melting, Materials Transactions *50*, 618–621 (2009).

S. Reber, W. Zimmermann & T. Kieliba, Zone melting recrystallization of silicon films for crystalline silicon thin-film solar cells, Solar Energy Materials and Solar Cell *65*, 409–416 (2001).

H. Zhang, J. Zhao, J. Xu, (. . .) B. Xu & B. Yang, Preparation of High-Purity Tin by Zone Melting Russian Journal of Non-Ferrous Metals *61*, 9–20 (2020).

15 Volatilization: Distillation and related processes

15.1 General

15.1.1 Historical development

Due to a lack of unambiguous sources, historical research has not yet reached a consensus on the origin of distillation processes. Partly, relevant knowledge is attributed to the Sumerians and Egyptians long before the turn of the millennium, partly the appearance of real distillation devices is assumed to be much later. What is certain, however, is that in the first centuries A.D. Chr. in Alexandria distillations were accomplished.

The knowledge of Hellenistic science was taken over by the Arabs, imparted to the Occident and substantially developed here in the Middle Ages; above all, the problem of obtaining alcohol had a stimulating effect.

The theoretical penetration of the processes in distillation columns and the construction of efficient apparatus for technology and laboratories did not take place until the 20th century. This extensive work was prompted by the large-scale distillations carried out in the petroleum industry.

15.1.2 Definitions – auxiliary phases (auxiliary substances)

Distillation is defined as the conversion of a liquid to the vapor state by heating and the subsequent condensation of the vapor by cooling.

The removal of gases from liquids or solids without condensation is not covered by this definition, but will be included in this chapter.

In a number of distillation processes, auxiliary substances are added to facilitate the exaggeration of the volatile compounds or to improve the separation effects. In the sense of the earlier definition (cf. ch. 6), these are not actually *auxiliary phases,* since the liquid or gaseous phase is not first formed anew by these substances. For the sake of simplicity, however, the term *auxiliary phases* will be retained.

15.1.3 Vapor pressure curves of pure substances – graphical and mathematical representation – Clausius-Clapeyron equation – Cox Diagrams

The dependence of the vapor pressure[1] on the temperature can be represented in a good approximation by the following equation:

1 Pressure is defined as the quotient force/area; depending on the unit of measurement for force and area, different expressions are obtained: 1 atmosphere (atm) corresponds to the mean air pressure of

https://doi.org/10.1515/9783111181417-015

$$p = 10^{\,\mathrm{A} - \frac{\mathrm{B}}{T}} \tag{1}$$

A and B are constants, p = vapor pressure, T = abs. temperature). This exponential equation is usually transformed for more convenient evaluation:

$$\log p = \mathrm{A} - \frac{\mathrm{B}}{T} \tag{2}$$

If log p is plotted against $\frac{1}{T}$ the result is a straight line (cf. Fig. 15.1).

Equations (1) and (2), respectively, follow from the *Clausius-Clapeyron equation* under simplifying assumptions (independence of the enthalpy of vaporization λ[2] from temperature; neglect of the liquid volume V_{fl} with respect to the gas volume V_{g}):

$$\frac{\mathrm{d}p}{\mathrm{d}T} = \frac{\lambda}{T \cdot \left(V_{\mathrm{g}} - V_{\mathrm{fl}}\right)}. \tag{3}$$

Table 15.1 and Fig. 15.1 show the vapor pressure of water as a function of temperature as an example of the behavior of liquids when heated.

The constants in eq. (2) are determined either from the pressures of a liquid measured at two different temperatures or – more precisely – by taking the vapor pressure curve over a larger temperature interval and calculating the relevant straight lines according to eq. (2).

Various methods have been devised to derive the entire vapor pressure curve from as few experimental values as possible. Of these, only the so-called *Cox diagram* should be mentioned, in which only one pressure-temperature value pair of the investigated substance is required (e.g. the boiling point at atmospheric pressure).

To draw the Cox diagram, a straight line at an angle of about 45° is placed in a coordinate grid. The abscissa represents the vapor pressures in logarithmic order. With the help of the arbitrarily drawn straight lines, the temperature values of a known vapor pressure curve are assigned to the pressures, so that the scale of the temperatures on the ordinate results from the position of the relevant straight lines. If the vapor pressure-temperature value pairs of other, chemically similar substances are entered into this coordinate network, straight lines are obtained which intersect at a point (cf. Fig. 15.2).

760 Torr, i.e., the pressure exerted by a 760 mm high column of mercury on its support. This is 1.033 kp/cm^2 (density of Hg = 13.5954 g/cm^3 at 0 °C). The physical unit of pressure is 1 bar = 10 N/cm^2 = 1.02 kp/cm^2; the unit of pressure in the SI system is 1 pascal (Pa), 1 Pa = 1 N/m^2, 10^5 Pa = 1 bar. In engineering, the technical atmosphere (at) = 1 kp/cm^2 is used. In Anglo-Saxon literature, the expression psi = pounds per square inch = 0.0703 at (15 psi = 1 at) is found. In the past, the unit of force was 1 dyn; 10^5 dyn = 1 newton (N).

2 The molar enthalpy of vaporization λ is defined as the amount of heat required to convert 1 mole of a liquid to the vapor state against atmospheric pressure.

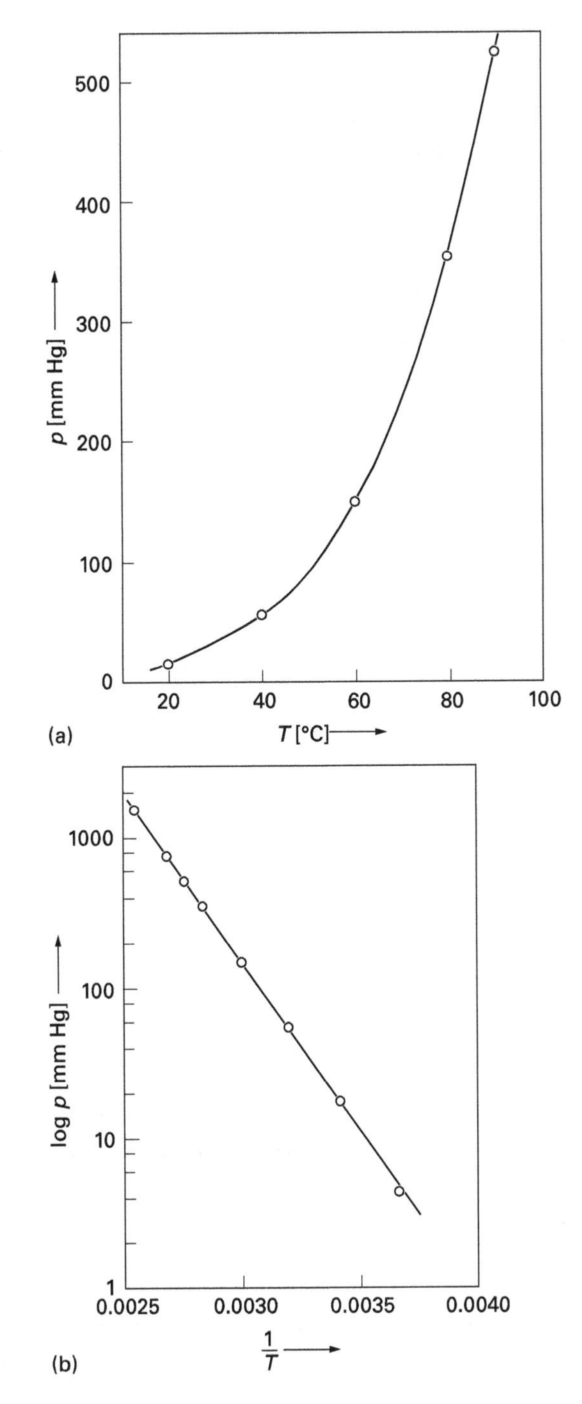

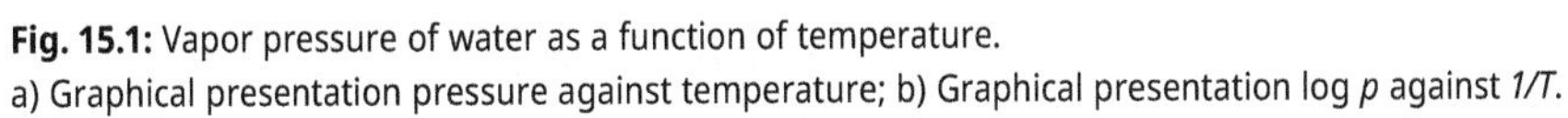

Fig. 15.1: Vapor pressure of water as a function of temperature.
a) Graphical presentation pressure against temperature; b) Graphical presentation log p against $1/T$.

Tab. 15.1: Vapor pressure of water as a function of temperature.

T (°C)	1/T	p (mm Hg)	log p
0	0,00366	4,6	0,6628
20	0,00341	17,4	1,2406
40	0,00320	55,3	1,7427
60	0,00300	149,2	2,1738
80	0,00283	355,1	2,5504
90	0,00276	525,8	2,7208
100	0,00268	760,0	2,8808
120	0,00255	1489,2	3,1729

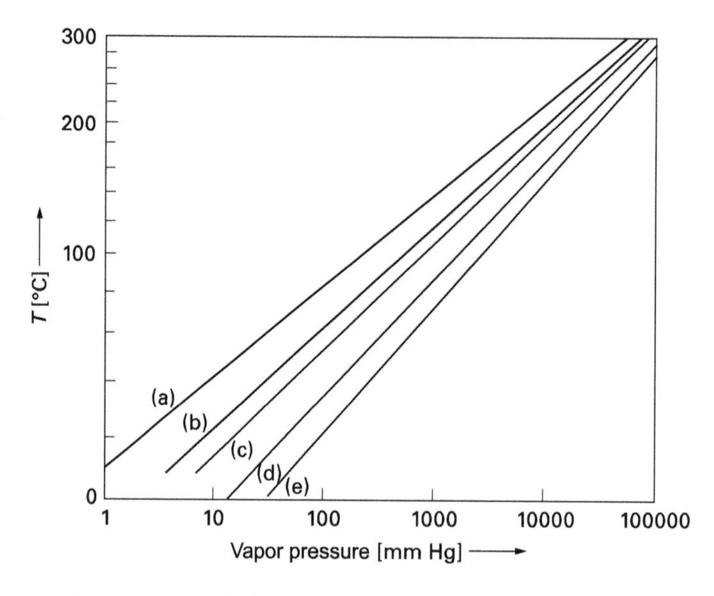

Fig. 15.2: Cox diagram for lower alcohols.
(a) i-Amyl alcohol; (b) i-butanol; (c) propanol; (d) ethanol; (e) methanol.

If this point is known for a particular class of substance, the vapor pressure curve of an unknown substance can be determined from a single vapor pressure-temperature pair of values and this point (provided that the substance belongs to the class of substances for which the Cox diagram was determined).

15.1.4 Vapor pressures of binary liquid mixtures – Raoult's law – vapor pressure diagrams – boiling diagrams – equilibrium diagrams

In the case of mixtures of two substances, in addition to temperature and pressure, the concentrations in the liquid and gaseous phases are variable. To describe the conditions prevailing in such systems, it is useful to use the concept of an *ideal* mixture, in which the interaction forces between the unequal molecules are as great as those between the equal molecules.

In the gaseous state, *Dalton's partial pressure law* applies to such mixtures:

$$P = p_1 + p_2 \tag{4}$$

(P = Total pressure; p_1 and p_2 = Partial pressures of components 1 and 2.)

Thus, the total pressure of a gas mixture in a given volume is composed additively of the pressures that the individual components would produce individually in the same volume.

Since according to the gas laws (with N for the number of moles and the gas constant R)

$$P \cdot V = N \cdot R \cdot T \text{ bzw. } P = \frac{N \cdot R \cdot T}{V} \tag{5}$$

and

$$p_1 = \frac{N_1 \cdot R \cdot T}{V}, \tag{5a}$$

applies:

$$\frac{p_1}{P} = \frac{N_1 \cdot R \cdot T/V}{N \cdot R \cdot T/V} = \frac{N_1}{N} = y_1 \tag{6}$$

It follows that the partial pressure p_1 of a component of a gas mixture divided by the total pressure P *is* equal to the mole fraction y_1 of this component (or: the partial pressure in % of the total pressure is equal to the concentration in mole %).

Vapor pressures of liquid mixtures. The vapor pressure of the components of liquid mixtures is determined not only by the temperature, but also by their concentration in the mixture. For ideal mixtures, Raoult's law applies at a fixed temperature:

$$p_1 = P_1 \cdot x_1. \tag{7}$$

(p_1 = Partial pressure of component 1;
P_1 = Pressure of pure component 1 at same temperature;
x_1 = Mole fraction of component 1 within liquid mixture).

Accordingly, the partial pressure of a component of an ideal mixture is equal to the product of the vapor pressure of the pure substance and its mole fraction in the liquid. For example, the vapor pressure of pure benzene at 100 °C is 1344 mm Hg, and the partial pressure of benzene in a mixture of 9 mol toluene and 1 mol benzene at the same temperature is 134 mm Hg.

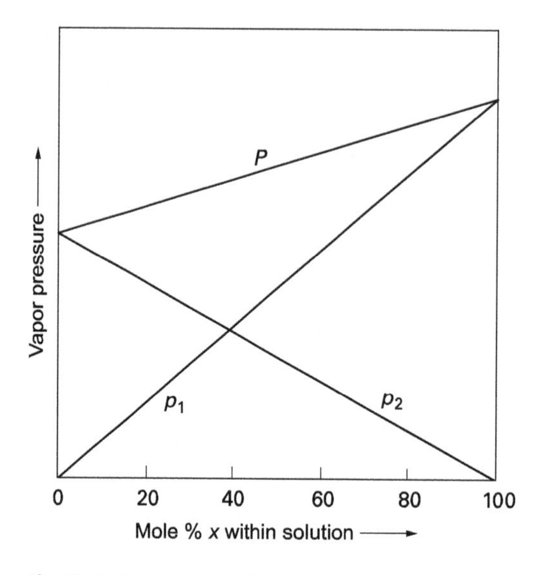

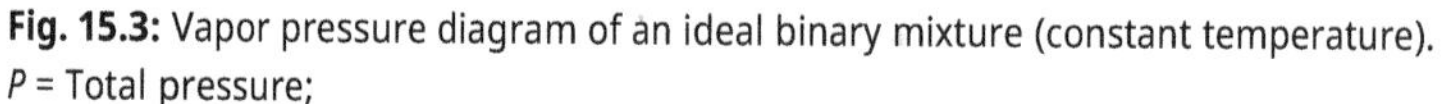

Fig. 15.3: Vapor pressure diagram of an ideal binary mixture (constant temperature).
P = Total pressure;
p_1 = partial pressure of component 1;
p_2 = partial pressure of component 2.

These relationships are represented graphically in the so-called vapor pressure diagram, in which the partial pressures of the components and the total pressure are plotted against the composition of the binary liquid mixture (Fig. 15.3). (Occasionally, the composition of the vapor phase is also plotted in the vapor pressure diagram). A vapor pressure diagram is valid only for a given temperature.

However, this method of representation is less suitable for overlooking the conditions during distillation, since here one usually works at constant pressure and variable temperature.

The behavior of binary liquid mixtures at variable temperature is represented by the so-called *boiling diagram* (*x-t-diagram*). This diagram shows the dependence of the boiling point of the liquid on its composition and the composition of the vapor at the respective boiling temperature (Fig. 15.4). Such a diagram is only valid for a certain pressure (e.g. 1 bar).

A mixture x_1 containing, say, 30 mol% of component x boils at temperature T_1; the associated vapor y_1 contains 64 mol% of component x.

The so-called *equilibrium diagram* or the *equilibrium curve x-y is* derived from the boiling diagram as a further representation. The composition of the vapor is plotted against the composition of the solution in a square coordinate system (Fig. 15.5, see also Fig. 6.3).

According to the diagram in Fig. 15.5, a liquid x_1 with, for example, 30 mol% x yields a vapor y_1 with 64 mol% x (usually the concentration of the more volatile component in

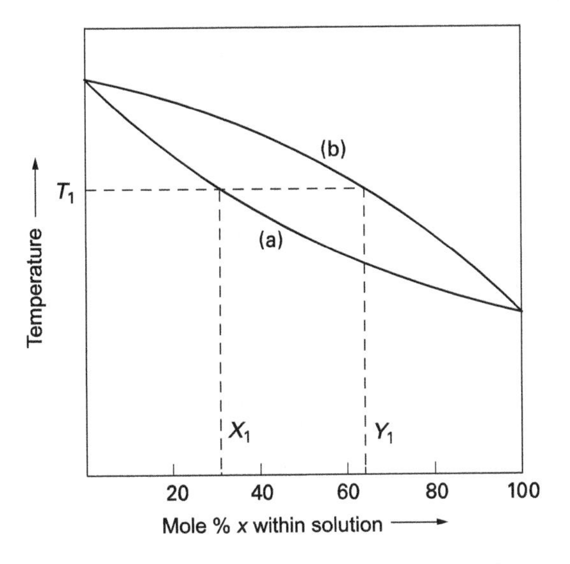

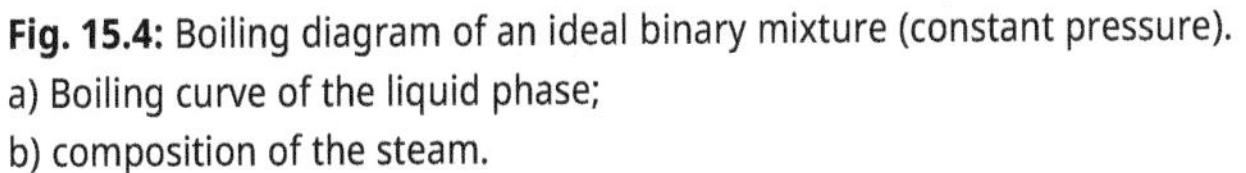
Fig. 15.4: Boiling diagram of an ideal binary mixture (constant pressure).
a) Boiling curve of the liquid phase;
b) composition of the steam.

the liquid is denoted by x, the concentration of the *same component* in the vapor by y; hence the expressions *x-t diagram* and *x-y diagram*).

Nearly ideal mixtures are formed by a number of chemically very similar compounds, e.g. benzene – toluene, n-hexane – n-heptane, 1,2-dibromoethane – 1,2-dibromopropane and others.

In the majority of cases, however, more or less large deviations from the ideal behavior occur, through which mutual attractive or repulsive forces of the different types of molecules become noticeable. In extreme cases, which are of considerable importance for distillations, even minima in the vapor pressure diagrams (and thus maxima in the boiling diagrams) and, conversely, maxima in the vapor pressure or minima in the boiling diagrams are observed. Figure 15.6 shows the vapor pressure diagram of a system with vapor pressure minimum.

Both partial pressure curves deviate downward from the values expected under ideal behavior (drawn in dashed lines).

The associated boiling diagram is shown in Fig. 15.7. A mixture x_1 with, for example, 20 mol% x in the liquid phase boils at the temperature T_1, the vapor has the composition y_1 = 10 mol% of the more volatile component x, and the distillate also has this composition at the start of distillation. Since the distillate contains less of component x than the associated liquid, x accumulates in the flask; the boiling point rises until point A is reached, at which flask contents and vapor have the same composition, and the remainder of the distillation takes place at constant temperature and constant distillate composition. Such a constant boiling mixture is called an *azeotrope*.

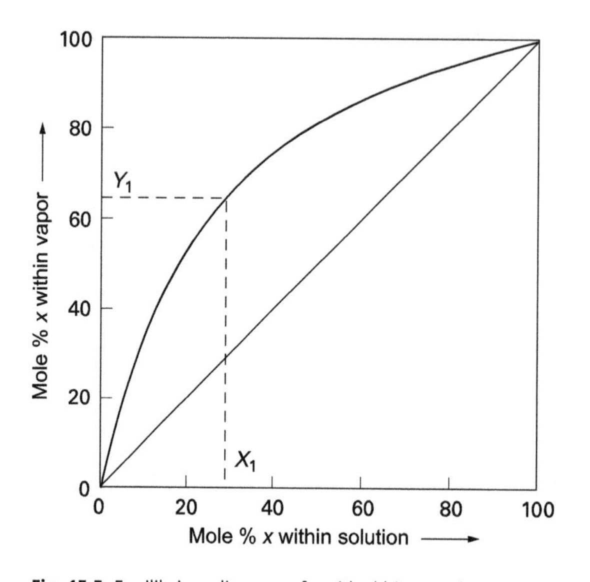

Fig. 15.5: Equilibrium diagram of an ideal binary mixture (constant pressure).

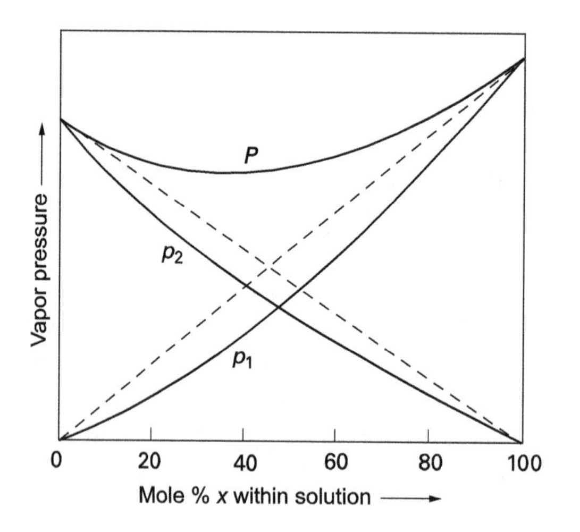

Fig. 15.6: Vapor pressure diagram of a binary system with vapor pressure minimum. P = Total pressure; p_1 = partial pressure of component 1 (= x); p_2 = partial pressure of component 2.

A mixture x_2 containing 65 mol% x, i.e., a composition that is on the other side of the azeotrope, boils at temperature T_2; the vapor y_2 contains 88 mol% x, so that the liquid in the flask is depleted of x during distillation and the boiling temperature rises until, again, point A is reached with the mixture boiling at a constant rate.

In such systems, therefore, only one component can ever be obtained pure by distillation; in addition, an azeotrope is formed which cannot be broken down into its

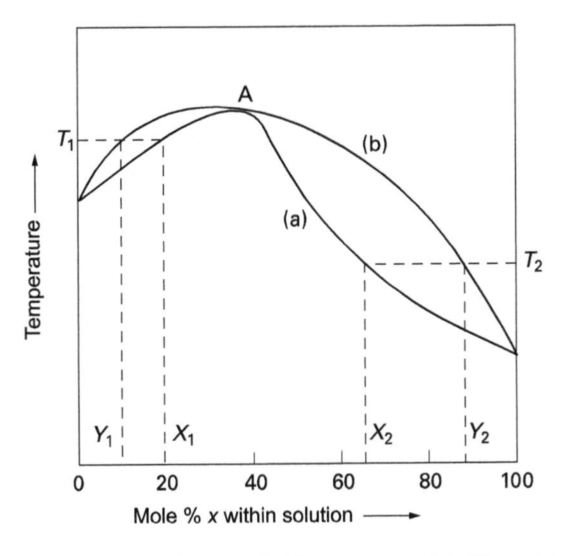

Fig. 15.7: Boiling diagram of a binary system with boiling point maximum (vapor pressure minimum).
a) Boiling curve of the liquid;
b) composition of the vapor. A = azeotrope.

components by distillation under the pressure once selected (unless the system is influenced by the addition of further substances).

The equilibrium diagram shows an s-shaped bent curve intersecting the diagonal line at the composition of the azeotrope (Fig. 15.8).

For systems with vapor pressure maximum, the partial pressure curves of the two components deviate upward from the ideal behavior (Fig. 15.9).

Accordingly, a minimum boiling temperature occurs in the boiling diagram (Fig. 15.10). For example, if a mixture x_1 with 10 mol% x *is* distilled, the boiling point is reached at the temperature T_1; the associated vapor has the composition y_1 = 25 mol% x. During distillation, component x accumulates in the liquid, the boiling point rises, and one finally arrives at pure component 2 (0% x). If one wants to enrich component x, the distillation would have to be repeated with the distillate. One would then start from the mixture y_1, which would give a vapor with about 41 mol % x *at* the beginning of the distillation (not plotted in Fig. 15.10). The composition of the distillate, with frequent repetition, moves closer and closer to the point A, which again represents an azeotrope, since here the liquid and vapor have the same composition. From then on, further separation by distillation is no longer possible.

The same applies if the initial mixture is on the other side of point A in composition. For example, the mixture x_2 with 90 mol% boils at the temperature T_2, and the vapor has the composition y_2 = 80 mol% x. During distillation, x is enriched in the flask and can eventually be obtained pure. To enrich component 2, the distillation would have to be repeated with the distillate.

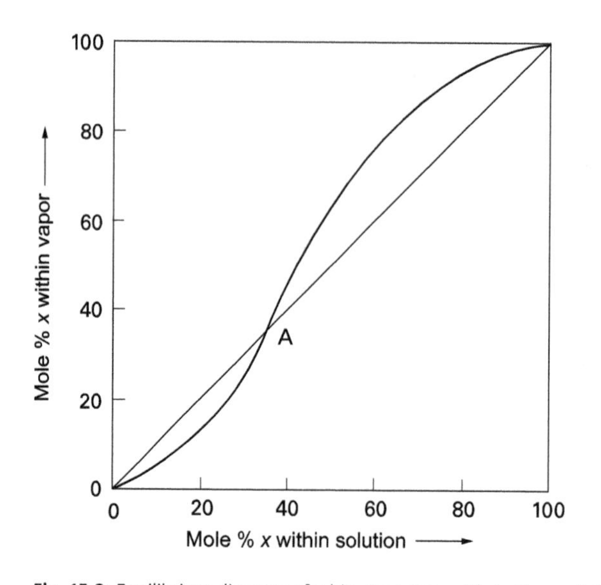

Fig. 15.8: Equilibrium diagram of a binary system with boiling point maximum.

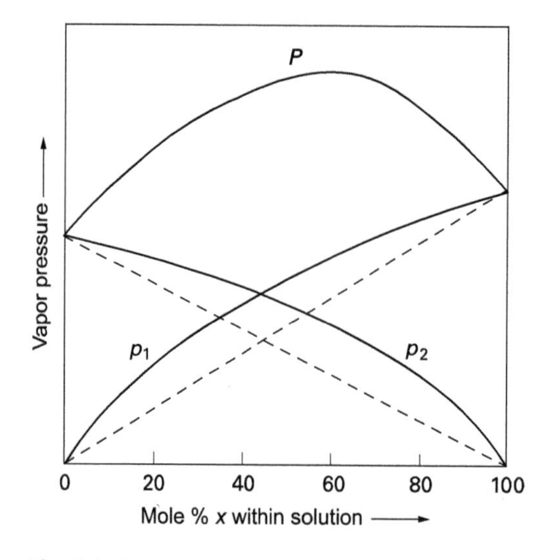

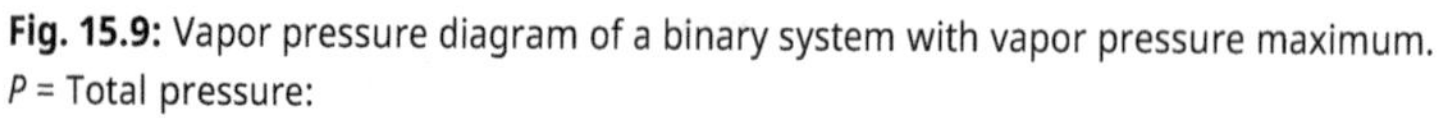

Fig. 15.9: Vapor pressure diagram of a binary system with vapor pressure maximum.
P = Total pressure:
p_1 = partial pressure of component 1 (= x);
p_2 = partial pressure of component 2.

Here, however, one approaches the point A more and more, until again the azeotrope passes.

Thus, even in the presence of such a mixture, only one component can ever be obtained pure, while the azeotrope is formed on the other side.

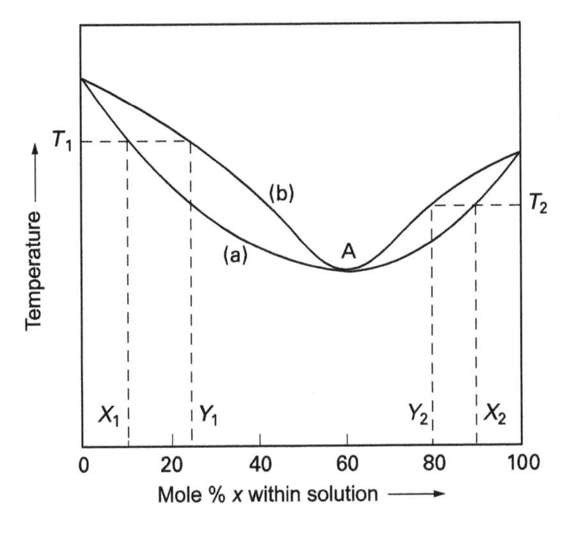

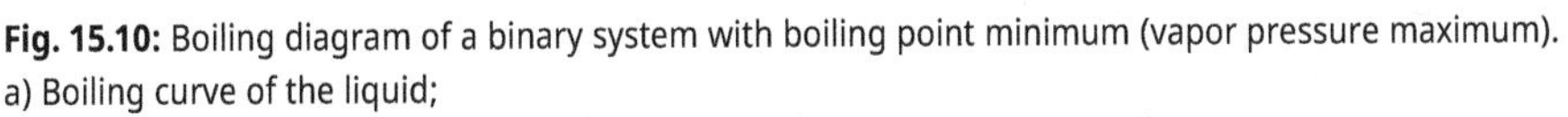

Fig. 15.10: Boiling diagram of a binary system with boiling point minimum (vapor pressure maximum).
a) Boiling curve of the liquid;
b) composition of the steam.
A = azeotrope.

The equilibrium curve is also s-shaped; it intersects the diagonal at the composition of the azeotrope (Fig. 15.11).

Finally, binary systems are mentioned in which the two liquids are practically insoluble in each other, thus forming two liquid phases.

The vapor pressure of each of the two substances is constant here at a given temperature and independent of the quantity ratio, and the total pressure is given by

Addition of the two pressures (Fig. 15.12). The mixture boils when the sum of the pressures reaches atmospheric pressure.

The boiling diagram is accordingly a straight line representing the constant boiling temperature (independent of the quantity ratio of the two liquids) (Fig. 15.13). The vapor composition would be plotted as a point on this straight line.

The equilibrium diagram is again a straight line, since the vapor composition is independent of the quantitative ratio of the liquids (Fig. 15.14). If a mixture of this type is distilled, the two components will constantly transition in the ratio of their vapor pressures. When one of the two liquids is completely distilled off, the boiling point jumps to the value of the other, and the latter distills off pure.

In general, the distillation of mutually insoluble liquids has no significance, since both can be separated mechanically without difficulty. However, the corresponding diagrams are given here, since they are authoritative for steam distillation and related processes (see below).

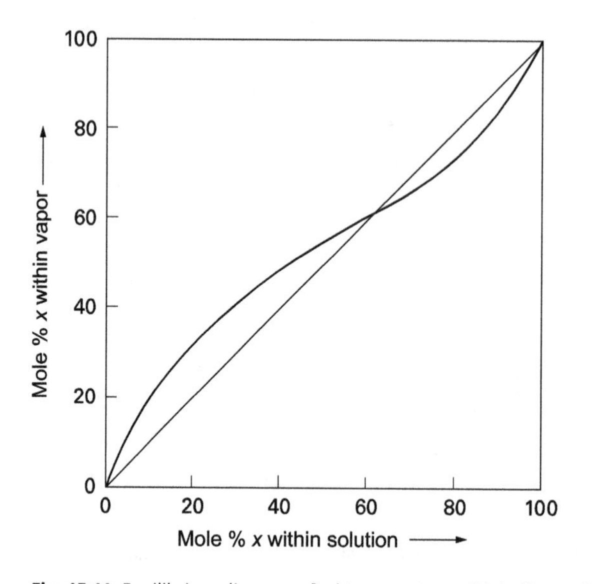

Fig. 15.11: Equilibrium diagram of a binary system with boiling point minimum.

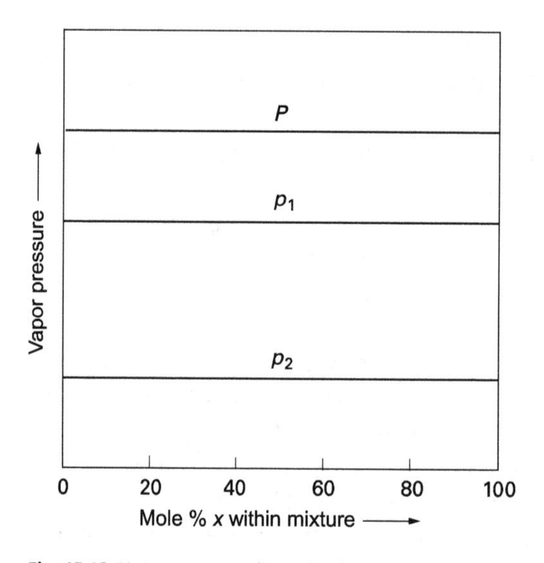

Fig. 15.12: Vapor pressure diagram of a binary system of two liquids insoluble in each other.
P = Total pressure;
p_1 = pressure of component 1;
p_2 = pressure of component 2.

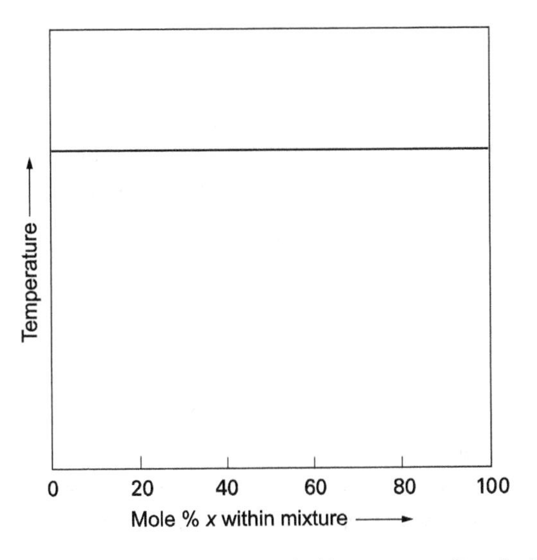

Fig. 15.13: Boiling point curve of a binary system of two liquids insoluble in each other.

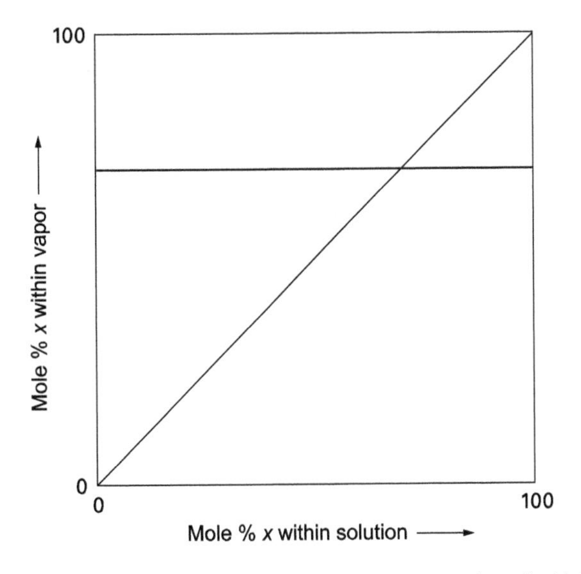

Fig. 15.14: Equilibrium diagram of a binary system of two liquids insoluble in each other.

15.1.5 Volatility – relative volatility (separation factor) – presentation of the result of distillations (distillation curves)

The *volatility* V_1 of a substance 1 is defined as the ratio of the partial pressure p_1 in the vapor to the mole fraction x_1 in the liquid:

$$V_1 = \frac{p_1}{x_1} \tag{8}$$

This quantity corresponds to the partition coefficient, since the partial pressure in the gas phase, like the mole fraction in the liquid, is an indication of concentration. The greater the volatility, the easier it is to remove a substance from the liquid phase.

Relative volatility (separation factor). The relative volatility β of a substance is the ratio of its volatility V_1 to the volatility V_2 of a second substance:

$$\beta = \frac{V_1}{V_2} = \frac{p_1/x_1}{p_2/x_2} = \frac{p_1/p_2}{x_1/x_2}. \tag{9}$$

The relative volatility therefore corresponds to the separation factor defined earlier.

The relative volatility can increase or decrease with increasing temperature; predominantly it decreases with increasing temperature, so that distillation should generally be carried out at the lowest possible temperature (vacuum distillation, see below).

Representation of the result of distillations (distillation curve). The result of a distillation is usually represented by the so-called *distillation curve*, in which a property of the distillate is plotted against the quantity transferred (Fig. 15.15). The distillation temperature is often chosen to characterize the distillate, but other quantities are also often used, e.g. density, refractive index, dielectric constant etc.

The distillation curve gives – if the temperature jumps between the individual substances are sufficiently pronounced and no azeotropes occur – an overview of the number and quantities of the components in the initial mixture.

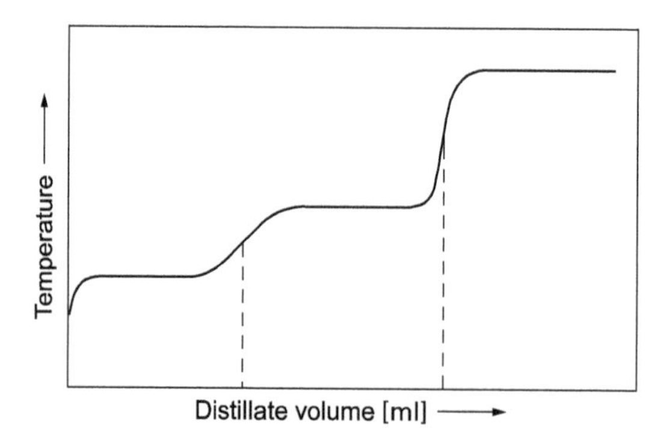

Fig. 15.15: Distillation curve.

Mixtures of numerous substances with closely spaced boiling points (e.g. hydrocarbons) often do not give uniform fractions during distillation; the distillation curve rises monotonically in such cases. However, by distilling under well-defined conditions (initial volume, equipment dimensions, distillation rate), it is possible to obtain curves suitable for comparison with similar samples or with standard mixtures. The procedure is referred to as *boiling analysis*.

Furthermore, the steepness of the transitions between two fractions indicates the efficiency of the separation: In the ideal case, no impure intermediate fraction occurs at all, and the temperature rise is vertical. The worse the separation, the flatter the curve in the transition region, and the greater the amount of intermediate fraction. As a measure of the steepness of the rise, one can use the angle formed by the tangent line at the inflection point with the abscissa (Fig. 15.16).

A more precise insight into the course of a distillation is obtained by plotting the composition of the distillate against its quantity, where a logarithmic scale is expediently chosen for the ordinate with the percentage composition (Fig. 15.17). However, the procedure requires quite a number of quantitative analyses of individual fractions.

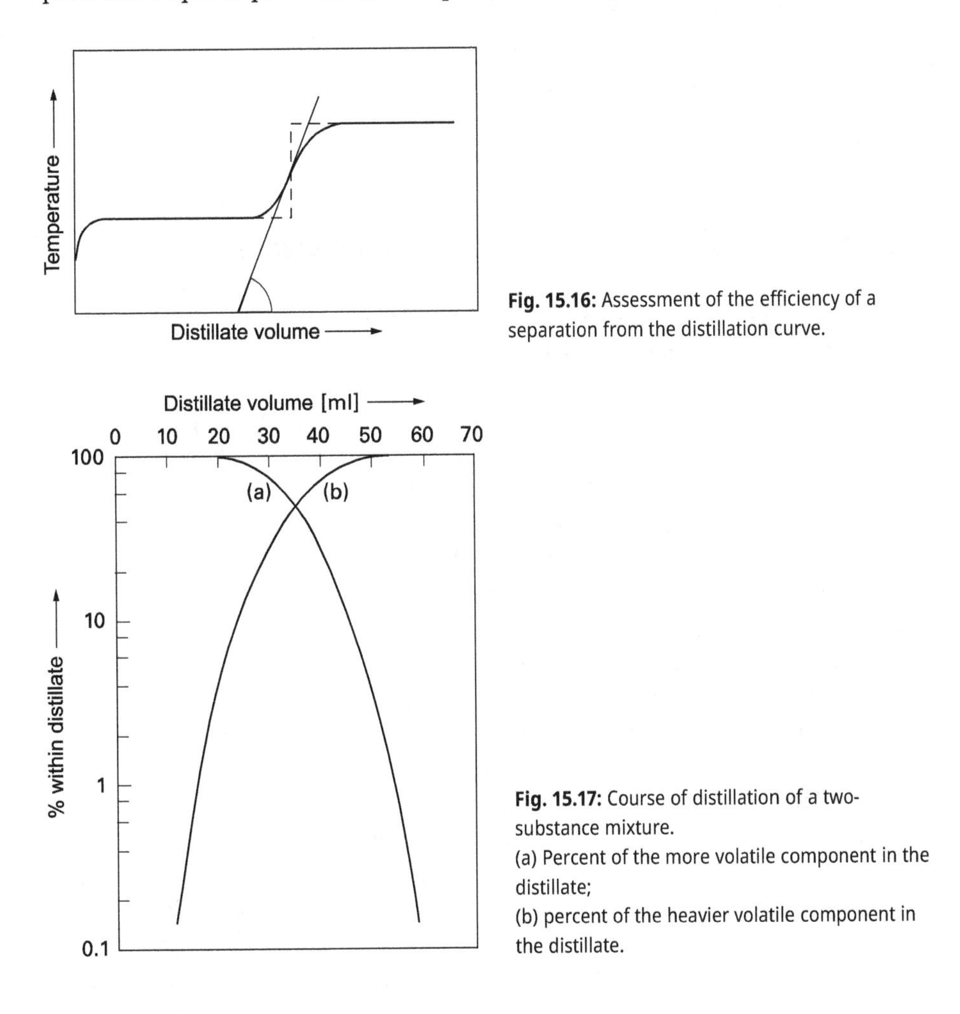

Fig. 15.16: Assessment of the efficiency of a separation from the distillation curve.

Fig. 15.17: Course of distillation of a two-substance mixture.
(a) Percent of the more volatile component in the distillate;
(b) percent of the heavier volatile component in the distillate.

Also rarely is a method used in which the quantity g transferred per degree of temperature change is plotted against the distillation temperature (Fig. 15.18). Such a diagram can be thought of as being derived from the ordinary distillation curve (Fig. 15.15) by swapping the coordinates and differentiating.

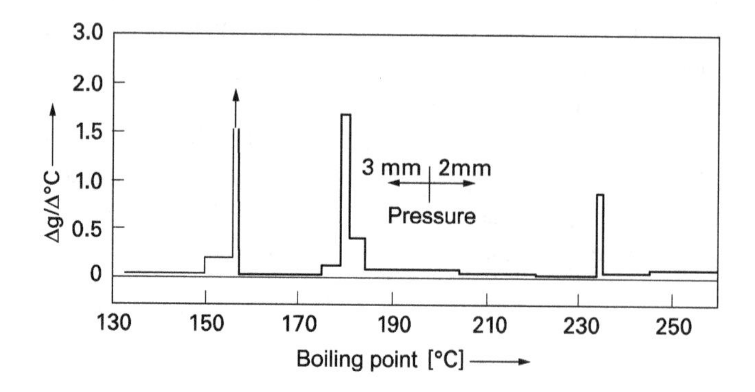

Fig. 15.18: Differential distillation curve.

15.2 Separations by one-time equilibrium adjustment

15.2.1 Head space analysis – extraction of dissolved gases (gases in metals and solutions)

In the so-called *head space analysis* first described by Schulek (1956), volatile constituents of an analytical sample are detected and determined by examining the gas above the sample. The material is filled into a sufficiently dimensioned vessel, sealed and the evaporation equilibrium is waited for; then a part of the gas phase is removed for further analysis, usually by gas chromatography.

The vapor pressure of the compounds under investigation can be increased by raising the temperature, by saturation with a salt such as Na_2SO_4, $(NH_4)_2SO_4$ or NaCl, and by chemical reactions (e.g. by esterification of alcohols or by silylation). In many cases, the volatile substances are enriched by freezing after removal.

In head space analysis, the simplest method is to fill the sample into a small serum vial or Erlenmeyer vial, which is sealed with a plastic or silicone septum. A portion of the supernatant gas is removed using the puncture cap technique already described in the section *Gas chromatography* with an injection syringe and needle (Fig. 15.19).

The method is mainly used for the qualitative analysis of aroma compounds in foods and beverages and for the investigation of volatile components in biological material (cf. Tab. 15.2). In quantitative evaluation, it must be taken into account that the separations are incomplete and the distribution coefficients are included in the calculation. Therefore, the method of the inner standard (addition of a substance with

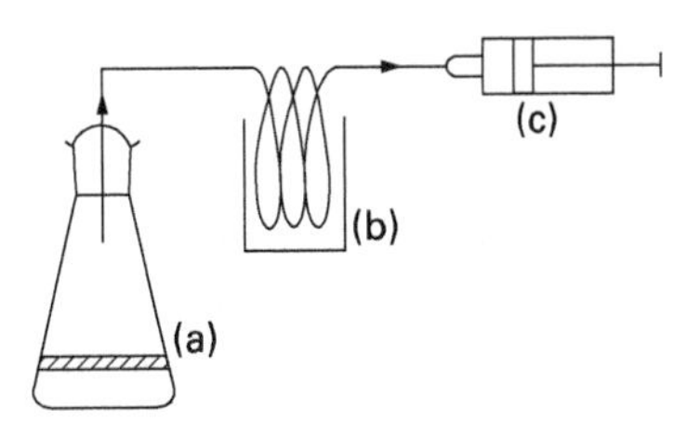

Fig. 15.19: Head space analysis with enrichment by freezing out.
a) Vessel with sample; b) PTFE capillary in cold bath; c) Syringe.

Tab. 15.2: Application of the head space analysis (examples).

Sample	Proven ingredients	Conditions
Vegetables, fruits	Flavorings	100 °C
Fruits	Flavorings	Room temperature
Milk	Flavorings	Na_2SO_4 addition; 50 °C argon atmosphere.
Fruit juice	Flavorings	Na_2SO_4 addition; 30 °C
Fermentation broths	Volatile components	NaCl or $(NH_4)_2SO_4$ addition; 10–36 °C
Biological material	Lower esters, aldehydes, mercaptans, alcohols	Na_2SO_4 addition; 60 °C
Blood	Ethanol	60 °C
Blood	Methanol, ethanol, i-propanol, acetone	Na_2SO_4 addition; 50 °C
Biological tissue	Alcohols	Esterification with HNO_2
Boiler feed water	H_2	–
High polymer plastics	Monomer residues	Room temperature

similar properties that can be easily detected separately from the actual analyte) is usually used.

The advantage of the method is that it is simple and fast to perform; larger series of samples can be examined automatically, and good accuracies have been achieved.

One source of error is the absorption of many organic compounds in septa made of polymer; one can avoid this interference by using an aluminum coating on one side of the septum.

Furthermore, errors occur if the equilibrium setting in the head space is not waited for and if the syringe is too cold during gas sampling, so that portions of the volatile components condense in the injection needle.

15.2.2 Suction of dissolved gases (gases in metals and solutions)

In the sense of the definition of distillation given above, processes in which a gas is extracted from a non-volatile substance do not belong to the actual distillation methods, since the evaporation product is not condensed. However, since these methods also involve separation by volatilization, they will also be discussed here. Two groups of methods can be distinguished: The volatilization of gases from metals and alloys and the volatilization from – mostly aqueous – solutions.

Gases in metals. Hydrogen is physically dissolved or only weakly bound in metals and can be expelled by heating the sample at temperatures below the melting point. As a rule, an equilibrium pressure is reached, and a distribution law of the form

$$\frac{c_1^2}{c_2} = \text{konst.,} \tag{10}$$

which is usually transformed to

$$V = a \cdot \sqrt{p}. \tag{11}$$

The solubility V of hydrogen in the metal (in ml/100 g at 0 °C and 760 mm Hg) is therefore proportional to the square root of the H_2 pressure (in mm Hg). The partition coefficient α is a temperature-dependent constant that must be determined experimentally for each metal and alloy (to do this, determine the hydrogen concentration in the metal at at least two different pressures and plot V vs. $\sqrt{p}$ is plotted. The slope of the straight line gives α).

If α is known, the hydrogen can be determined by pressure measurement at a specific temperature, taking into account the proportion remaining in the metal. One thus works according to the method of incomplete separation with the aid of the partition coefficient (cf. Part 1, Chap. 4). Another possibility is the complete removal of the gas from the metal, which can be done by repeated extraction or more simply by heating above the melting point.

The elements oxygen and nitrogen, which are also counted among the *gases* in metals, are bound as oxides and nitrides, respectively, and must first be released by chemical reactions before they can be volatilized. The following reactions are mainly used for this purpose

1. Dissolution of the metal in acids, whereby nitride nitrogen is converted into ammonium salt (so-called *Kjeldahl digestion*);
2. heating in H_2 atmosphere, forming water from oxides, and
3. heating with carbon to very high temperatures, converting oxides to CO and nitrides to N_2 (hydrogen escapes as H_2).

The Kjeldahl method is to be counted among the digestion methods and will not be discussed here. The oxygen determination by reaction with hydrogen, which goes back to Ledebur, is only suitable for the reduction of oxides of non-noble elements; SiO_2 and

Al_2O_3, for example, are not noticeably attacked at temperatures up to about 1000 °C. The reaction, on the other hand, is used to determine the oxygen content of Pb, Cu, Bi, W, and others. An apparatus in which bismuth metal is heated statically under H_2 atmosphere is reproduced in Fig. 15.20; the water formed is absorbed in a magnesium perchlorate layer, and the decrease in the amount of hydrogen is measured with the aid of the burette. More commonly, the sample is heated in flowing hydrogen, but this procedure will be discussed later, since it involves the operation of one-sided repetition.

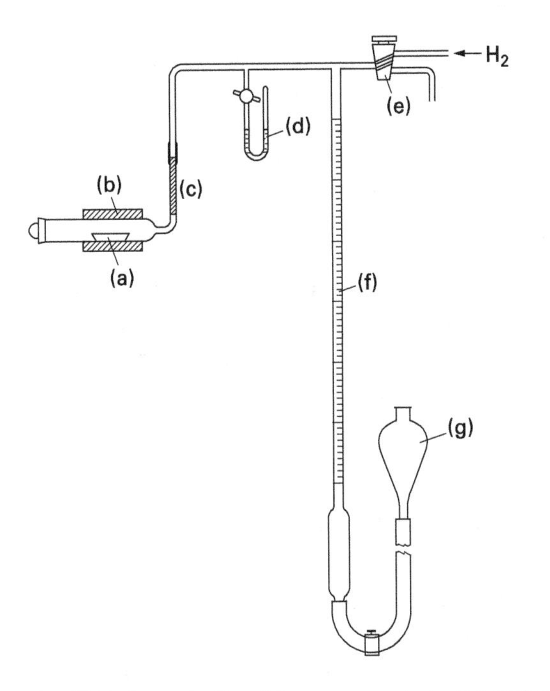

Fig. 15.20: Determination of the oxygen content of bismuth metal.
a) Shuttle with sample;
b) oven;
c) magnesium perchlorate layer;
d) pressure gauge;
e) three-way stopcock for evacuating and admitting hydrogen;
f) microburette;
g) mercury storage vessel.

Of greater importance than the reaction with hydrogen is the reaction with carbon, which was probably first used by Tucker (1881) for oxygen determination. Since all oxidic crucible materials also react to form CO at the very high temperatures required for this, work is carried out in graphite crucibles according to a proposal by Walker (1912), which thus serve simultaneously as a container and carbon source.

A simple apparatus for carrying out this reaction to determine the oxide content of copper is shown in Fig. 15.21; the sample is placed in a graphite boat in a tube which can be evacuated, and the vessel is shut off and heated after evacuation by closing a stopcock. Finally, analysis is performed by measuring the pressure of the evolved carbon monoxide.

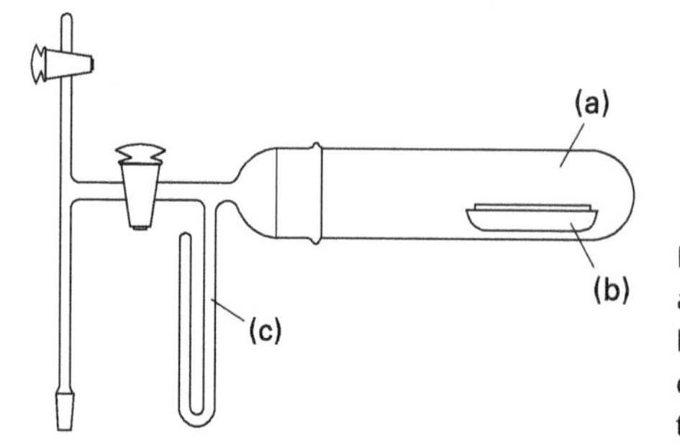

Fig. 15.21: Determination of oxygen in copper.
a) Evacuable vessel;
b) graphite boat with sample;
c) ground joint with pressure gauge and connection to the pump.

For the investigation of many base metals, however, much more complex methods, the so-called *vacuum melting methods,* must be used. In these, the evolved gases are removed from the vessel by a high vacuum pump and transferred to the measuring section of the instrument (the term *hot extraction* often found in the literature should be avoided, as there is no extraction in the usual sense).

For heating the samples, induction furnaces are probably used without exception, with which the required temperatures of 1500–2200 °C are achieved without difficulty. Eddy currents are generated in the graphite crucible and in the sample by a high-frequency coil located outside the vessel, which causes the heating. Non-conducting substances, such as the quartz wall, on the other hand, hardly absorb any energy. A major advantage of this heating method is the ability to cool the vessel wall, thus avoiding any thermal stress on the vessel.

The graphite crucible is usually surrounded by a layer of graphite powder to contain the heat radiation. It can be suspended from a tantalum wire or placed in a cup-shaped stand (Fig. 15.22).

Since a prolonged degassing of the apparatus is required before each determination, several samples are usually inserted simultaneously and brought into the crucible one after the other with the aid of a magnetic insertion device.

One source of error in the process is the partial distillation of the metal samples at the high temperatures required for reaction with carbon; the fractions distilled off settle in the colder parts of the apparatus and reabsorb the released gases to a considerable extent. The effect is referred to as the *getter effect.*

To eliminate this interference, the crucible is covered during the melting period. In addition, it has proved advantageous to throw the sample into a bath of a high-boiling metal which has already been pre-melted in the crucible (*bath method*). On the one hand, this lowers the partial pressure of the metal to be investigated, and on the other hand, the temperature can often be lowered, which has the same effect. Furthermore, the viscosity of the melt is lowered and thus the time required for the gases to diffuse

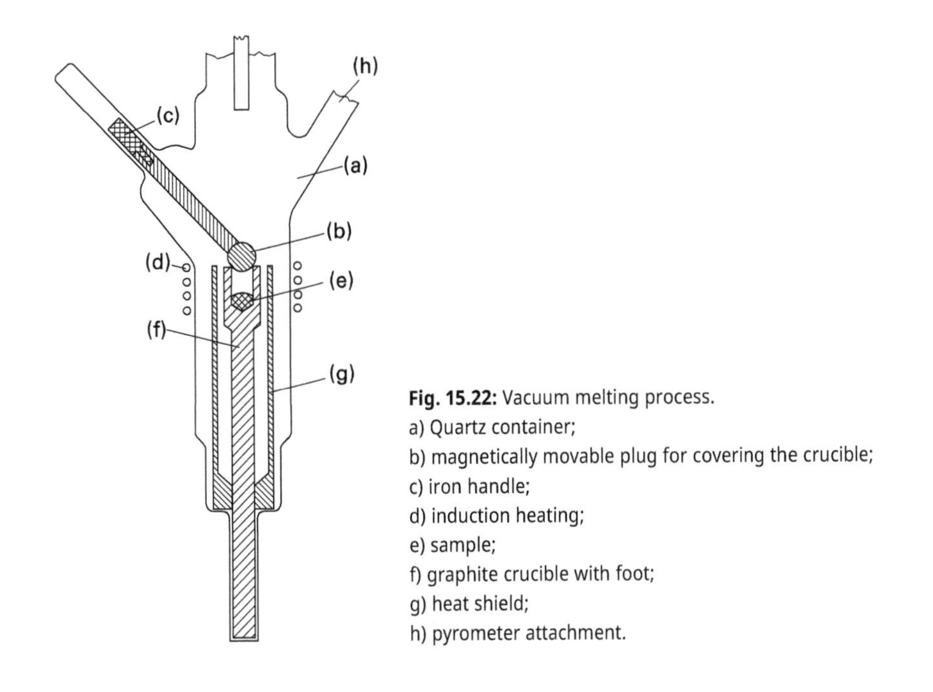

Fig. 15.22: Vacuum melting process.
a) Quartz container;
b) magnetically movable plug for covering the crucible;
c) iron handle;
d) induction heating;
e) sample;
f) graphite crucible with foot;
g) heat shield;
h) pyrometer attachment.

out is shortened. The metals used in the bath include carbonaceous iron, tin and platinum, the latter of which has particularly favorable properties.

A variant of the bath method is to wrap the analytical sample in a metal foil or sheet or to enclose it in a metal capsule; the sample is then thrown into the high-heat crucible with the wrapping (so-called *flux method*). In this way, splashing of the metal can be prevented, which easily occurs when the sample is not wrapped and thrown into the hot bath.

Some examples of the application of the vacuum melting process are given in Tab. 15.3.

Gases can be removed from liquids by a method given by van Slyke (1924). The liquid to be examined (e.g., blood) is admitted into a pipette-like vessel filled with mercury from a funnel placed above it (Fig. 15.23). Then, by lowering the storage vessel for the mercury, an empty space of about 50 ml capacity is created above the liquid and the whole is shaken until the dissolved gases are expelled. Finally, compress it again to a certain volume (e.g., 2 ml) by raising the mercury level and determine the amount by pressure measurement.

Tab. 15.3: Vacuum melting methods for the determination of hydrogen, oxygen and nitrogen in metals and alloys (application examples).

Metal	Determined elements	Temperature (°C)	Bath	Flux
Ag	O	1400	–	–
B, Cr, Cu, Si, Fe	O, H, N	1500–1600	Fe	
Be	O	1850	Pt	Pt
Cr	O	1600	Fe + Sn	Al + Fe
Cu	O	1150	–	–
Fe	O	1600–2000	–	–
Ferrosilicon	O	1750	Sn + C	Ni
Ga, In	O	1700	–	Pt
Mo, Ta, Ti, W	O	2000	–	–
Mo	O	1750–1800	Fe + Sn	
Nb	O	1650	Fe	–
Pd	O	2100	–	–
Pu, U, Zr	O	1900	Pt	–
Si	O	1700	Fe + Sn	
Ta	O, H, N	1860	Pt	–
Th	O, H, N	1900	Pt	–
Ti	O	1900	Sn	Sn
U	O, H, N	1820	Pt	–
UC	O, H, N	2000	Pt	–
V	O	1900	Pt	–
W	O	1950	Pt	–
Zr	O, H, N	1860	Pt	–

15.3 Separations by one-sided repetition without auxiliary substance

15.3.1 Drying

Drying of moist samples and precipitates is mainly done by heating to about 105–130 °C in a drying oven. Furthermore, small blocks made of aluminum with a wide bore to accommodate crucibles have proven effective, which are heated to the desired temperature with a Bunsen burner; an inert gas can be introduced through a nozzle attached to the lower end.

Furthermore, drying in the so-called *drying gun* is very effective, in which the water is removed from the sample in a vacuum and absorbed in P_2O_5. The glass vessel containing the material to be dried can be brought to a certain temperature either by the vapor of a boiling organic liquid or by introducing it into a drying oven.

Drying of substances containing water very tightly bound is usually carried out in electric ovens at higher temperatures (up to about 1100 °C).

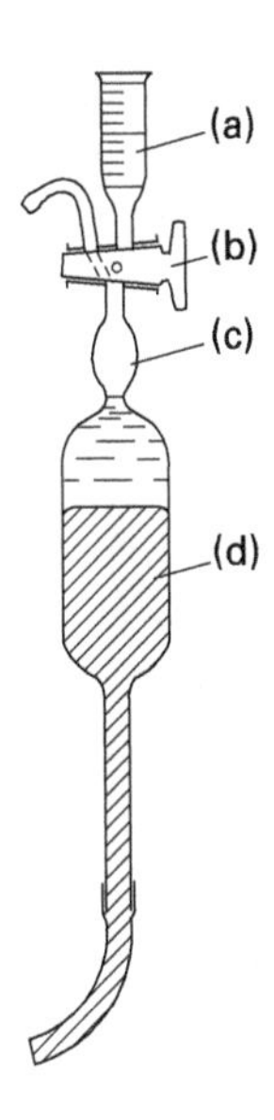

Fig. 15.23: Determination of dissolved gases in body fluids.
a) Graduated filling funnel for the sample;
b) three-way valve;
c) measuring volume;
d) mercury (barrier liquid).

For the sake of completeness, drying in desiccators at room temperature over various absorbents for water should also be mentioned.

15.3.2 Solvent evaporation – overhead distillation

Larger quantities of water are usually removed by evaporation over a free flame, on a water bath or sand bath. Also favorable are the so-called surface radiators, in which the heat of a heating coil acts on the surface of the liquid from above; the evaporation process is thus much faster than in the case of evaporation on the water bath, without losses due to splashing.

Overhead distillation (usually simply referred to as *distillation*) is widely used to separate and purify organic liquids. It works according to the principle of one-sided repetition, since the evaporation equilibrium, which is continuously disturbed by the condensation of the vapor, is repeatedly established in the vapor space. According to what was said in Chap. 1, the process yields effective separations only if the yield on one side of the scheme is close to 100%; this is the case if the residue remaining in the flask is non-volatile, i.e. it remains quantitatively at each equilibrium of the volatile component.

Overhead distillation does not have a high separation capacity and is rarely useful when liquid mixtures are present. Only if the individual components have very different boiling points can satisfactory separation be achieved by distilling at gradually increased temperature and collecting the distillate in portions. This process is known as *fractional distillation*.

The usual devices for laboratory preparations usually consist of a flask, an intermediate piece with a thermometer and a cooler. Devices in which the flask with the liquid to be distilled lies at an angle in a heating bath and is continuously rotated during heating so that the contents of the flask are constantly in motion have also proved successful. The advantage is faster distillation as a result of better heat transfer from the heating bath to the liquid, and vacuum distillation can also be carried out largely without splashing and without foaming. Distillation apparatuses of this type are referred to as *rotary evaporators* (Rotavapor®).

Low-boiling liquids can be distilled in arrangements such as those shown in Fig. 15.24. The initial mixture is frozen in a tube with a ground joint a, this is attached to the apparatus and the whole is evacuated; then the vessel a is allowed to thaw slowly and the vapor formed is precipitated in the deep-frozen vessel b at e.g. –180 °C (the term *cryodiffusion* is found in the literature for this process; however, it is a distillation process).

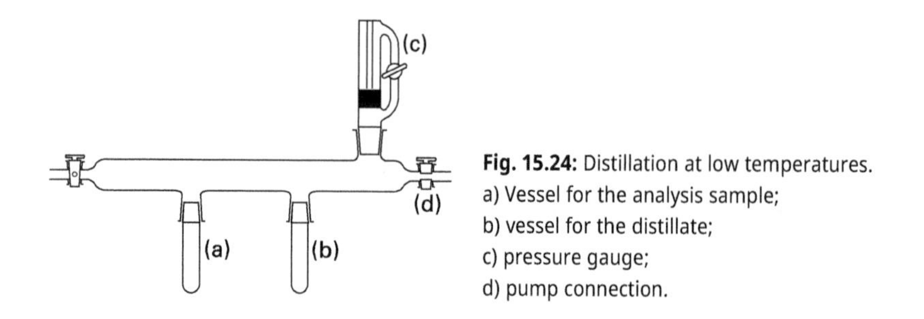

Fig. 15.24: Distillation at low temperatures.
a) Vessel for the analysis sample;
b) vessel for the distillate;
c) pressure gauge;
d) pump connection.

The usual distillation devices are no longer suitable for substance quantities of less than about 5 ml, and numerous modifications have therefore been developed for separations on a smaller scale. Of these, an example in which the receiver can be changed by moving a plunger will be given (Fig. 15.25). Quantities of about 0.05 ml (1 drop) can be distilled from a flat support to a cooled slide above.

Faults due to splashing. In all distillations in which the liquid mixture is heated to boiling, a fault occurs which often considerably influences the separations:

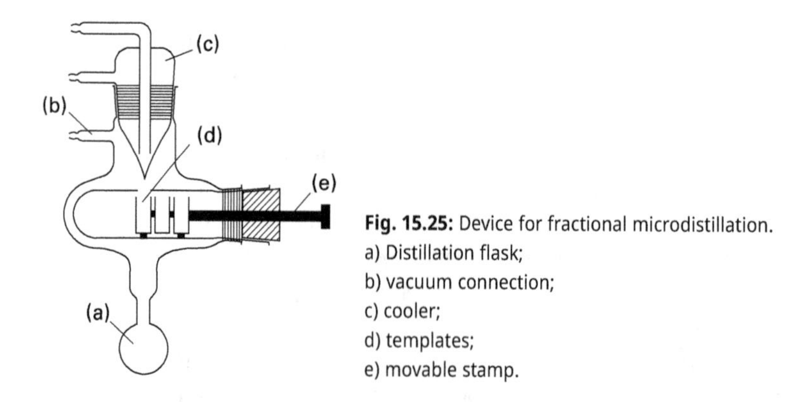

Fig. 15.25: Device for fractional microdistillation.
a) Distillation flask;
b) vacuum connection;
c) cooler;
d) templates;
e) movable stamp.

As the vapor bubbles break through the liquid surface, fine mist droplets are formed by spraying, which are carried by the vapor stream into the condenser and thereby into the distillate. The extent of this disturbance depends on the distillation rate, the apparatus dimensions, and the surface tension of the liquid; in general, it can be expected that, on the order of magnitude, about 0.1% of the liquid will enter the distillate in this way. Various mist eliminators have been proposed to eliminate this error, but their effectiveness is limited.

15.3.3 Short path distillation

The so-called *short-path distillation* or *molecular distillation* process goes back to the work of Caldwell (1909), Brönstedt (1921) and Burch (1928). Distillation is carried out under a high vacuum (approx. 10^{-3} mm Hg) and the condenser is placed so close to the liquid that the mean free path of the molecules in the vapor corresponds in size to the distance between the condenser and the liquid surface. Of the numerous apparatus arrangements described in the literature, an example is given in Fig. 15.26.

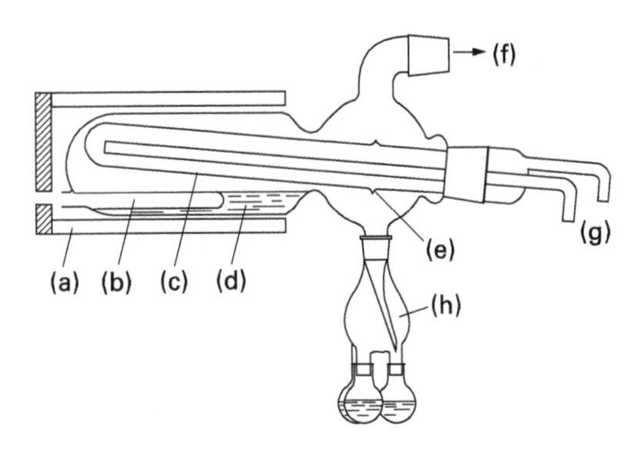

Fig. 15.26: Short path distillation.
a) Oven;
b) thermocouple insert;
c) cooler;
d) substance;
e) ring bead;
f) vacuum connection;
g) cooling water inflow and outflow;
h) submission.

Compared to all other distillation processes, Short Path Distillation has a fundamental peculiarity: Not only differences in vapor pressures but also in molecular weights are decisive for the separations. Therefore, substances with identical vapor pressures can also be separated with this method, provided only the molecular weights are different.

The main field of application of the method is the preparative recovery or purification of high-boiling organic compounds which would already decompose during distillation under normal pressure or in ordinary vacuum (a few Torr). In special cases, however, the method can also be used for analytical separations.

At this point, the distillation of metals should be mentioned, which is usually also carried out under high vacuum in order to lower the distillation temperature as much as possible and to exclude the effect of atmospheric oxygen (there is, of course, no danger of decomposition as with organic compounds). In analytical chemistry, distillation of alkali metals, cadmium and lead has occasionally been used (Tab. 15.4). Phosphorus and tellurium can also be removed from some metals in this way; vacuum is not required for the distillation of mercury.

Tab. 15.4: High-vacuum distillation of metals (examples).

Source material	Distilled off	Conditions
Li + Li_2O	Li	700 °C; $7 \cdot 10^{-6}$ Torr
Brass	Zn	800–850 °C; 10^{-4} Torr
Al-Zn alloys	Zn	850 °C; $3 \cdot 10^{-3}$ Torr
Sn-Zn alloys	Zn	850 °C; ≤ 0.05 Torr
Cu-Sn-Cd alloys	Cd	800 °C; ≤ 0.05 Torr
Cu-Pb alloys	Pb	850 °C; ≤ 0.05 Torr
Sn-Pb alloys	Pb	1000 °C; ≤ 0.05 Torr

15.3.4 Microdiffusion

Volatile substances dissolved in liquids often have a considerable partial pressure even at ordinary or moderately elevated temperature (e.g., ammonia above aqueous solution). They can be quantitatively removed from the liquid if the vapor-solution equilibrium is continually disturbed by absorption of the component present in the vapor.

Separations according to this principle were first carried out by Schlösing (1851); the process was later elaborated mainly by Conway (1957) and called *microdiffusion* (the term was chosen because the vapor diffuses from the liquid to the absorbent; however, it is a volatilization process).

The method of operation may be illustrated by the example of the separation of ammonia from an ammonium salt solution: The inner container of a modified Petri dish about 6 cm in diameter contains dilute sulfuric acid as absorbent (Fig. 15.27). The analytical solution is poured into the outer ring-shaped part of the dish. The vessel is then covered with a flat ground glass plate so that only a narrow gap remains open at the edge. Through this gap, strong KOH or soda solution is poured from a pipette into the outer ring, the cover plate is slid over the opening as quickly as possible and the reagent and analysis solution are mixed by carefully swirling them around. The dish is then left to stand until all of the NH_3 has been absorbed by the absorbent solution.

The duration of such a separation depends, apart from the partial pressure of the volatile substance, essentially on the layer thickness of the solution and also on the dimensions of the apparatus. The volatilization can be accelerated by rocking the vessel and by moderately increasing the temperature; salt additions in high concentrations are also favorable for increasing the volatility. In the above apparatus, a highly volatile substance such as NH_3 diffused at least 99.5% into the inner cup from about 1 ml of solution in 60 min. Heavier volatile substances, e.g. propionic acid, require several hours.

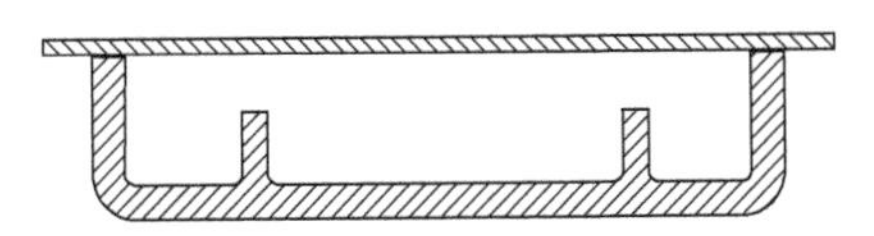

Fig. 15.27: Microdiffusion apparatus.

Numerous other devices have been proposed for separations according to this principle, of which only the device for blood alcohol determination given by Widmark (1922) will be mentioned (Fig. 15.28).

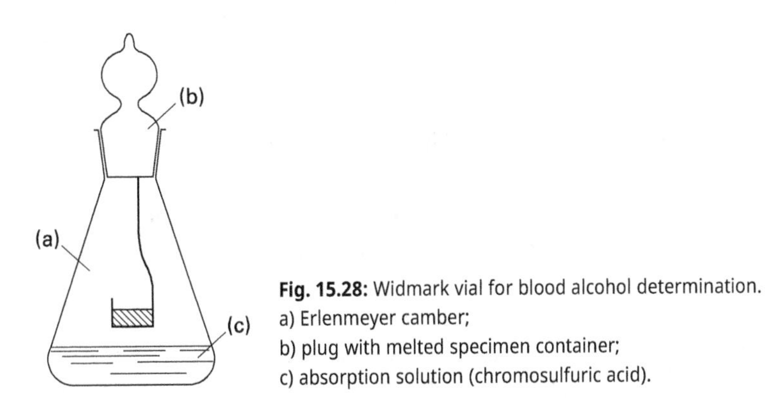

Fig. 15.28: Widmark vial for blood alcohol determination.
a) Erlenmeyer camber;
b) plug with melted specimen container;
c) absorption solution (chromosulfuric acid).

By its nature, the method is only suitable for highly volatile substances; however, the range of applications can be extended, since many substances with vapor pressures that are too low can be converted into more suitable compounds. For example, one can use NO_3^- and NO_2^- by reduction, acid amides by hydrolysis into NH_3, many nitrogen-containing organic substances by Kjeldahl decomposition also into NH_3, others by enzyme reactions or oxidation into CO_2 and many more.

Volatilization by microdiffusion is mainly used to separate trace components down to the nanogram range; the lower limit is given by the sensitivity of the determination method. On the other hand, one should not exceed a few milligrams of substance by this method. Some application examples are given in Tab. 15.5.

Tab. 15.5: Microdiffusion (application examples).

To be separated	Added reagent	Absorption solution
NH_4^+, NH_3	KOH, K_2CO_3 et al.	H_2SO_4, HCl, boric acid
CO_3^{2-}, CO_2	H_2SO_4	$Ba(OH)_2$
CN^-, HCN	H_2SO_4	NaOH
F^-, $(CH_3)_3SiF$	Hexamethyldisiloxane	NaOH/isopropanol
N_3^-, HN_3	H_2SO_4	$AgNO_3$
S^{2-}, H_2S	H_2SO_4	NaOH
Cl^-, Cl_2	$KMnO_4$	KI; Fast Green FCF
Br^-, Br_2	$K_2Cr_2O_7 + H_2SO_4$	AI
I^-, I_2	$K_2Cr_2O_7 + H_2SO_4$	KI
CO	H_2SO_4	$PdCl_2$
Volatile amines	NaOH; K_2CO_3	Various acida
Fatty acids $C_1 - C_4$	$H_2SO_4 + Na_2SO_4$	Na citrate
CH_3OH, i-propanol	K_2CO_3 [x)]	H_2SO_4
C_2H_5OH	K_2CO_3 [x)]	$K_2Cr_2O_7 + H_2SO_4$
Phenols	H_2SO_4	NaOH
Formaldehyde	–	Chromotropic acid
Acetaldehyde	–	Semicarbazide + HCl
Acetone	–	$NaHSO_3$; 2,4-dinitrophenylhydrazine

[x)] to increase the volatility.

15.4 Separations by one-sided repetition with auxiliary substance

15.4.1 Distillation in the stream of a non-condensed gas: Gases in metals – expulsion of gases from solutions with an auxiliary gas – vacuum distillation

The vacuum melting method for the determination of gases in metals described in the previous section can be simplified considerably in terms of apparatus according to a suggestion by Singer (1940) by expelling the gases released by heating in the graphite crucible from the apparatus with a stream of inert gas (usually argon, more rarely helium or nitrogen) (Fig. 15.29). This avoids the need for the costly high-vacuum apparatus, but requires very careful purification of the inert gas. The method known as the *carrier gas method* is therefore relatively rarely used; some examples are given in Tab. 15.6. In addition to the determination of gases in metals, the method has served to determine the nitrogen content of rocks and the oxygen content of organic compounds and metal oxides.

Expulsion of gases from solutions with an auxiliary gas. For the rapid expulsion of gases from solutions, a stream of an auxiliary gas is often used, from which the component sought is removed in a subsequent absorption device.

In the simplest case, the propellant gas is evolved in the solution itself; for example, in Gutzeit's arsenic detection method, the H_2 gas evolved from zinc and hydrochloric acid carries the arsine hydrogen produced by reduction out of the reaction vial.

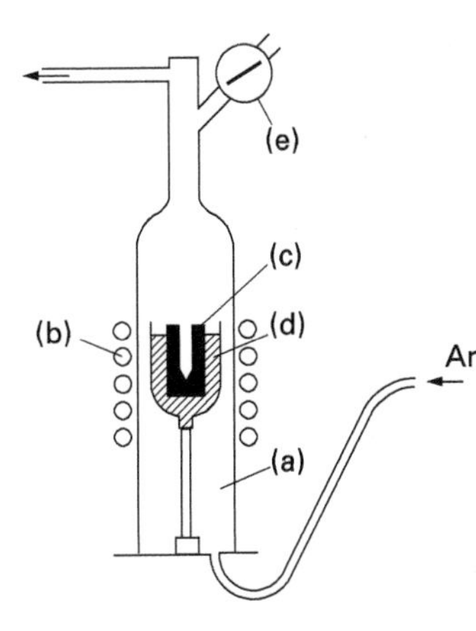

Fig. 15.29: Carrier gas method for the determination of gases in metals.
a) Quartz vessel;
b) induction coil;
c) crucible;
d) heat shield (graphite powder);
e) specimen injection.

Tab. 15.6: Carrier gas process (application examples).

Metal	Analyte	Temp. (°C)	Bath	Flux
Be	O	2600	Ni	Cu
Fe	O	1800	Pt	–
Mn	O	1650	Sn	–
U	N	1950	Pt	–
V	H	1950	Pt	–
Y	O	2100–2200	–	Pt
Zr	O	2100	–	Pt

Devices in which an external gas flow is passed through the solution are more versatile. If volatilization is carried out at room temperature, it is sufficient to pass the gas stream – as finely distributed as possible – through the analysis solution.

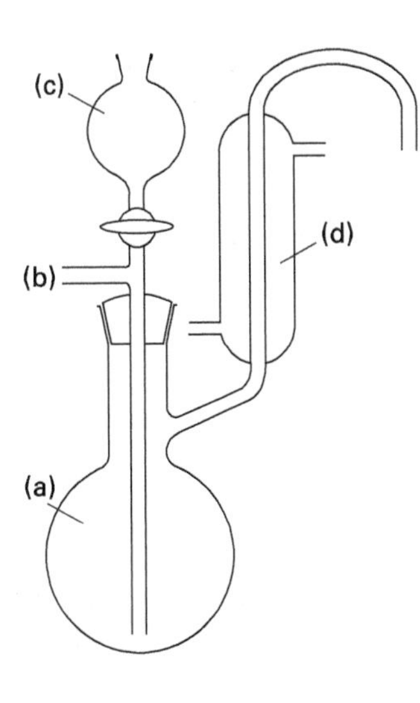

Fig. 15.30: Expulsion of gases from solutions with an auxiliary gas.
a) Piston with specimen;
b) gas supply;
c) dropping funnel;
d) reflux condenser.

Devices in which the sample can be heated usually consist of a flask with gas introduction, a dropping funnel for feeding reagent solutions, and a reflux condenser in which vaporized solvent is retained (Fig. 15.30); devices operating on this principle are described in numerous variations in the literature.

Some application examples for the method are given in Tab. 15.7.

Tab. 15.7: Expulsion of gases from solutions with an auxiliary gas (application examples).

Substance	Reagent additive	Auxiliary gas	Exaggerated
Carbonates	HCl	Air	CO_2
Arsenate	Zn + HCl	H_2	AsH_3
Sulfides (in metals)	H_3PO_4	CO_2	H_2S
Dimethyl ether	HI	CO_2	CH_3I
Blood	–	Air	C_2H_5OH
Food (+ H_2O)	HCl	N_2	SO_2
Food	–	Air	NH_3

Vacuum distillation. In vacuum distillation commonly used in the laboratory, usually only the vacuum of a water jet pump is applied, whereby pressures of about 15–20 Torr (depending on the water temperature) can be achieved. The boiling point of liquids is generally lowered by about 50 °C as a result.

Since in vacuum distillation with small quantities of substances, it is difficult to avoid bumping or splashing of the mixture, a fine stream of air from a long capillary is used to bubble through the liquid. In this way, the liquid is kept in constant motion and the evaporated compounds are forced into the condenser.

15.4.2 Distillation in the stream of a subsequently condensed gas (*codistillation*): Steam distillation – distillation from aqueous solutions – distillation in the vapor stream of organic auxiliary liquids

If an auxiliary liquid is added to the mixture of substances to be separated, the vapors of which carry along the component to be separated, the term *condistillation* is used according to a suggestion by Rassow (the term *codistillation* is more common).

Probably the best known of these methods is *steam distillation*, which is used to volatilize substances that have a low but nevertheless noticeable vapor pressure at the boiling point of water. A stream of water vapor is passed through the analysis sample (or through a mixture of water and analysis sample), which carries the volatile components in accordance with their vapor pressures; the vapors are condensed in a cooler.

In the simplest case, the analysis sample is insoluble in water; according to the vapor pressure diagram of systems of components insoluble in each other (cf. Fig. 15.12), the partial pressures add up, so that for a total pressure of 760 Torr, the distillation temperature must always be below 100 °C.

The process leads to rapid separation of volatile components even if their partial pressures are only low, since the vapor densities of the various components in the gas phase behave like their molecular weights. The vapor contains substance 1 to be separated with molecular weight M_1 and component 2 (water, molecular weight 18) in the ratio of their pressures at the boiling point of the mixture. At a total pressure of 760 Torr, it thus consists of p_1 /760 of substance 1 and p_2 /760 water. However, the weight amounts of a_1 and a_2 in the vapor are $p_1 \cdot M_1$ /760 and $p_2 \cdot 18/760$, the weight ratio of substance to be separated and water thus being)

$$\frac{a_1}{a_2} = \frac{p_1 M_1}{p_2 \cdot 18} = \frac{p_1}{p_2} \cdot \frac{M_1}{18} \tag{12}$$

As an example of such a distillation, nitrobenzene is to be overdistilled by steam distillation. At the boiling point of the system (about 99 °C), the vapor pressure of nitrobenzene is 20 torr, and that of water is 740 torr; thus, the vapor pressures behave as 20/740 = 0.027, but the amounts by weight in the vapor behave as 20/740 · 123/18 = 0.185. This ratio is higher than the partial pressure ratio by a factor of M_1 /18 = 123/18 = 6.8. Thus, with little water, a rather large amount of nitrobenzene is carried over.

In general, the higher the molecular weight of the overdistilled substance, the more favorable the method.

The method is also used to separate substances that are present in solution. The same considerations apply to the weight ratios in the vapor; however, in contrast to the behavior of systems of components that are insoluble in one another, the partial pressure of the compound to be separated decreases with decreasing concentration, so that much larger quantities of water are usually required for quantitative overdriving. This undesirable effect can be more or less compensated for by adding salts in high concentration to the starting solution (to increase the partial pressure) or non-volatile compounds, e.g. sulfuric acid (to increase the boiling point).

When carrying out steam distillations, a vessel is required for developing the steam; this is passed through the flask with the substance mixture (Fig. 15.31); for analytical purposes with smaller sample quantities, more compact designs are also used (Fig. 15.32).

Steam distillation is mainly used for the preparative purification of organic compounds; in analytical chemistry, this method is mainly important for the isolation of volatile natural products.

The advantage is the gentle separation with low equipment requirements. On the other hand, the separation effect is low, since mixtures can only be separated into volatile and non-volatile components.

Distillation from aqueous solutions. Distillation from aqueous solutions is in principle a steam distillation, except that the steam used to exaggerate volatile substances is generated in the solution itself.

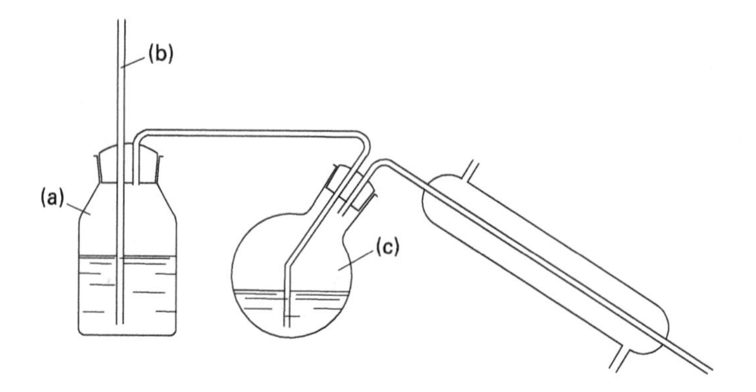

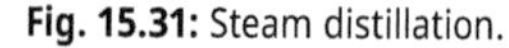

Fig. 15.31: Steam distillation.
a) Vessel with water; b) riser tube; c) flask with substance mixture.

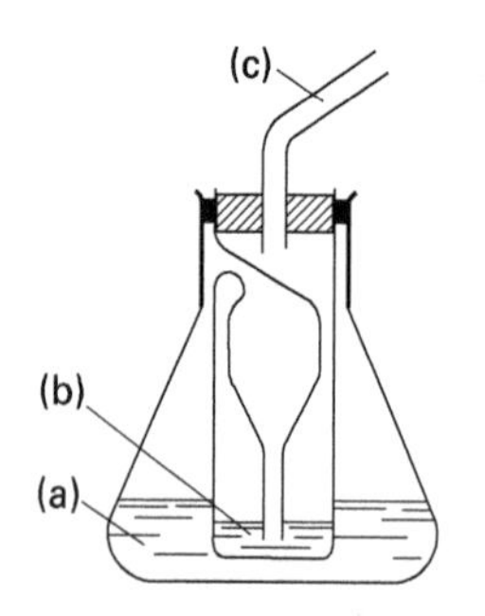

Fig. 15.32: Steam distillation for the determination of volatile acids in wine.[3]
(a) Water; (b) sample; (c) steam discharge to cooler.

Devices are used as in simple overhead distillation, but numerous variants differing in details have been described. To accelerate such distillations, a gas stream is often additionally passed through the solution.

The process has considerable importance for inorganic analytical chemistry, since only a few inorganic compounds can be distilled off from aqueous solutions, for which high selectivity is therefore achieved. The starting solutions must usually have a certain composition so that the volatile compound is formed in sufficient concentration; e.g., to distill arsenic(III) chloride, one needs a high HCl concentration (approx. 20–25%, cf. Fig. 15.33), to distill OsO_4 and RuO_4, one needs a high oxidation potential, and so on. Often the boiling point of the solution is raised by adding H_2SO_4, occasionally a gas is passed through the solution to cause certain reactions; e.g. CrO_2Cl_2 is distilled with the aid of an HCl stream (cf. Tab. 15.8).

The effectiveness of the procedure is somewhat compromised by the aforementioned overspray of liquid droplets.

3 The idea originated with M. Pozzi-Escot (1904).

Distillation in the vapor stream of organic auxiliary liquids. For various reasons, water vapor is substituted by vapors of organic liquids for the exaggeration of volatile substances: First, water vapor is not suitable when distilling off water itself; second, the temperature during distillation can be raised by using high-boiling liquids; and finally, in special cases, the vapors can cause special chemical reactions (e.g., the esterification of boric acid by methanol).

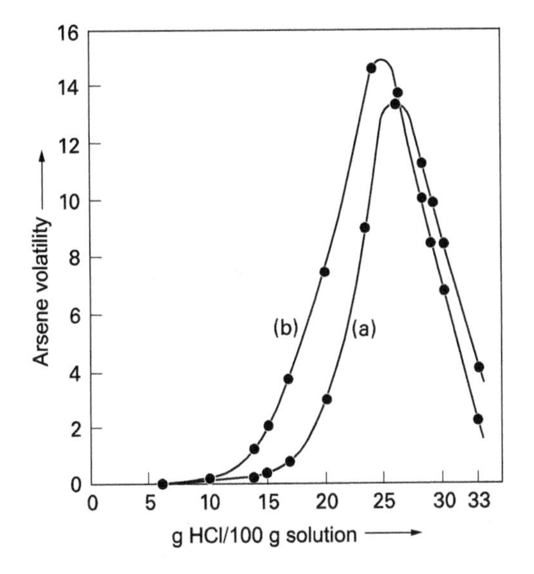

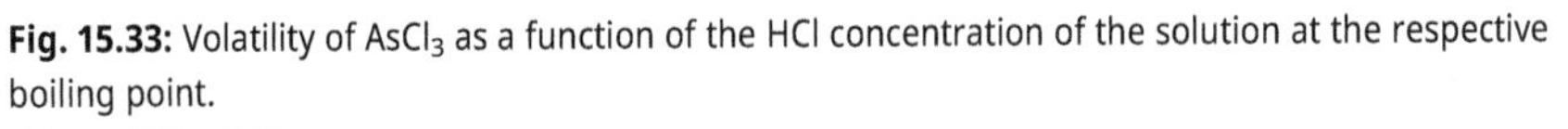

Fig. 15.33: Volatility of $AsCl_3$ as a function of the HCl concentration of the solution at the respective boiling point.
a) Pure HCl solution;
b) HCl solution with 13% H_2SO_4 addition.

Tab. 15.8: Distillation of inorganic compounds from aqueous solutions (examples).

Element to be separated	Solution composition	Distilled compound
As	Strong hydrochloric acid	$AsCl_3$
Ge	Strong hydrochloric acid	$GeCl_4$
Sb	Strong hydrochloric acid	$SbCl_3$
Se	HBr + H_2SO_4	$SeBr_4$
Sn	HBr + HCl + H_2SO_4	$SnBr_4$
F	$HClO_4$ + Silicate	H_2SiF_6 (SiF_4)
Cr	$HClO_4$ + HCl gas stream	CrO_2Cl_2
Os	5 M NHO_3	OsO_4
Ru	Strong perchloric acid	RuO_4
S	H_3PO_4 or H_2SO_4	H_2S
N	Strongly alkaline	NH_3

In the practical procedure, the auxiliary liquid is usually added to the sample and the mixture is distilled off overhead. Circulation methods are also frequently used (see below).

Xylene and various chlorinated hydrocarbons are particularly suitable as auxiliary liquids for water determination; e.g. tetrachloroethene is added in the well-known method according to Tausz and Rumm (1926). The mixture condensed in the cooler separates into two phases, so that the water can be determined by simple volume measurement.

In this way, water is determined primarily in fatty or greasy mixtures. The method fails if water-soluble and volatile substances other than water (e.g. lower alcohols) are present in the starting material, also if the water is very tightly bound in the sample, and finally if water is present only in traces. If the distillation temperature is to be raised, higher boiling liquids such as ethylene glycol (bp. + 197 °C), tetralin (bp. + 206 °C), various fractions of aliphatic hydrocarbons and others can be added.

Boric acid is distilled as a methyl ester from strongly acidic solutions containing methanol, which must contain dehydrating agents to shift the unfavorable esterification equilibrium; sulfuric acid is usually added.

15.4.3 Azeotropic distillation

As explained above, a mixture of two substances cannot be separated by distillation if an azeotrope occurs. In so-called *azeotropic distillation*, a ternary azeotrope is generated in such systems by adding a suitable third substance, with the aid of which the desired substance can be obtained pure.

Probably the best-known example of this process is the production of absolute alcohol. While only a mixture with a maximum of 95% ethanol can be obtained from aqueous ethanol solutions by distillation, a ternary azeotrope with 74% benzene, 18.5% ethanol and 7.5% water is obtained after the addition of benzene, with the aid of which the water can be completely removed from the starting mixture.

Such methods have gained great importance in technology, but so far play only a minor role in analytical chemistry.

15.4.4 Closed loop procedure

Procedures in which the gas phase is circulated have been devised using both gases and liquids as auxiliary substances. The main advantage of conducting gases in a *closed loop* is the small amount of extraneous material that is passed over the solid or liquid analytical sample; this allows the blank values to be reduced, which is particularly important when separating trace amounts of H_2O or CO_2. The principle of operation is shown in Fig. 15.34.

This requires a circulating pump for the gas, which is usually designed as a peristaltic squeeze pump; in the determination of H_2 in aluminum, argon was circulated as an auxiliary gas with a steel diaphragm pump, and the resulting H_2-Ar mixture could be conveniently analyzed by thermal conductivity measurement.

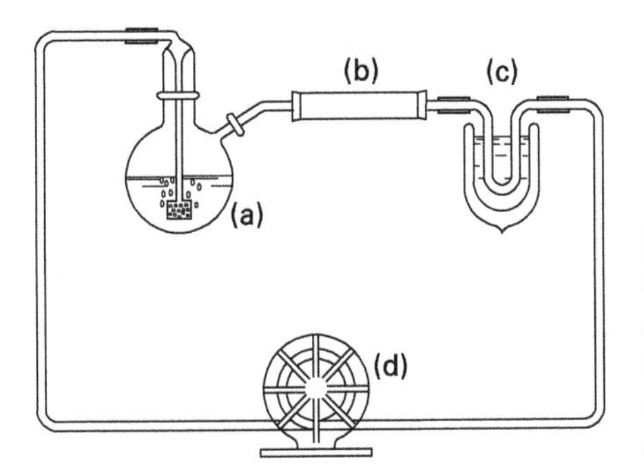

Fig. 15.34: Expulsion of gases from solutions by recirculation of an auxiliary gas.
a) Flask with analysis solution;
b) absorption vessel for water vapor;
c) freeze-out loop;
d) circulating pump.

However, the main application of distillative circulation methods is probably in the determination of water with the aid of organic liquids. The apparatus used is particularly simple, because the distillate separates into two phases, of which the organic phase can be returned directly to the flask. Of the numerous designs in which auxiliary liquids are used, some of which are specifically lighter and some heavier than water, only one example is

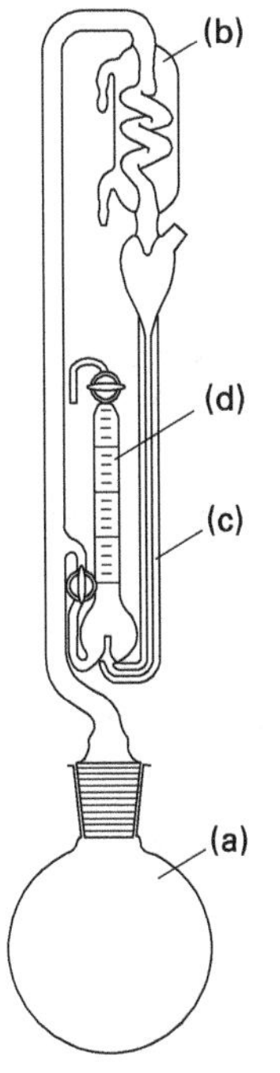

Fig. 15.35: Water determination by circulation distillation using the tetrachloroethane method.
a) Distillation flask;
b) flow-through cooler;
c) capillary;
d) collection vessel for the distilled water.

given (Fig. 15.35). An important detail here is the use of a flow-through cooler from which the precipitated water droplets are flushed quantitatively into the collection vessel.

The advantage of the process compared to distillation without circulation is the small amount of circulating liquid, which can also dissolve only a correspondingly small amount of water that escapes determination.

In other applications of *closed loop distillation*, the separated component of the analytical sample remains dissolved in the distillate; in such behavior, the auxiliary liquid must first be separated from the substance in question before it can be returned to the distillation flask. This separation is usually achieved by a second distillation under modified conditions.

For example, when separating boric acid, the methyl ester is distilled into a collector with methanolic NaOH, from which the methanol is distilled back into the flask with the analysis sample with saponification of the ester (Fig. 15.36).

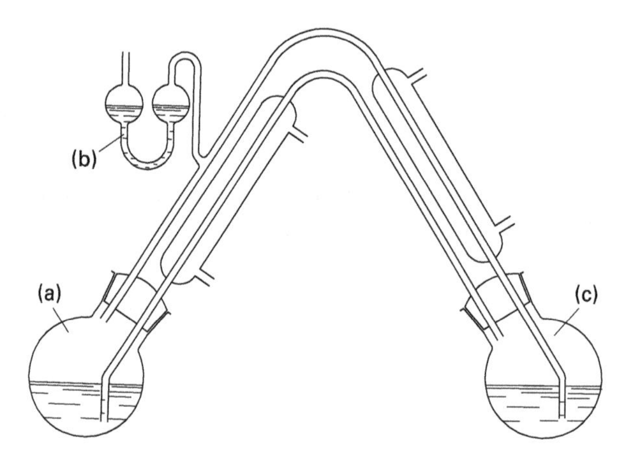

Fig. 15.36: Closed loop distillation for the separation of boron as boric acid methyl ester.
a) Strongly acidic borate-methanol solution; b) pressure equalization vessel with absorbing solution; c) methanolic NaOH solution.

In principle, HF can be separated from aqueous solutions as H_2SiF_6, with aqueous NaOH solution in the receiver from which the water is distilled off again.

15.5 Countercurrent method without auxiliary substance

15.5.1 General – use of boiling and equilibrium diagrams – theoretical bottoms (separation stages) – reflux ratio – operating contents

The distillation methods discussed so far are all not very effective; they only allow the separation of substances with widely differing boiling points, but fail in the presence of substance mixtures with small differences in volatility (or low separation factors). To improve the efficiency, the separation must also be repeated during distillation. This is done in so-called distillation columns, which, however, operate according to a principle that differs from the other separation processes in columns.

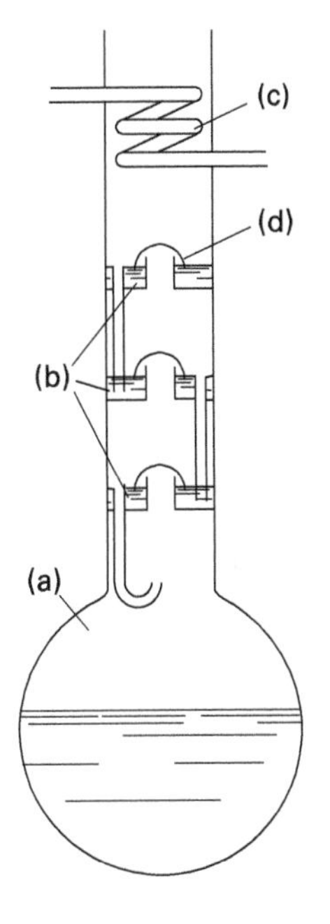

Fig. 15.37: Distillation column, (*bubble tray tower*) (principle).
a) Flask with initial mixture;
b) plates;
c) reflux condenser;
d) bell.

The liquid mixture to be separated is contained in a flask on which a tall top, the column, is placed (Fig. 15.37). This contains box-shaped compartments for holding liquid, the so-called *trays* or *plates*, which are provided with an overflow and with a central opening for the passage of the vapor. The openings are covered with bell-shaped lids so that rising steam is forced to bubble through the liquid on the plate. A reflux condenser is located on the top of the column.

The substance mixture in the flask may consist of two ideally behaving liquids with slightly different boiling points. When heated, the developed vapor initially rises to the first plate and is condensed here. As the whole system heats up, the liquid of the first plate evaporates in its turn, is condensed on the next higher plate and so on, until the vapor reaches the reflux condenser. If the flask is heated further, a continuous countercurrent of rising vapor and descending liquid is formed; after some time, vapor-liquid equilibria will be established along the column, i.e., the composition of the liquid on each tray is in equilibrium with the vapor rising from it. The distillate dripping from the condenser into the column is called *reflux* and the mode of operation in which all the distillate flows back into the column is called *total reflux distillation*.

If the boiling diagram of the present two-substance mixture is known, the concentration ratios within the column can be given – assuming that vapor-liquid equilibrium has been fully established on each plate.

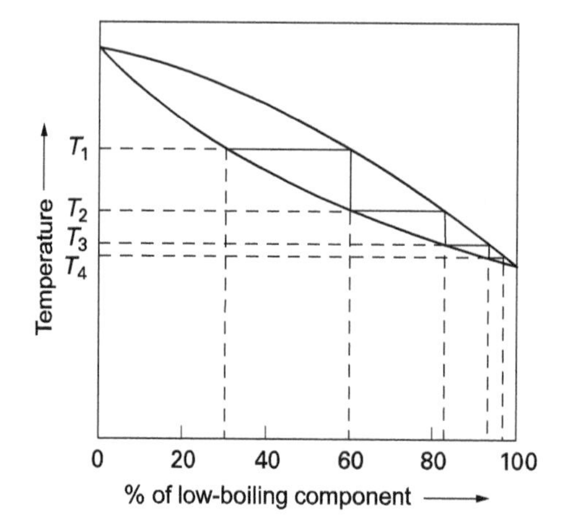

Fig. 15.38: Graphical determination of the concentration ratios in a bubble tray tower according to the boiling diagram.

The initial mixture may contain 30% of the more volatile component (see Fig. 15.38). It boils at the temperature T_1, the vapor and thus also the liquid on the first (lowest) stage contain 60% of the more volatile component. This latter mixture boils at T_2 and yields a vapor (and a liquid on the second plate) with 83% of the light volatile component; this mixture in turn boils at T_3 and yields a vapor and a liquid on the third plate with 92.5% of the light volatile component, which boils at T_4; finally, the final product with 97% purity condenses at the cooler.

A corresponding graphical evaluation can be performed using the equilibrium diagram (Fig. 15.39); an ascending staircase curve is obtained, from which, however, the boiling temperatures of the individual fractions are not apparent.

In this example, there are 4 equilibrium settings between the two phases, i.e. 4 separation plates with only three trays in the column. The difference is due to the fact that the evaporation equilibrium has already been established once below the column in the flask. In the literature, the number of separation plates is often equated with the number of trays, i.e. the one equilibrium setting below the column is not taken into account. For columns with a larger number of plates, the difference can be neglected.

Theoretical plate. In practice, the ideal case of complete equilibrium setting taken as a basis above is never achieved in every evaporation process, the efficiency of each tray and thus also that of the entire column is consequently lower than would be expected according to theory.

When evaluating a distillation column, one now takes as a basis the separation performance actually achieved and calculates how many times the equilibrium would have had to be established in order to achieve a separation performance found. Such

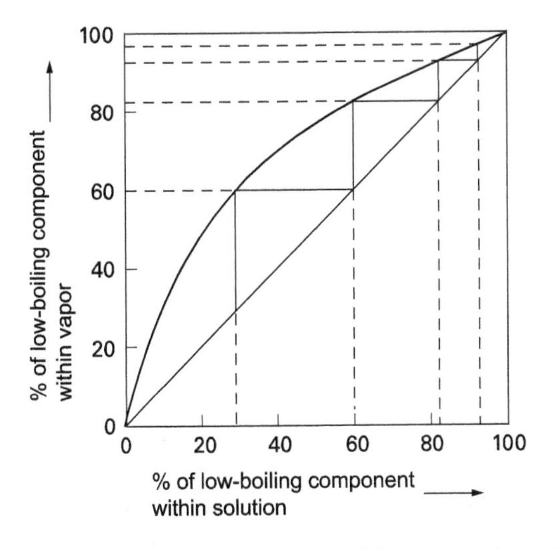

Fig. 15.39: Graphical determination of the concentration ratios in a bell tray tower according to the equilibrium diagram.

a calculated equilibrium setting is called a *theoretical plate*. According to what has been said, the number of theoretical plates of a distillation column is always smaller than the number of plates installed in reality.

The term *height of a theoretical plate* (column length divided by the number of theoretical plates) is derived from this term in accordance with the considerations for other separation processes (*height of a separation plate*). In the Anglo-Saxon language area, the term *HETP* = *height equivalent to a theoretical plate* can be found for this, which was first used by Peters. The terms *theoretical plate* and *separation plate* will be used synonymously in the following.

Reflux ratio. Total reflux distillation is of no practical significance, since no distillate is obtained and consequently no separation of the mixture takes place. However, this mode of operation is the basis for theoretical derivations of the ratios in the column, it is further used in the determination of the number of theoretical trays (see below).

In separations by distillation in a column, a part of the liquid condensed at the head is removed and the rest is allowed to flow back. The ratio of the quantity flowing back in the unit of time (e.g. in ml/min) to the quantity removed is the so-called *reflux ratio*. Accordingly, the total reflux mode of operation means the reflux ratio ∞.

Due to the distillate withdrawal, the column no longer operates under optimum conditions, so the purity of the distillate decreases. The more distillate is withdrawn compared to the reflux and the lower the reflux ratio, the poorer the separation. The influence of the reflux ratio can be determined quantitatively according to a graphical method given by McCabe and Thiele (1925), but the rather complicated considerations will not be dealt with here, since their importance lies mainly in the field of technical distillation.

Operating hold-up. During distillation, a portion of the liquid, called the working hold-up, is in the column itself. Since this fraction increases the amount of transition fraction between two separate components, it has a detrimental effect on the result of distillation, and attempts are made to design the equipment in such a way that the operating hold-up becomes as small as possible.

15.5.2 Determining the number of theoretical plates of a distillation column

As shown in Fig. 15.39, the number of theoretical plates required to obtain a head fraction of a certain purity can be determined by a step curve. The preconditions are that the equilibrium diagram and the initial concentration in the flask are known and that distillation is carried out under total reflux.

This observation can be used inversely to determine the number of separation plates: A mixture of two substances, the equilibrium diagram of which is known, is placed in the flask (e.g. benzene – toluene 50:50) and distilled under total reflux until equilibria have been established in the entire system; this usually takes about 1–2 h. The number of plates is then determined. Then a small amount of distillate and flask contents are taken and analyzed, and the step curve is plotted on the equilibrium diagram according to the results of the analyses. The number of separation plates can be read out immediately.

To simplify the procedure, diagrams have been worked out in which a suitable measurand, e.g. the refractive index, is determined in the distillate and in the flask contents. The two values are entered on the curve of the diagram and the number of separation plates is read out from the ordinate as the difference between two numbers (see Fig. 15.40).

An approximate overview of the numbers of separation plates required for separation at different boiling point differences of liquids is given in Tab. 15.9.

The number of theoretical plates of a distillation column is not a very precisely determinable quantity; on the one hand, the reproducibility of the value when repeating the test with the same test mixture is not good, since unavoidable irregularities during distillation influence the result somewhat, and on the other hand, different values are obtained when using different test mixtures, since the rate of mass transfer between the two phases can change. Despite these shortcomings, the number of theoretical plates is quite useful for an initial assessment of the efficiency of a column and for comparing different columns.

15.5.3 Column types – return control – vacuum distillation – cryogenic distillation

The column with individual trays shown in Fig. 15.37 is rarely used in analytical chemistry, since such devices are difficult to make from glass and also have a large operating volume.

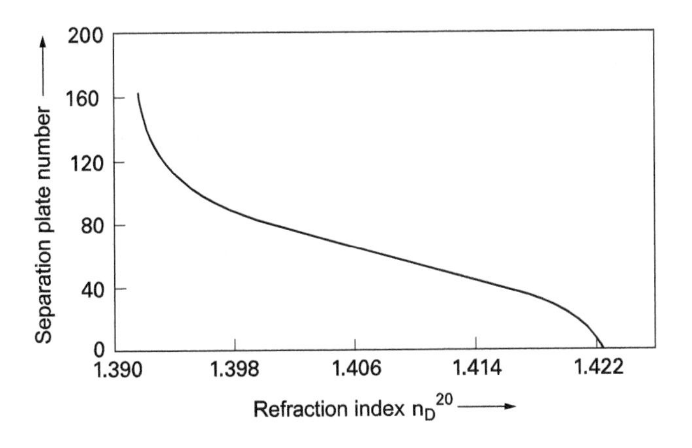

Fig. 15.40: Determination of the number of separation plates of a distillation column from the refractive index of the test mixture 2,2,4-trimethylpentane – methylcyclohexane.

Tab. 15.9: Required numbers of separation plates for binary mixtures of hydrocarbons with different boiling point differences.

Boiling point difference °C at 760 mm Hg	Required number of separation plates
7,0	20
5,0	30
3,0	55
1,5	100

More common are the so-called *packed columns*, in which a continuous layer of irregularly shaped packing is present in place of the individual trays. Their function is to distribute the downflowing reflux as finely as possible so that the liquid presents a large surface area to the rising vapor. The packing materials used include *Raschig rings*, saddles or wire spirals, the latter being particularly suitable for laboratory columns. When filling the columns, care must be taken that the packing is arranged irregularly so that no channels can form within the layer. Furthermore, the distillate should not flow down the tube wall unhindered.

For packed columns, as for all other types of column, care must be taken to ensure that the heat losses in the column are low. They are therefore usually provided with a jacket consisting either of a layer of insulating material, possibly with additional heating, or of an evacuated, internally mirrored vacuum jacket.

Since the operating contents of packed columns are still quite large, they are, like tray columns, not very suitable for distilling small quantities of substances. It is therefore necessary to try to reduce the extent of this disturbing factor. This is done by omitting the packing, but care must be taken to ensure that the rising vapor and reflux in the empty column come into sufficient contact with each other. To achieve this, the so-called *annular gap bubble columns* and columns with rotating inserts have been developed.

The annular gap bubble columns first given by Craig (1936) contain a somewhat thinner fused tube inside the column, so that only a narrow annular space of about 1–2 mm width remains between the two tube walls. As a result, an intensive mass transfer takes place between the liquid flowing down the walls as a thin film and the vapor (Fig. 15.41).

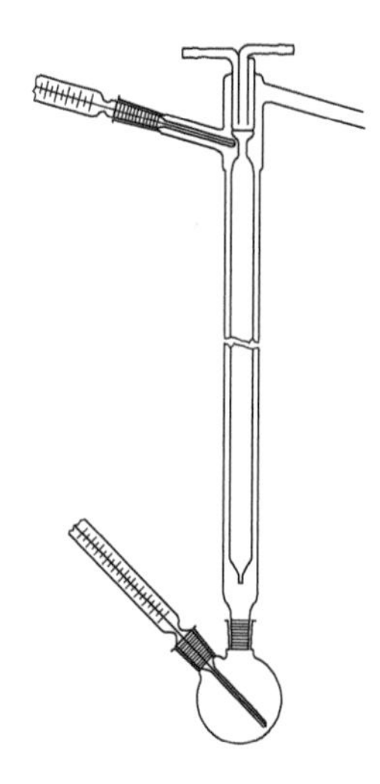

Fig. 15.41: Annular gap bubble column.

Columns with rotating inserts are designed as *spinning belt columns* and as *rotary drum columns*. In the spinning belt columns first described by Lesesne and Lochte (1938), there is a rapidly rotating metal or PTFE belt in the separation column through which the liquid droplets of the reflux are finely sprayed and the gas phase is swirled so that intensive contact takes place.

The belt is flat in the simplest case, but spirally wound or star-shaped designs have also been described. The speed should be between 1000 and 2000 min^{-1}, an increase brings no improvement, but possibly even a deterioration of the separation effect.

In the case of the rotary roll columns, the inner insert of an annular gap column is rotated, resulting in a combination of a rotary belt and annular gap column. The separation effect is very good with a small operating content.

Reflux control. Of the numerous proposals for controlling the reflux (return flow) ratio, only a column head with a swiveling funnel mounted below the reflux condenser in such a way that the distillate runs down into it may be mentioned (Fig. 15.42). At certain time intervals, which can be set as desired, an iron rod attached to the funnel is pulled to one side by a magnet mounted outside the column and the distillate is directed into a discharge pipe. The reflux ratio is given by the duration for which the magnet is switched on and off.

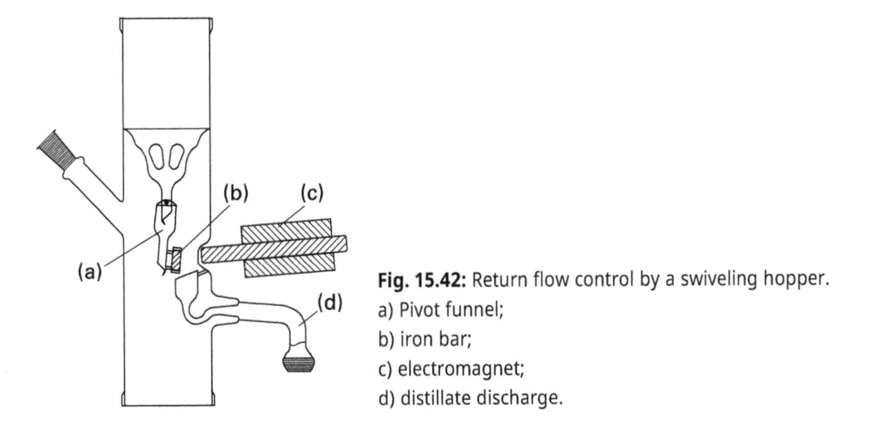

Fig. 15.42: Return flow control by a swiveling hopper.
a) Pivot funnel;
b) iron bar;
c) electromagnet;
d) distillate discharge.

Vacuum distillation. The column types described above can also be operated under vacuum. As already mentioned, the relative volatility usually increases when the boiling temperature is lowered, so that the application of vacuum brings an improvement in the separations. Figure 15.43 shows the effect of pressure on an equilibrium diagram, and Fig. 15.44 shows the same effect in a different representation (change in relative volatility with pressure).

Cryogenic distillation. Special distillation equipment has been developed for the separation of low-boiling gases, which permits largely automatic operation. The main difficulties here are sampling and condensation of the distillate, which is therefore sometimes taken in gaseous form.

15.6 Countercurrent method with auxiliary substance

15.6.1 Extractive distillation

In so-called *extractive distillation*, a higher-boiling auxiliary liquid is allowed to flow against the vapor rising in the column; if this liquid interacts particularly strongly with some of the compounds in the vapor, their volatilities are selectively reduced and the separation factors with respect to the unaffected components are increased. The process is cumbersome, since the added auxiliary liquid must subsequently be removed by another method; it is of importance mainly for technical purposes.

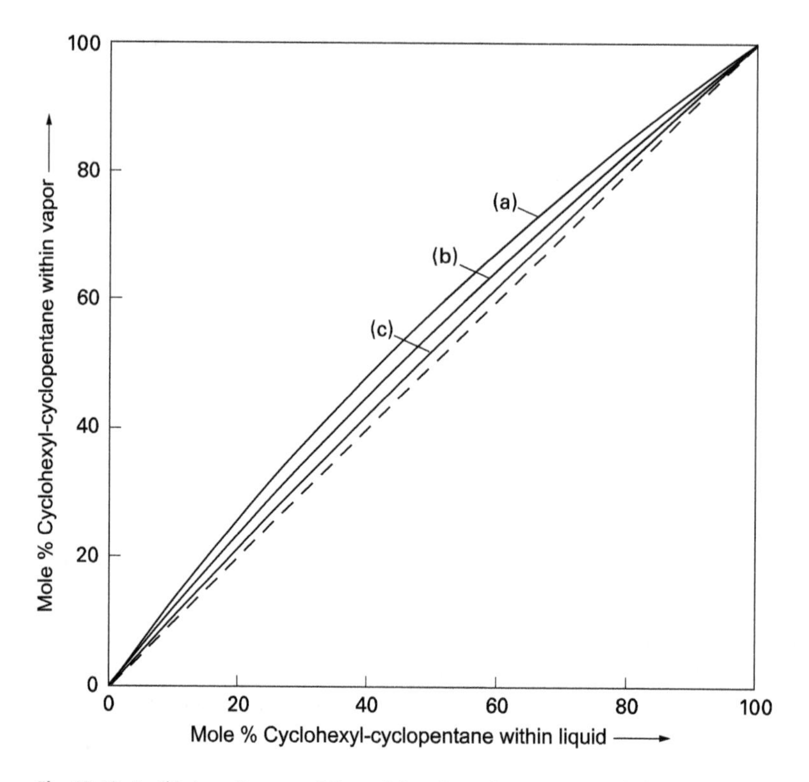

Fig. 15.43: Equilibrium diagram of the cyclohexyl – cyclopentane – n-dodecane system at different pressures.
(a) 20 mm Hg; (b) 100 mm Hg; (c) 400 mm Hg.

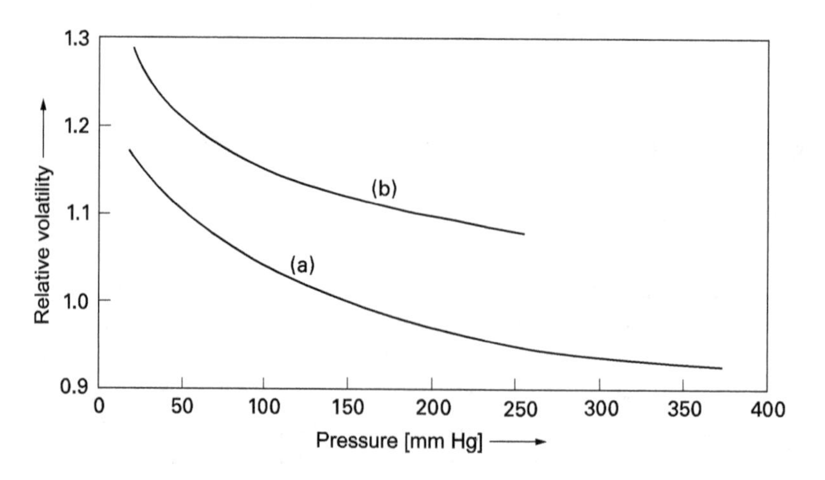

Fig. 15.44: Relative volatilities as a function of pressure.
a) System n-tridecane – dicyclohexyl; b) system cyclohexyl – cyclopentane – n-dodecane.

15.6.2 Azeotropic distillation

The azeotropic distillation process already mentioned can also be carried out in countercurrent to improve efficacy. However, it has also found hardly any analytical applications and should only be mentioned for the sake of completeness.

15.6.3 Inserting auxiliary substances

If a third liquid is added to a mixture of two substances, the boiling point of which lies between the boiling points of the substances to be separated and which does not interact preferentially with either of them, the separation is improved (Fig. 15.45). The process is referred to as *amplified distillation* in Anglo-Saxon literature.

The favorable effect of the addition in this case is based on the fact that the disturbing influence of the column contents is eliminated; during the transition from the lower-boiling to the higher-boiling component, the auxiliary liquid itself assumes the role of the contents.

When separating a large number of compounds, it is useful to add a mixture of substances whose boiling range covers that of the substances to be separated; for example, the separation of amine mixtures can be improved by adding suitable hydrocarbon fractions.

A similar procedure in principle is used if an auxiliary liquid is added to the starting mixture, the boiling point of which is higher than that of each component of the mixture. In this way, even the last residues of the sample can be overdistilled, the removal of which from the column is otherwise not readily possible.

This method is also cumbersome, so despite the often significant improvement in separations, its use is limited.

15.7 Efficacy and scope of the method

Simple overhead distillation for the separation of substances with widely differing boiling points is an indispensable tool, especially for drying analytical samples and removing solvents. In inorganic analysis, the method is of considerable importance because of its often high selectivity. The determination of gaseous constituents after volatilization from solutions, melts or solids is also a frequently used method which in many cases cannot be replaced by any other.

Column distillation has been developed into a very effective process in recent decades. Tab. 15.10 gives an overview of the separation levels achievable with laboratory equipment; however, the values depend not only on the design of the apparatus, but also on the distillation speed and – to a lesser extent – on the type of samples. The data in Tab. 15.10 should therefore only be regarded as approximate values.

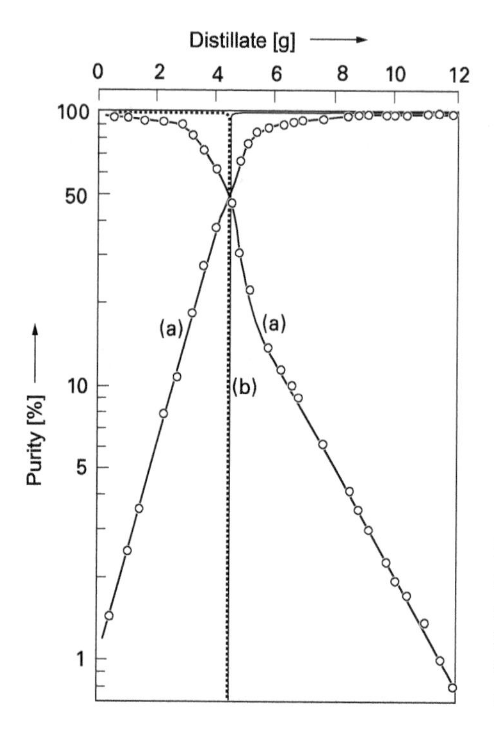

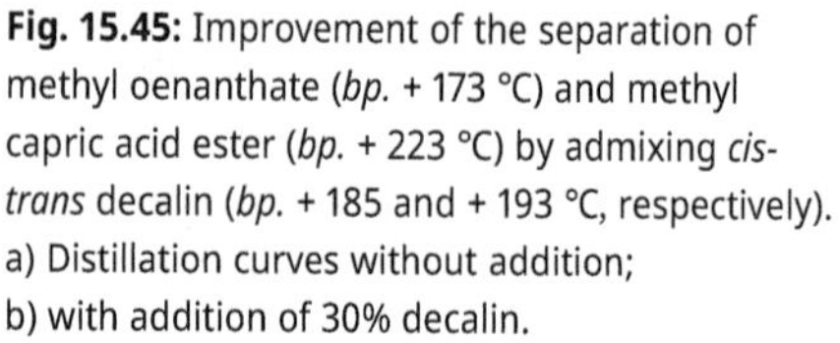

Fig. 15.45: Improvement of the separation of methyl oenanthate (*bp.* + 173 °C) and methyl capric acid ester (*bp.* + 223 °C) by admixing *cis-trans* decalin (*bp.* + 185 and + 193 °C, respectively).
a) Distillation curves without addition;
b) with addition of 30% decalin.

In analytical work, columns of about 1 m, at most of about 2 m length are usually used; thus, normally 30–50 or about 50–100 separation plates are available. An increase in this number is limited by increasing operating contents, increasing pressure drop in the column and increasing distillation times. For these reasons, columns with several hundred plates, manufactured for special purposes, are rarely used in analytical practice.

Tab. 15.10: Separation plate heights of different distillation columns.

Type	Plate height (cm)
Packed column with wire spirals	3–7
Packed column with special packing (*Heligrid*®)	0,9–1,9
Rotating belt column	2–3
Rotating roller column	0,9–1,7
Annular gap column	0,5–3

While separations by column distillation are carried out to a large extent in preparative and technical chemistry, the importance of this method for analytical chemistry has declined considerably since the introduction of gas chromatography. The main disadvantages are the long duration of a separation and the occurrence of intermediate fractions due to the operating contents of the columns; the above-mentioned method of *amplified distillation* for eliminating intermediate fractions is cumbersome, so that it has not been widely used.

Despite these disadvantages, column distillation is likely to continue to be used to a limited extent for analytical purposes, since it permits the use of larger quantities of substances and thus seems particularly suitable for the enrichment of trace impurities.

General literature

Fundamentals of distillation can be found in all physical chemistry books.

More specific introductions

J.-P. Duroudier, Distillation, Elsevier, Amsterdam (2016).
A. Górak & Z. Olujić, Distillation: Equipment and Processes, Elsevier, Amsterdam (2014).
A. Kiss, Advanced Distillation Technologies: Design, Control and Applications, J. Wiley & Sons, New York (2013).
E. Mahrous & M. Farag, Trends and applications of molecular distillation in Pharmaceutical and Food Industries. Separation and Purification Reviews 51, 300–317 (2022).
S. Moldoveanu & V. David, Modern Sample Preparation for Chromatography, Elsevier, Amsterdam (2015).
R. Saidur, E. Elcevvadi, S. Mekhilef, & H. Mohammed, An overview of different distillation methods for small scale applications, Renewable & Sustainable Energy Reviews *15*, 4756–4764 (2011).

Literature to the text

Azeotropic distillation

M. Doherty & J. Knapp, Distillation, azeotropic and extractive, in: Kirk-Othmer Separation Technology, Wiley & Sons, Hoboken *1*, 918–984 (2008).
W. Luyben & I.-L. Chien, Design and Control of Distillation Systems for Separating Azeotropes, John Wiley & Sons, Hoboken (2010).
A. Pereiro, J. Araujo, J. Esperanca, I. Marrucho, and L. Rebelo, Ionic liquids in separations of azeotropic systems – A review, Journal of Chemical Thermodynamics *46*, 2–28 (2012).
J. Stichlmair, H. Klein & S. Rehfeldt, Distillation: Principles and Practice, J. Wiley & Sons, New York (2021).

Head Space Analysis

S. Countryman & J. Bonilla, Analysis of Residual Solvents by Head Space Gas Chromatography (Book Chapter), in: Handbook of Analysis of Oligonucleotides and Related Products, 331–350 (2011).
R. Kumar, R. Varma, S. Sen & S. Oruganti, Head-space miniaturization techniques (Book Chapter), in: Emerging Freshwater Pollutants: Analysis, Fate and Regulations, 95–116 (2022).

N. Snow & G. Slack, Head-space analysis in modern gas chromatography, TrAC, Trends in Analytical Chemistry *21*, 608–617 (2002).

T. Wampler, Analysis of food volatiles using headspace-gas chromatographic techniques, Food Science and Technology (New York, NY, United States) *115* (Flavor, Fragrance, and Odor Analysis), 25–54 (2002).

Rotating belt column

L. Bittorf, K. Pathak & N. Kockmann, Spinning Band Distillation Column-Rotating Element Design and Vacuum Operation, Industrial and Engineering Chemistry Research *60*, 10854–1086 (2021).

Gases in metals

T. Kannengiesser & N. Tiersch, Comparative study between hot extraction methods and mercury method-A national round robin test Welding in the World *54*, R108–R114 (2010).

H. Kipphardt, T. Dudzus, K. Meier, (. . .), M. Hedrich & R. Matschat, Measurement of oxygen and nitrogen in high purity metals used as national standards for elemental analysis in Germany by classical carrier gas hot extraction (HE) and HE after activation with photons, Materials Transactions *43*, 98–100 (2002).

C. Kramer, P. Ried, S. Mahn, (. . .), N. Langhammer & H. Kipphardt, Design and application of a versatile gas calibration for non-metal determination by carrier gas hot extraction, Analytical Methods *7*, 5468–5475(2015).

Y.-J. Zhu, The status quo and prospect of gas analysis in metal, Yejin Fenxi/Metallurgical Analysis *34*, 19–23 (2014).

Co-distillation

F. Ferreira, M. Carneiro, D. Vaitsman, F. Pontes, M. Monteiro, L. da Silva, & A. Alcover Neto, Matrix-elimination with steam distillation for determination of short-chain fatty acids in hypersaline waters from pre-salt layer by ion-exclusion chromatography, Journal of Chromatography A *1223*, 79–83 (2012).

J. Masteli, The essential oil co-distillation by superheated vapour of organic solvents from aromatic plants, Flavour and Fragrance Journal *16*, 370–373 (2001).

H. Muhamad, P. Abdullah, Pauzi T. Al & S. Chian, Optimization of the sweep co-distillation clean-up method for the determination of organochlorine pesticide residues in palm oil, Journal of Oil Palm Research *16*, 30–36 (2004).

F. Peng, L. Sheng, B. Liu, H. Tong & S. Liu, Comparison of different extraction methods: Steam distillation, simultaneous distillation and extraction and headspace co-distillation, used for the analysis of the volatile components in aged flue-cured tobacco leaves Journal of Chromatography A *1040*, 1–17 (2004).

B. Rao, P. Kaul, K. Singh, G. Mallavarapu & S. Ramesh, nfluence of co-distillation with weed biomass on yield and chemical composition of rose-scented geranium (pelargonium species) oil Journal of Essential Oil Research *17*, 41–43 (2005).

N. Sahraoui, M. Vian, I. Bornard, C. Boutekedjiret, and F. Chemat, Improved microwave steam distillation apparatus for isolation of essential oils, Journal of Chromatography A *1210*, 229–233 (2008).

C. Zhang, R. Eganhouse, J. Pontolillo, I. Cozzarelli, & Y. Wang, Determination of nonylphenol isomers in landfill leachate and municipal wastewater using steam distillation extraction coupled with comprehensive two-dimensional gas chromatography/time-of-flight mass spectrometry, Journal of Chromatography A *1230*, 110–116 (2012).

Short Path Distillation

D. Bethge, Short path andmolecular distillation (Book Chapter), in: Vacuum Technology in the Chemical Industry, pp. 281–293 (2014).

F. Pudel, P. Benecke, K. Vosmann, & K. Matthäus, 3-MCPD- and glycidyl esters can be mitigated in vegetable oils by use of short path, European Journal of Lipid Science and Technology *118*, 396–405 (2016).

L. Vázquez & C. Akoh, Fractionation of short and medium chain fatty acid ethyl esters from a blend of oils via ethanolysis and short-path distillation, JAOCS, Journal of the American Oil Chemists' Society *87*, 917–928 (2010).

X. Xu, Short-path distillation for lipid processing, Healthful Lipids, 127–144 (2005).

Microdistillation

H. Ağalar, B. Temiz & B. Demirci, HS-SPME and microdistillation isolation methods for the volatile compounds of bergamot fruits cultivated in Turkey, Natural Volatiles and Essential Oils *6*, 21–25 (2019).

R. King, B. May, D. Davies & A. Bird, Measurement of phenol and p-cresol in urine and feces using vacuum microdistillation and high-performance liquid chromatography, Analytical Biochemistry *384*, 27–33 (2009).

M. Kurkcuoglu, N. Sargin & K. Baser, Composition of volatiles obtained from spices by microdistillation, Chemistry of Natural Compounds *39*, 355–357 (2003).

K. Lam, E. Cao, E. Sorensen & A. Gavriilidis, Development of multistage distillation in a microfluidic chip, Lab on a Chip *11*, 1311–1317 (2011).

A. Ziogas, G. Kolb, H.-J. Kost & V. Hessel, Development of high performance micro rectification equipment for analytical and preparative applications, Chemie Ingenieur Technik *83*, 465–478 (2011).

Microdiffusion

M. Heiling, J. Arrillaga, R. Hood-Nowotny & X. Videla, Preparation of ammonium-15N nitrate-15N Communications in Soil Science and Plant Analysis 37, 337–346 (2006).

G. Roda, S. Arnoldi, M. Cas, (. . .), R. Froldi & V. Gambaro, Determination of cyanide by microdiffusion technique coupled to spectrophotometry and GC/NPD and propofol by fast GC/MS-TOF in a case of poisoning, Journal of Analytical Toxicology *42*, e51–e57 (2018).

Annular gap column

W. Fischer, Fully automatic distillation. Annular gap column for careful separation of small amounts, Chemie-Anlagen + Verfahren *34*, 104 (2001).

16 Volatilization: Sublimation

16.1 General

16.1.1 Historical

Sublimations were already carried out in antiquity; for example, the Greek physician Dioscorides mentions the extraction of mercury by sublimation of cinnabar in the 1st century AD.

16.1.2 Definitions – p, T diagrams – auxiliary gases

Sublimation is defined as the transition of a solid substance into the gas phase (without the appearance of a liquid) and the subsequent condensation of the gas. The condensed components are the *sublimate* (for the substance to be sublimed, the term *sublimand* has been proposed, but like the term *desublimation* for condensation, it has not become common).

A substance can be sublimed if its vapor pressure already reaches *noticeable* values at temperatures below the melting point, so that it can be vaporized at a sufficient rate for practical purposes. However, the sublimation rate depends not only on the sublimation pressure, but is significantly influenced by the particle size of the substance and the speed at which the gas is carried away from the surface.

The vapor pressure curve of a subliming substance is shown in Fig. 16.1. Both the liquid and the solid phase exhibit a temperature-dependent vapor pressure; the two curves intersect at the melting point, at which liquid, solid phase and vapor are in invariant equilibrium. Substances whose vapor pressure already reaches the value of 760 Torr below the melting point cannot be melted at normal pressure.

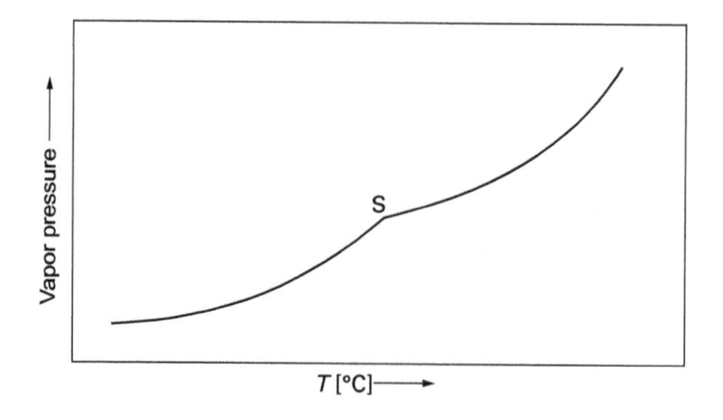

Fig. 16.1: Vapor pressure curve of a sublimating substance.
S = Melting point.

https://doi.org/10.1515/9783111181417-016

To accelerate sublimations, which are slow due to low vapor pressure of the substance in question, one can work in a vacuum or pass an inert auxiliary gas over the sample, which carries the vapor away.

A somewhat different type of sublimation occurs when an auxiliary gas is applied which reacts with the sample and produces the volatile compound in the first place. For example, in the above-mentioned mercury recovery by *sublimation* of the sulfide, the vaporizing element is set free only by the action of air (or of added iron filings). Such volatilization processes will also be discussed below.

16.2 Separations by one-sided repetition

16.2.1 Sublimation under normal pressure – vacuum sublimation – freeze drying

In sublimations for analytical purposes, the method of one-sided repetition is used exclusively; the evaporation equilibrium is continuously disturbed – just as in overhead distillation – by condensation of the evaporate.

A very simple sublimation method consists of heating the substance on one slide; the sublimate is collected on the underside of a second slide held over the sample. The considerable losses in this method can be reduced by placing the sample in a small glass ring between the two slides, but even then some of the sublimate is lost.

Quantitative work usually uses closed apparatus with cold fingers that can be evacuated so that substances with low sublimation pressure can still be volatilized at useful rates; it also reduces or eliminates decomposition phenomena in sensitive samples because the process occurs at a lower temperature.

A vertical version of such an apparatus is shown in Fig. 16.2; the cooling finger can be filled with a cooling mixture or cooled with flowing water by means of an insert (not shown).

The disadvantages of such devices are that parts of the sublimate can fall off the cooling finger into the non-volatile residue, and further that, especially when working rapidly, non-volatile components are mechanically entrained by the sublimate in fine distribution. Thus, even in a series of quantitative separations, somewhat too high values were found on average for the volatilized substances.

The sublimate can be prevented from falling back by arranging the cooler or cooling finger horizontally; furthermore, the vapors can be allowed to pass through a filter plate or a glass frit, which at the same time eliminates the entrainment effect (cf. Fig. 16.3).

Freeze-drying (lyophilization). Even at quite low temperatures (down to approx. −30 °C), ice still has sufficient vapor pressure so that it can be removed by sublimation. One applies this behavior in the so-called *freeze-drying*. Aqueous solutions are frozen at about −20 to −30 °C in a flask, and the ice is then removed by pumping under vacuum. When working quantitatively, it is recommended to pass the vapor

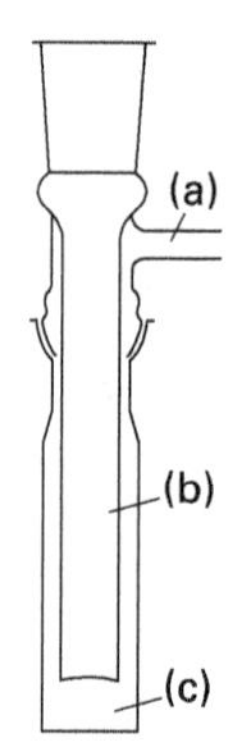

Fig. 16.2: Sublimation unit with cooling finger.
a) Vacuum port; b) cold trap; c) sample room.

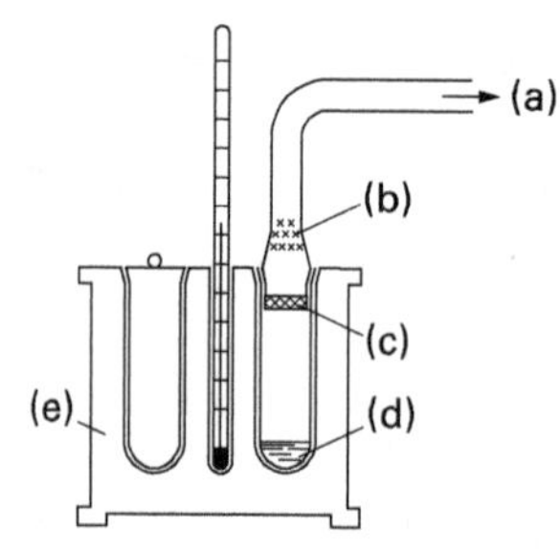

Fig. 16.3: Device for fractional sublimation.
a) Vacuum connection; b) sublimate;
c) glass frit; d) sample; e) heating block.

through a filter to avoid losses due to the entrainment effect. In trace analysis, for example, pesticide residues in water can thus be transferred to a more permanent form of sample storage. *Lyophilization* processes have become particularly important in pharmaceuticals and food technology. The preservation of proteins and other labile bioorganic substances is often achieved by transferring them to a solvent-free dry state by freeze-drying. This can also be carried out as an aerosol spray process.

16.2.2 Fractionated sublimation – resublimation

If different sublimable substances are present in a mixture at the same time, *fractional sublimation* can be used to achieve more or less extensive separations. First, the most volatile component is sublimed off, then the substrate is changed and sublimation is repeated while increasing the temperature (see Fig. 16.3). However, the process is only useful in the case of large differences in the sublimation prints.

Resublimation is the repeated sublimation of a sublimate for further purification. The process is hardly ever used because the effect is usually small.

16.2.3 Sublimation in the stream of an auxiliary gas – chlorination – evaporation analysis – pyrohydrolysis

Sublimation in the stream of an inert auxiliary gas offers few advantages over vacuum sublimation, especially since it can be difficult to condense the last sublimate residues from the gas stream. The process is therefore little used; as an example, consider the purification of iodine by sublimation with steam.

Sublimation with the aid of reactive gases has gained somewhat greater importance; of the various processes, chlorination, evaporation analysis (thermodesorption analysis) and pyrohydrolysis will be described. These methods have already been discussed as transport reactions in section 3.3. Here, the analyte becomes transportable through the auxiliary phase.

Treatment with chlorine at moderately elevated temperatures leads to the formation of volatile chlorides in a number of metals, some nonmetals, and various sulfides, arsenides, selenides, and tellurides. Some metals that react too violently with Cl_2 are better reacted with gaseous HCl (e.g. Al, Be); furthermore, CCl_4, C_3Cl_8 or Cl_2 + C can be used to convert oxides such as TiO_2, Nb_2O_5, Ta_2O_5 and others into the likewise volatile chlorides (cf. Tab. 16.1).

These reactions can simplify the analysis of some complexly composed sulfide and arsenide ores (e.g., pale ores) and some alloys. Chlorination processes have also found application in the isolation of inclusions in metals and alloys.

Tab. 16.1: Sublimation under chlorination (examples).

Analysis sample	Reagent	Desublimated
Fe, Si, Ti, Ga, Sn, Sb, alloys	Cl_2	Metal chlorides
Al, Be, Sn	HCl	$AlCl_3$; $BeCl_2$; $SnCl_2$
P, As	Cl_2	PCl_5, $AsCl_3$
Sulfides, selenides, arsenides, tellurides	Cl_2; S_2Cl_2	S, As, Se, Te (as chlorides), metal chlorides
TiO_2, Nb_2O_5, Ta_2O_5	CCl_4; C_3Cl_8; Cl_2 + C	$TiCl_4$, $NbCl_5$, $TaCl_5$

Evaporation analysis. In so-called *evaporation analysis*, elements or compounds are expelled from solid samples with a gas stream at high temperatures and condensed on a cooled surface. Hydrogen is usually used as the propellant gas, by which As, Bi, Cd, Ge, Hg, In, Pb, Te, Tl, Sb, and Zn (as well as various low-volatility elements, which are not of interest in this context, however) are released from their compounds. Furthermore, oxides can be volatilized in the O_2 stream, sulfides in the H_2S stream, chlorides in the HCl gas stream, and others.

The apparatus is a quartz tube extended on one side to form a narrow nozzle, which holds the boat with the analysis sample (Fig. 16.4). At a short distance from the nozzle is a cooling finger, which can be covered with a thin cap made of aluminum. If the size ratios and the distance between the nozzle and the cooling finger are selected favorably, practically quantitative condensation of the evaporated substances can be achieved.

The condensate can be detached from the cooling finger or determined directly on it by spectroscopy.

The application range of the method lies mainly in the determination of traces of volatile metals in a non-volatile matrix; more rarely, compounds are volatilized (cf. Tab. 16.2). For the separation of fission products (As, Cd, Ge, Mo, Ru, Sb, Te, Zn) from irradiated uranium samples, an apparatus with induction heating usable up to 1500 °C was described (Fig. 16.4).

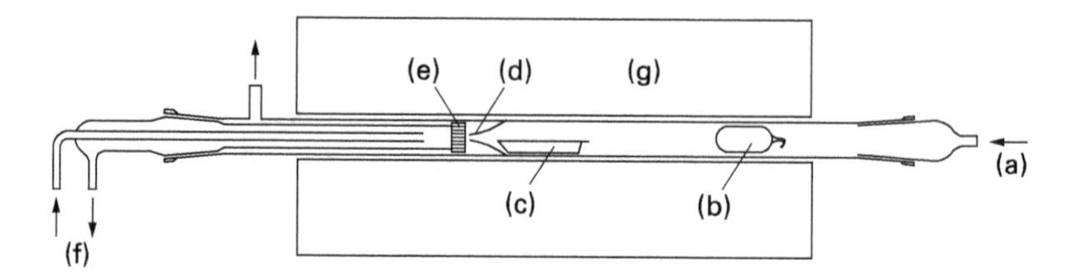

Fig. 16.4: Apparatus for evaporation analysis.
a) Gas feed; b) diffusion body; c) boat with sample; d) nozzle; e) cooling finger with cap; (f) cooling; (g) furnace.

Pyrohydrolysis. If solid inorganic analysis samples are treated with superheated water vapor at temperatures of about 800–1000 °C, various volatile acids are set free from their salts and separated.

The reactions can often be accelerated by the addition of non-volatile oxides (acid anhydrides) such as U_3O_8, WO_3, SiO_2 or V_2O_5; these combine with the bases formed by the hydrolysis reaction.

The apparatus used to perform analytical *pyrohydrolysis* consists of a boiling flask for the water, furnaces for superheating the steam and for heating the substance, and a cooler in which the volatilized components are precipitated together with the water vapor. Quartz is usually chosen as the material because of the high temperatures involved (Fig. 16.5).

Pyrohydrolysis is mainly used in the determination of fluorine in fluorides and silicates, e.g. glasses, ceramic bodies, ashes, etc.; it is a convenient and effective separation method for this element, but may fail if the fluorine is present only in traces, since residues may remain trapped inside solid particles. The method is also recommended for decomposing borides and removing boric acid from glasses; however, the temperature must be raised to 1100–1400 °C.

Tab. 16.2: Evaporation analysis (application examples).

Sample	Evaporated substance	Gas flow	Temperature (°C)	Time (min)	Additive compound
Burn off	Zn	H_2	1100	30	Coal
Pyrite, Cu ores	Zn	H_2	1100	30	C + Fe
Al alloy	Zn	H_2	1100	60	–
Ga, In	Zn	H_2	1000	30	–
Graphite, burn off	Cd	H_2	800	20–30	–
Copper	Pb	H_2	1150	20–35	–
Li-Glass	Tl	H_2	950–1000	30–50	–
Brownstone	Tl	O_2	950–1000	30–50	–
ZnO	Tl	O_2	950–1000	30–50	–
Bi, Pb, Zn	Tl	O_2	950–1000	30	Al_2O_3
Bauxite	Be	$N_2 + H_2O$	1050	40	CeF_3

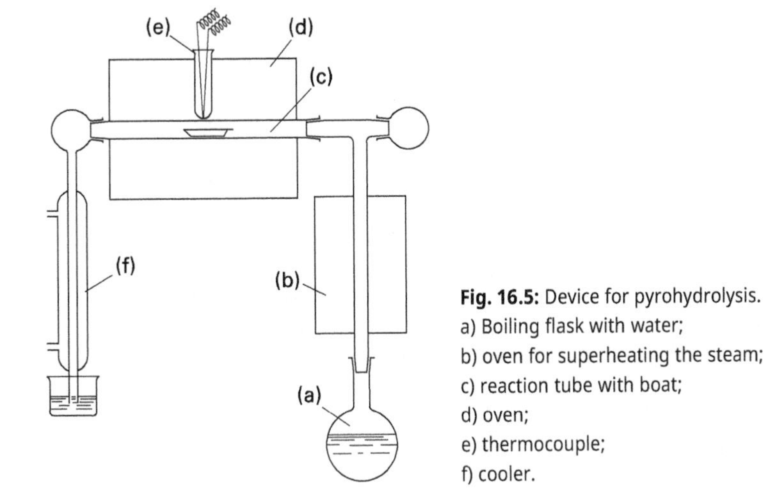

Fig. 16.5: Device for pyrohydrolysis.
a) Boiling flask with water;
b) oven for superheating the steam;
c) reaction tube with boat;
d) oven;
e) thermocouple;
f) cooler.

16.2.4 Efficacy and scope of the method

The sublimation process often allows the effective separation of individual volatile components from complicated mixtures; the method also frequently performs well in the enrichment of trace elements.

In inorganic analysis, the main application is probably in sublimation using a reacting gas, especially in the chlorination processes. In the field of organic analysis, sublimation methods are used to detect and isolate small amounts of volatile compounds; after condensation of the sublimate on a slide, microscopic examination of the crystals can often be used to identify amounts of substances in the µg range and below. Thin-layer chromatographically separated substances can also often be recovered pure in this way; the stain containing the compound to be detected is scraped off the plate and the substance and adsorbent are separated in a microsublimation apparatus (e.g., on a Kofler bench). Other applications include the isolation of natural products from plants and of toxic compounds from biological material.

If we disregard the sources of error already mentioned (entrainment effect and falling back of the sublimate), which can be eliminated by instrumental measures, the most important disturbance in all sublimation methods is that the volatilization need not be quantitative, because the subliming substance must diffuse out of the interior of solid granules; this process can take a more or less long time, depending on the temperature and size of the particles. Therefore, the analytical sample should be very finely powdered for all sublimations.

General literature

H. Brinkmann, C. Kelting, S. Makarov, O. Tsaryova, G. Schnurpfeil, D. Woehrle & D. Schlettwein, Fluorinated phthalocyanines as molecular semiconductor thin films, in: C. Woell, Organic Electronics, John Wiley / VCH Weinheim 37–60 (2009).

J. Green, Sublimation (Book Chapter), in: Encyclopedia of Analytical Science, pp. 410–415 (2004).

M. Gronbach, L. Kraußer, T. Broese, C. Oppermann & U. Kragl, Sublimation for enrichment and identification of marker compounds in fruits, Food Analytical Methods *14*, 1087–1098 (2021).

A. Gupta, Short review on controlled nucleation, International Journal of Drug Development & Research *4*, 35–40 (2012).

J. Han, D. Nelson, E. Sorochinsky & V. Soloshonok, Self-disproportionation of enantiomers via sublimation; new and truly green dimension in optical purification, Current Organic Synthesis 8, 310–317 (2011).

J.-L. Pi, C.-F. You & C.-H. Chung, Micro-sublimation separation of boron in rock samples for isotopic measurement by MC-ICPMS Journal of Analytical Atomic Spectrometry *29*, 861–867 (2014).

Literature to the text

Error during sublimation

R. Rouseff, Analytical methods to determine volatile sulfur compounds in foods and beverages, ACS Symposium Series *826* (Heteroatomic Aroma Compounds), 2–24 (2002).

Freeze drying (lyophilization)

G. Adams, The principles of freeze-drying, Methods in Molecular Biology 368 (Cryopreservation and Freeze-Drying Protocols), 15–38 (2007).

A. Gupta, Annealing is one of the best method for rapid drying in lyophilization technique, Internationale Pharmaceutica Sciencia *2*, 78–84 (2012).

Y.-F. Maa & R. Costantino, Spray freeze-drying of biopharmaceuticals: applications and stability considerations, Biotechnology: Pharmaceutical Aspects *2* (Lyophilization of Biopharmaceuticals), 519–561 (2004).
C. Ó'Fágáin, Storage and lyophilisation of pure proteins, Methods in Molecular Biology *681*, 179–202 (2011).
L. Qian & H. Zhang, Controlled freezing and freeze drying: a versatile route for porous and micro-/nano-structured materials, Journal of Chemical Technology and Biotechnology *86*, 172–184 (2011).
I. Roy & M. Gupta, Freeze-drying of proteins: some emerging concerns, Biotechnology and Applied Biochemistry *39*, 165–177 (2004).
S. Shukla, Freeze drying process: a review, International Journal of Pharmaceutical Sciences and Research *2*, 3061–3068 (2011).

Pyrohydrolysis

F. Antes, M. Dos Santos, R. Loureno Guimarães, (. . .), E. Moraes Flores & V. Dressler, Heavy crude oil sample preparation by pyrohydrolysis for further chlorine determination, Analytical Methods *3*, 288–293 (2011).
S. Jeyakumar, V. Raut, and K. Ramakumar, Simultaneous determination of trace amounts of borate, chloride and fluoride in nuclear fuels employing ion chromatography (IC) after their extraction by pyrohydrolysis, Talanta *76*, 1246–1251 (2008).
P. Mello, J. Barin, F. Duarte, C. Bizzi, L. Diehl, E. Muller, I. Edson, and E. Flores, Analytical methods for the determination of halogens in bioanalytical sciences: A review, Analytical and Bioanalytical Chemistry *405*, 7615–7642 (2013).
R. Sawant, M. Mahajan, D. Shah, U. Thakur & K. Ramakumar, Pyrohydrolytic separation technique for fluoride and chloride from radioactive liquid wastes, Journal of Radioanalytical and Nuclear Chemistry *287*, 423–426 (2011).
C. Senger, K. Anschau, L. Baumann, (. . .), P. Mello & E. Muller, Eco-friendly sample preparation method for silicon carbide using pyrohydrolysis for subsequent determination of tungsten by ICP-MS, Microchemical Journal *171*, 106781 (2021).
D. Wu, H. Deng, W. Wang & H. Xiao, Catalytic spectrophotometric determination of iodine in coal by pyrohydrolysis decomposition, Analytica Chimica Acta *601*, 183–188 (2007).

Evaporation analysis

D. Allan, J. Daly, & J. Liggat, Thermal volatilization analysis of TDI-based flexible polyurethane foam, Polymer Degradation and Stability *98*, 535–541 (2013).
S. Chattopadhyay, H. Tobias & P. Ziemann, A method for measuring vapor pressures of low-volatility organic aerosol compounds using a thermal desorption particle beam mass spectrometer Analytical Chemistry *73*, 3797–3803 (2001).
X. Liu, J. Zhao, Y. Liu & R. Yang, Volatile components changes during thermal aging of nitrile rubber by flash evaporation of Py-GC/MS, Journal of Analytical and Applied Pyrolysis *113, 3367*, 193–201 (2015).
T. Matsueda, Y. Hanada, Y. Yao, T. Tanizaki, T. Kuroiwa, M. Moriguchi & K. Tobiishi, Thermo-desorption analysis of dioxins and related compounds in flue gas, Organohalogen Compounds *60*, 521–524 (2003).

17 Condensation

17.1 General

Condensation is the transition from the gas phase to the liquid or solid form; for condensations at low temperatures, the term *freezing out* is also found.

Condensation has already been mentioned as a component of distillation and sublimation processes; here we will discuss methods in which the actual separation occurs during condensation (rather than through the preceding volatilization).

Separations by condensation are mainly carried out by the method of single equilibration, and a method with systematic repetition in a separation tube has also been described.

17.2 Separations by one-time equilibrium adjustment

17.2.1 Freezing out of a component from a static gas volume – fractional condensation

Separations of gas mixtures can be achieved when there are large differences in the boiling points of the components by condensing the more volatile components. For example, small helium contents in neon can be determined by cooling a flask containing the gas mixture to the temperature of liquid hydrogen (20.4 K); the neon precipitates in solid form, in which it has only a low vapor pressure, and the helium is enriched in the gas phase.

In the presence of gas mixtures with several components, fractional condensation can possibly be carried out by freezing out at different temperatures.

17.2.2 Condensation in the temperature gradient

Fractional condensation is more often carried out in a tube provided with a temperature gradient, at the hot end of which the mixture of substances to be separated is in solid or liquid form. Depending on the condensation temperatures, the various components precipitate in individual zones at a greater or lesser distance from the hot end.

The separation efficiency of such processes is quite low; although the initial part of the condensate is sharply defined, after the colder side the condensate smears over a longer distance, so that zones which are not very far apart tend to overlap (cf. Fig. 17.1). The method is therefore only suitable for separating substances with large vapor pressure differences.

https://doi.org/10.1515/9783111181417-017

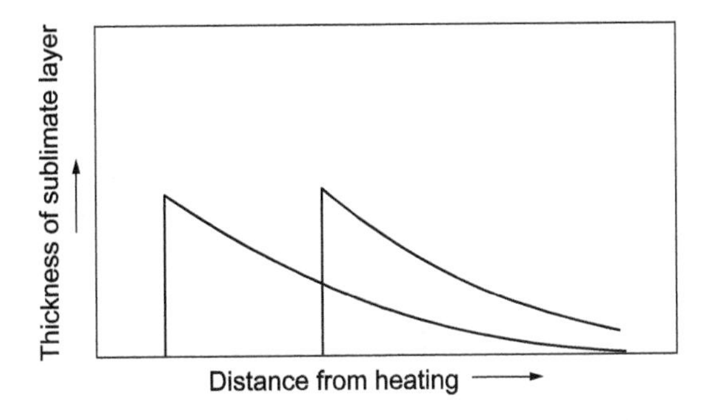

Fig. 17.1: Condensation of two substances in a temperature gradient.

As an example of application, let us show the separation of some metal chlorides volatilized in a quartz tube at about 1000 °C; this is surrounded by a copper jacket which is cooled at one end (Fig. 17.2). The chlorides of iridium, platinum and tungsten settle in well-separated zones.

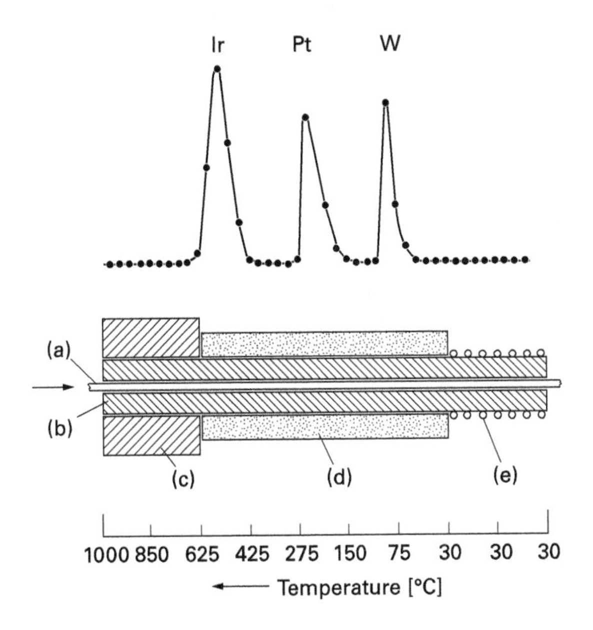

Fig. 17.2: Condensation in the temperature gradient.
Top: Separation result.
Below: Apparatus.
a) Quartz tube; b) copper tube; c) furnace; d) insulation; e) cooling coil.

17.2.3 Freezing out components from a gas stream

To remove condensable components from flowing gases, cold traps are used which are described in various embodiments. The simplest are U-tubes and cooling coils which are immersed in a cold bath (cf. Fig. 15.19). Since all condensation processes involve the formation of aerosols which are difficult to knock down, efforts have been made to improve the effectiveness of cold traps by filter inserts, filling with adsorbents, circular guidance of the gas flow, baffles or by extended spaces with lower flow velocity, etc.

The difficulty of quantitative separation by condensation increases with decreasing concentration of the substance in the gas.

As already mentioned in the section on *Gas chromatography*, the quantitative condensation of trace components is a task which has not yet been solved satisfactorily; by filling the cold traps with adsorbents, separations of about 90–95% are achieved. The most effective method has proved to be to condense the entire carrier gas together with the components it contains in the analysis sample; however, this means that the originally intended separation of gas and condensate is abandoned and the method is limited to the special case of gas chromatography. Similar in principle are the methods already mentioned, in which a condensable auxiliary substance is mixed with the gas stream.

17.3 Systematic repetition of condensations

Condensation in a temperature gradient can be extended to the method of systematic repetition in the following way: place the boat with the analysis substance in a longer tube through which an inert gas flows and draw a relatively short temperature gradient in the direction of flow along the tube (Fig. 17.3).

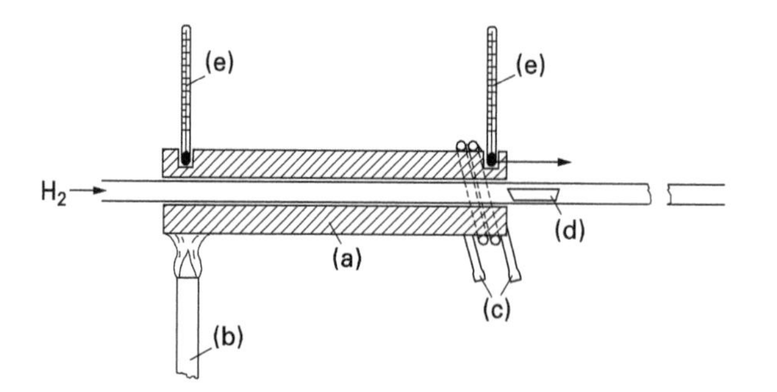

Fig. 17.3: Systematic repetition of condensations (*horizontal distillation*).
a) Sliding aluminum body; b) Bunsen burner; c) cooling coil; d) boat with substance; e) thermometer.

When the sample enters a sufficiently hot zone of the furnace, it is vaporized, and the vaporized portions settle in various colder parts of the tube depending on the condensation temperature, are vaporized again as the temperature gradient advances, and so on.

With this arrangement developed by Jantzen, which is called *horizontal distillation* but will be counted here among the condensation methods, the formation of sharper zones with higher purity is obtained than with simple condensation in the temperature gradient.

Literature to the text

J. Christian, D. Petti, R. Kirkham, & R. Bennett, Advances in sublimation separation of technetium fromlow-specific-activity molybdenum-99, Industrial & Engineering Chemistry Research *39*, 3157–3168 (2000).

H. Eun, Y. Cho, T. Lee, (. . .), G. Park & H. Lee, An improvement study on the closed chamber distillation system for recovery of renewable salts from salt wastes containing radioactive rare earth compounds, Journal of Radioanalytical and Nuclear Chemistry *295*, 345–350 (2013).

E. Karlsson, J. Neuhausen, R. Eichler, (. . .), A. Vögele & A. Türler, Thermochromatographic behavior of iodine in 316L stainless steel columns when evaporated from lead–bismuth eutectic, Journal of Radioanalytical and Nuclear Chemistry *328*, 691–699 (2021).

B. Zhuikov, Kinetic Approach to calculating the peak shape in thermochromatography of ultra small amounts of substances, Radiochemistry *63*, 446–453 (2021).

I. Zvára, Vacuum thermochromatography: Diffusion approximation, evaluation of desorption energy, Journal of Radioanalytical and Nuclear Chemistry *299*, 1847–1857 (2014).

Part III: Separations due to different migration rates in one phase

18 Introduction

The complete separation of two substances by different migration rates shown in the introduction to Part 1 (section 2.2), like the complete separation by distribution between two immiscible phases, represents an ideal case which cannot be realized in practice. The zones of the migrating compounds basically widen in the direction of movement, so that a more or less strong overlapping occurs.

The effectiveness of such a separation process therefore depends, on the one hand, on the length of the separation distance and the difference in migration velocities, and, on the other hand, on the extent of the disturbing zone broadening. Here, too, a separation factor can be defined: One cuts out of the separation distance the sections with substances A and B and determines in each of these sections the quantity ratio m_A/m_B. The separation factor β is then

$$\beta = \frac{m_\mathrm{A}/m_\mathrm{B}\,\text{in section 1}}{m_\mathrm{A}/m_\mathrm{B}\,\text{in section 2}}. \tag{1}$$

The migration of originally stationary particles must be caused by a force which accelerates the masses concerned. A constant velocity is then either achieved by ending the action of the force, or it results from the action of a counterforce which increases with the velocity (e.g. friction).

In the case of separations due to different migration rates, the following procedures can be distinguished:

- One-dimensional methods;
- countercurrent method and
- two-dimensional processes.

The one-dimensional processes are given by the migration in a gaseous or liquid medium in *one* direction (cf. 1st part, Fig. 2.2).

Countercurrent processes occur when the migration of the particles takes place in a liquid flowing in the opposite direction to the direction of migration. This has the same effect on the separation as an extension of a given separation distance.

In the two-dimensional methods, two crossed force fields (simultaneously or successively) act on the particles. Like the crossing of two immiscible phases, these methods permit the continuous separation of substance mixtures, even if they contain more than two components (cf. Fig 18.1). The migration velocities need only be different in one direction of movement.

Similarly, two-dimensional separations result when the particles are entrained in a moving medium in one direction and a force field acts perpendicular (orthogonal) to the direction of motion.

https://doi.org/10.1515/9783111181417-018

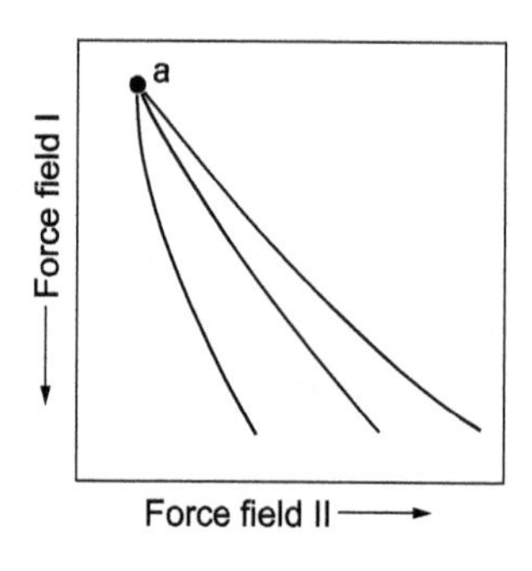

Fig. 18.1: Continuous separation in crossed force fields. a = substance introduction.

19 Migration of charge carriers in electric and magnetic fields (mass spectrometry)

19.1 Historical development

The separation of gaseous ions by electric fields was first achieved by Thompson (1911). The method was substantially improved by Dempster (1918) and by Aston (1919) and used to determine the isotopic composition of numerous elements. Another important advance was made by Mattauch and Herzog (1934) by developing the so-called *double-focusing* instruments, and finally *time-of-flight mass spectrometry* was elaborated by Stephens (1946) and by Cameron and Eggers (1948). Highest resolution mass spectrometers based on *FT ion cyclotron resonance* in an ion trap were developed by Carnisarow and Marshall (1974).

Preparative mass spectrometry gained particular importance through the parallel use of hundreds of calutron mass spectrometers during the Second World War in the so-called *Manhattan* project (production of the atomic bomb in the USA). This was the first time that ^{235}U was separated from ^{238}U by preparative methods.

Today, mass spectrometry represents the most important technique for the highly selective analytical separation of complex elemental and molecular ions. In combination with sensitive detection techniques, the identification of 10^{-15} g (absolute) is routinely achieved. Combined with a preceding chromatographic separation technique (e.g. GC, LC), it is currently the most effective method for separation, identification and quantification. Currently, the use of isotope-labeled analytes for traceable calibration or the measurement of natural isotope distributions is gaining increasing importance.

19.2 General

19.2.1 Definitions

Mass spectrometry is the term used to describe a process in which ions are accelerated in a gaseous state and separated according to their masses (more precisely: according to their mass/charge ratios) by suitable arrangements of electric and magnetic fields. The separated ions are either collected on a detector plate and detected here or measured electrically as an ion current. In the former case, one speaks of *mass spectrography*, in the latter (which is of primary importance for analytical chemistry) of *mass spectrometry*. If the substance to be investigated is in solid, not readily vaporizable form, the method is called *solid-state mass spectrometry*.

https://doi.org/10.1515/9783111181417-019

19.2.2 Principle structure of a mass spectrometer

According to the above, a mass spectrometer basically consists of three parts: First of all, the device for generating the ions, the so-called *ion source*, then the separating device and finally the receiver with the auxiliary means for detecting and quantitatively measuring the ion flow.

In the following, only the most important ionization and separation processes will be discussed.

19.2.3 Ionization process

Of the numerous methods for ionizing gases, only those most commonly used in analytical applications of mass spectrometry are described here; these are electron impact ionization, field ionization, and ionization in an electric spark, electron plasma, or by electrostatic discharging. Selective ionization is achieved by photoionization processes.

Electron impact (ES) – Ionization. Electrons can be knocked out of gas molecules hit by an electron beam at low pressure, forming positive ions. The experimental procedure is shown in principle in Fig. 19.1: An electron beam c emerges from the filament a through the aperture b into the collision space, where it meets a fine gas stream supplied from a nozzle. The electrons leave the shock chamber through a lower aperture and are collected in the cage d. One directs the formed gas ions out of the collision space by a weak *pulling voltage* between the plates e_1 (negatively charged) and e_2 (positive) and then accelerates them against the entrance slit of the spectrometer by a voltage of a few thousand volts applied to the plate f. The gas ions are then accelerated to the entrance slit of the spectrometer.

The gas pressure in the shock chamber is about 10^{-5} Torr, and ion currents of about 10^{-8} to 10^{-9} A are achieved with such arrays. Non-ionized atoms and molecules must be removed by continuous pump-down.

The substance feed through a nozzle shown in Fig. 19.1 is suitable for gases and relatively easily vaporizable liquids. In front of the nozzle is a reservoir in which a gas pressure of 0.01–1 Torr is maintained. The reservoir can usually be heated up to about 350 °C, so that sufficient pressure can be achieved even in the presence of compounds with higher boiling points.

Even higher boiling compounds are heated in a small oven placed directly in front of the joint space.

An important variable in electron impact ionization is the voltage between the filament and the collector, which determines the energy of the ionizing electrons. With sufficient energy, on the one hand several ions can be formed from a given electron, and on the other hand a gas atom or molecule can be ionized several times.

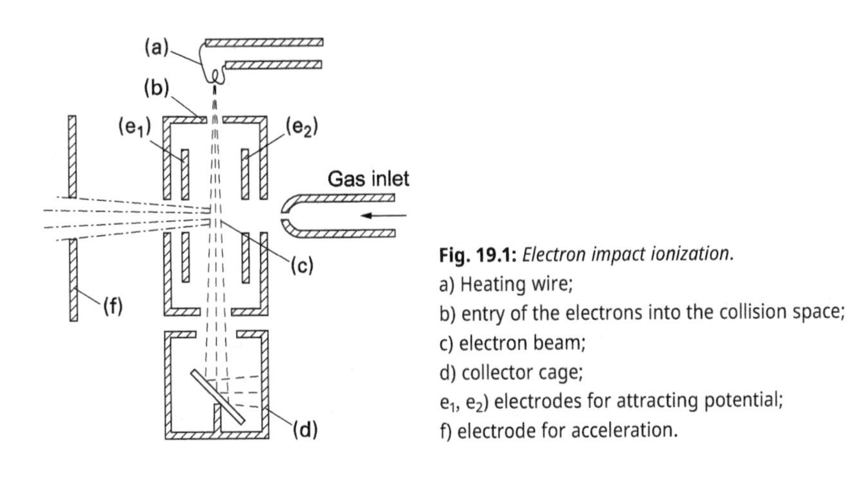

Fig. 19.1: *Electron impact ionization.*
a) Heating wire;
b) entry of the electrons into the collision space;
c) electron beam;
d) collector cage;
e_1, e_2) electrodes for attracting potential;
f) electrode for acceleration.

The number of pairs of charge carriers produced by an electron on a path of 1 cm in the gas is called *differential ionization*; in order to be able to compare this in different gases, it is related to a gas pressure of 1 Torr. Figure 19.2 shows the differential ionization of mercury as a function of the electron energy.

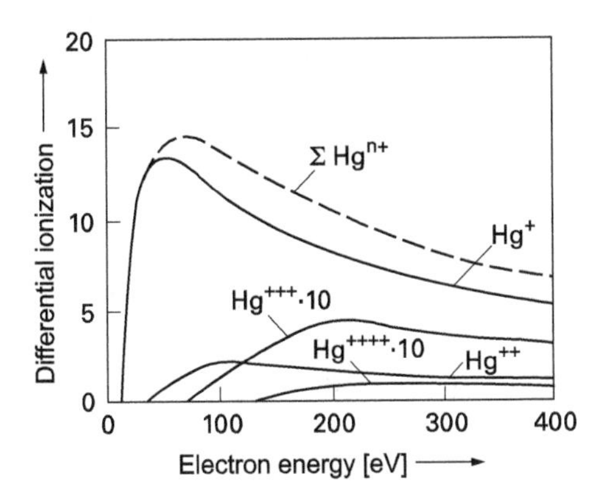

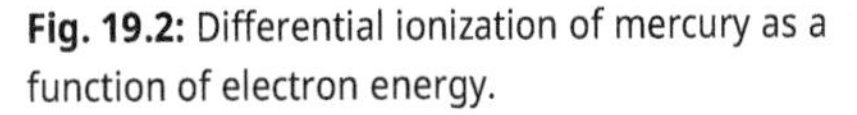

Fig. 19.2: Differential ionization of mercury as a function of electron energy.

As you can see, a certain minimum energy of the ionizing electrons is required for each ion before it can appear. This potential is therefore called the *appearance potential*. With increasing electron energy, the differential ionization increases to a maximum and then slowly decreases again. The maximum for most ions is between about 50 and 100 eV; one therefore usually excites with energies of about 70 eV.

Figure 19.2 further shows that multiple positively charged ions occur relatively rarely, and the higher the charge, the less.

Electron impact ionization can be carried out not only with atoms such as Hg or Ar, etc., but also – which is of much greater importance for analytical chemistry – with inorganic and organic molecules. In this process, the latter usually not only give ions corresponding to the unchanged molecule (except for the loss of one charge), but are additionally split into a more or less large number of fragments; this process is called *fragmentation*. The masses of the fragments and the relative amounts in which they are formed are readily reproducible under constant ionization conditions, so they can be used to characterize the compounds in question. In the case of larger organic molecules, the mass of the molecular ion often does not occur at all, but only fragments are formed; in such cases, mass spectrometric molecular weight determination is not readily possible.

Fast Atom Bombardment (FAB) ionization. If high-energy (approx. 16 keV) ionized atoms (e.g. Cs^+) are used instead of electrons, *soft ionization* of complex molecules is possible. However, the substance to be charged must be in a low-volatility matrix (e.g. nitrobenzyl alcohol, glycerol or thioglycerol). The analyte molecules are torn from the surface and charged by secondary ion formation.

Inductively coupled plasma (ICP) ionization. *ICP* as an ionization technique has gained acceptance in recent years, especially in elemental analysis and more recently for nanoparticle characterization. The reason for this is the efficient ion formation in a gas ion plasma. In this process, argon, for example, is heated to (electronic) temperatures up to 12000 K in a plasma formation excited by high frequency (free electrons and Ar^+). The Ar ion plasma is combined with the sample in a collision chamber in the form of an aerosol. In this process, the energy is transferred to the submicron sample particles and they are ionized. Sample aerosol formation can also be accomplished by laser ablation. This provides the possibility of scanning samples and creating two-dimensional distributions.

Secondary ion beam (SI) ionization. If lighter element or cluster ions (O_2^+, Cs^+, Ga^+, Ar^+, Bi^+; SF_5^+, Au_3^+, Bi_3^+, Bi_2^{3+}) with an energy of 0.2–25 keV are used for bombardment, neutral, positively and negatively charged particles are produced. The neutral particles are lost to the SI for analysis. This technique is particularly used for planar ablation of material surfaces and subsequent mass spectrometric depth profile analysis (*SIMS*).

Chemical ionization (CI). If a molecular weight determination is desired, it is possible to convert reactant gases (e.g. methane, or higher alkanes, ammonia, water) mixed in excess to the analyte molecules in the electron beam (150 eV) to transiently unstable protonation species:

$$CH_4 + e^- \longrightarrow \cdot CH_4^+ + 2e^- \longrightarrow CH_3^+ + H\cdot$$
$$\cdot CH_4^+ + CH_4 \longrightarrow CH_5^+ + \cdot CH_3$$
$$\cdot CH_4^+ + CH_4 \longrightarrow C_2H_5^+ + H_2 + H\cdot$$

The methyl radical cations formed react with methane in the plasma to form unstable carbonium ions CH_5^+ resp. $C_2H_5^+$. These can serve to protonate an analyte molecule.

In the next step, positively charged analyte molecule ions are formed simply by charge exchange.

$$M + CH_5^+ \longrightarrow CH_4 + [M+H]^+$$
$$AH + CH_3^+ \longrightarrow CH_4 + A^+$$

Since the proton affinity decreases in the sequence $CH_5^+ > C_2H_5^+ > H_3O^+ > C_4H_9^+ > NH_4^+$ decreases, only little energy is released during proton transfer and therefore molecular fragmentation is avoided. The ions formed in this process are dominated by the molecular ion, allowing molar mass determination. The extent of fragmentation can be controlled by the choice of gases and pressure conditions.

Negative chemical ionization (NICI). The thermalized electrons produced by the interaction of the reactant gas molecules with the accelerated electrons can be attached to halogen-containing molecules, and thus to compounds with high electron affinity. This effect occurs during ionization with low-energy electrons (0.1–1 eV), where the energy must be in a narrow range that depends on the type of ionized molecule.

Another variant is ionization at atmospheric pressure (*atmospheric pressure ionization, APCI).* In this process, nitrogen is charged unipolar to the gaseous analyte molecules in a corona. The nitrogen radical cation formed in this process can form both positively and negatively charged fragment ions through further gas phase reactions.

Electrospray Ionization (ESI). *Electrospray ionization* is now widely used. This technique is particularly suitable in combination with liquid chromatography. A conductive eluent is forced through a capillary, which is opposed by a corona needle as an electrode. A cone of liquid is formed (so-called *Taylor cone*), which is dominated at the tip by charges of the same name in the liquid near the corona needle. This leads to the spontaneous formation of a finely dispersed aerosol of running medium with analyte. Due to the Kelvin effect, the volatile fraction evaporates abruptly forming nm-sized ion clusters. The composition of the running medium as well as the polarity of the corona needle with applied high voltage in the higher kV range determines the fragmentation. In general, only low fragmentation is achieved. Fenn received the Nobel Prize for this in 2002.

Desorption Electrospray Ionization (DESI). The use of the charged aerosol jet for surface desorption of molecules adsorbed there leads to equally gentle formation of charged analyte ions and goes back to work by Cooks (2004).

Field Ionization (FI). As shown by Müller (1951), molecules can form positive ions in extremely strong electric fields with the loss of an electron. The required field strengths are obtained by opposing a very fine tip of about 0.1 μm radius and a plate a few mm apart; apply voltages of about 10 kV to the two parts, obtaining field strengths of the order of 10^8 V/cm.

Since ion yields are low with this arrangement, a fine wire or cutting edge is used in place of the tip, according to a suggestion by Beckey (1962).

Compared with electron impact ionization, *field ionization* has the advantage that, with suitable field strength, even complicated organic molecules give only a few or no fragments; the ion corresponding to the molecular mass accordingly occurs with high intensity. The method is therefore particularly suitable for molecular weight determination.

Matrix-assisted laser desorption ionization (MALDI). When high-intensity laser light is used for sample ionization, effective photon absorption is critical for ion formation. In this process, the analyte molecules are previously mixed with a crystalline material (the matrix, e.g., sinapic acid, 2,5-dihydroxybenzoic acid, α-cyanohydroxycinnamic acid) that absorbs well at the laser wavelength. Usually, a nitrogen laser with a wavelength of 337 nm is used. The processes involved are complex. Due to the intense energy transfer of the laser pulse (a few ns pulse duration), the matrix at the surface is abruptly transformed into an ultra-fine charged aerosol, which partially loses the matrix and leaves a charged analyte molecule. Particularly gentle ionizations of large biomolecules are thus possible.

Spark Ionization. High-melting inorganic compounds for which the vapor pressure of approx. 10^{-5} Torr required in the ion source is practically unattainable can be ionized by sparking. The substance is put into the form of thin rods, the mass is made electrically conductive – if necessary – by adding graphite, and a spark is made between the electrodes in the ion source in a high vacuum. This requires very high voltages. Sometimes a strong laser pulse can be applied for sparking by multi-discharging processes.

This method can be used to ionize practically all inorganic substances; more highly charged ions are also formed to a greater extent. A disadvantage is that the spark discharges are not constant over time and that the resulting ions have very different kinetic energies, which makes it difficult to separate them by mass (see below).

Photoionization (PI). Gaseous components can be converted into gas ions by the process of *photoionization* (< 20 eV). A distinction is made between *single-photon ionization (SPI)* and *resonance-enhanced multiphoton ionization (REMPI).* Photoionization allows soft, subtance-selective ionization with adequate wavelength selection. REMPI requires appropriately tunable laser light sources. The high photon density leads to nonlinear photon absorption and ionization.

19.2.4 Acceleration of ions in the electric field – energy dispersion

The basis of the separation of ions of different mass in the mass spectrometer is their different acceleration in the electric field. If an ion with the charge e (e = elementary charge = 1.60 · 10^{-19} coulomb) passes through a potential V, it absorbs the energy $e \cdot V$, *which* must be equal to the kinetic energy acquired:

$$e \cdot V = {}^{1}\!/_{2}\, m \cdot v^2, \tag{1}$$

or:

$$v = \sqrt{\frac{2\, e \cdot V}{m}}. \tag{1a}$$

According to eq. (1a), ions with the same charge but different masses as well as ions with different charges but the same masses are accelerated to different velocities. A special case is when charge and mass are doubled (or tripled etc.) at the same time. The ion with mass 2 m and charge 2e receives the same velocity when passing through a field as the ion with mass m and charge e; the ratio of mass to charge (the so-called *m/e ratio*) is decisive. At the accelerating voltages of about 2000 V commonly used in mass spectrometry, speeds of about 10^4–10^5 m/s are achieved by singly charged ions of average mass.

Energy dispersion. According to eq. (1a), ions of the same charge and mass coming from an ion source should have exactly the same final velocities after passing through the accelerating field. In fact, however, this is not or only approximately the case; even in an ion beam consisting of a uniform type of ions, different velocities occur. This *energy dispersion* results on the one hand from the fact that the gas molecules enter the ion source at different velocities, and on the other hand from the fact that ionization takes place at locations with somewhat different potentials.

In electron impact and field ionization, the energy dispersion is so small compared to the total energy that it can normally be neglected. The ions produced in this way are thus practically monoenergetic. However, the situation is different in spark ionization; as already mentioned, ions with very different initial energies are produced here, so that the velocities after acceleration are quite non-uniform.

19.3 One-dimensional mass spectrometric separations (time-of-flight mass spectrometers)

Time-of-flight MS. In a *time-of-flight mass spectrometer (TOF-MS)*, an ion packet generated in an electron impact ion source is accelerated by an extremely short voltage impact and then allowed to pass through a field-free section of up to 1 m in length. Here the ions separate according to their velocities so that they enter the collector one after the other at different *m/e ratios* (Fig. 19.3).

An actual separation of the ion types striking at a time interval of about 10–100 µs is not possible; however, after amplification, the individual pulses can be made

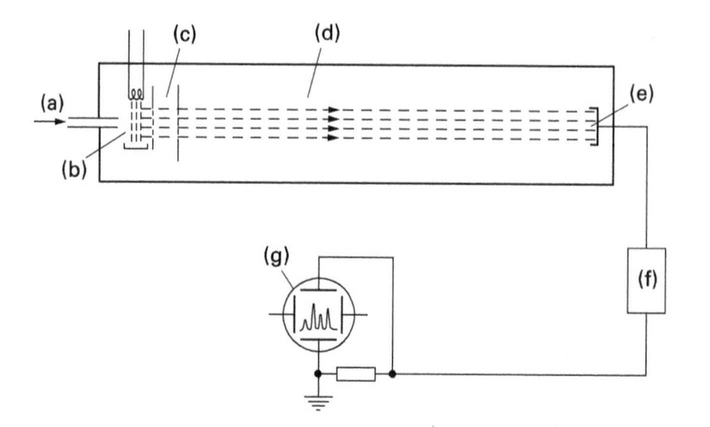

Fig. 19.3: Time-of-flight mass spectrometer (schematic).
(a) Gaseous sample inlet; (b) ion source; (c) acceleration path; (d) run path; (e) collector; (f) amplifier; (g) read out.

visible, e.g. as a standing image with the aid of an oscilloscope running synchronously with the acceleration pulses. The shorter the voltage pulses used for acceleration, the better the separation of ions of different masses.

Ion Mobility Spectrometry, IMS. *IMS* represents a largely similar separation and detection technique for charged gas ions. Already Langevin (1903) detected the presence of various gas ions in ambient air (at normal pressure and temperature), and shortly thereafter published the still valid considerations on the movement of gas ions in electric fields. Today, IMS is used as a fast gas sensor, but also increasingly as a possibility to separate chiral or complex biomolecules.

An essential feature of IMS is that the separation of charged gas ions, which are formed, for example, by impact ionization of neutral molecules with β-radiation or photoionization, takes place at ambient conditions. This eliminates the need for a vacuum.

However, the separation principle is analogous to TOF-MS in that the generated analyte gas ions are moved in a *drift tube* against a flowing gas in an electric field and thus reach the detector one after the other according to their electrical mobility. Figure 19.4 shows the basic structure of an IMS.

As can be seen, it consists of a reaction chamber to which the gaseous sample is fed, the subsequent drift chamber, and the ion detector. In the reaction chamber, charged analyte ions are generated from neutral analyte gas molecules by direct ionization or addition of charge carriers such as electrons or otherwise generated ion clusters. These are abruptly released into the drift space by reversing the polarity of a grid which separates the reaction space from the drift space. In the drift space, the analyte ions are accelerated by sequentially following ring electrodes to fly to an ion

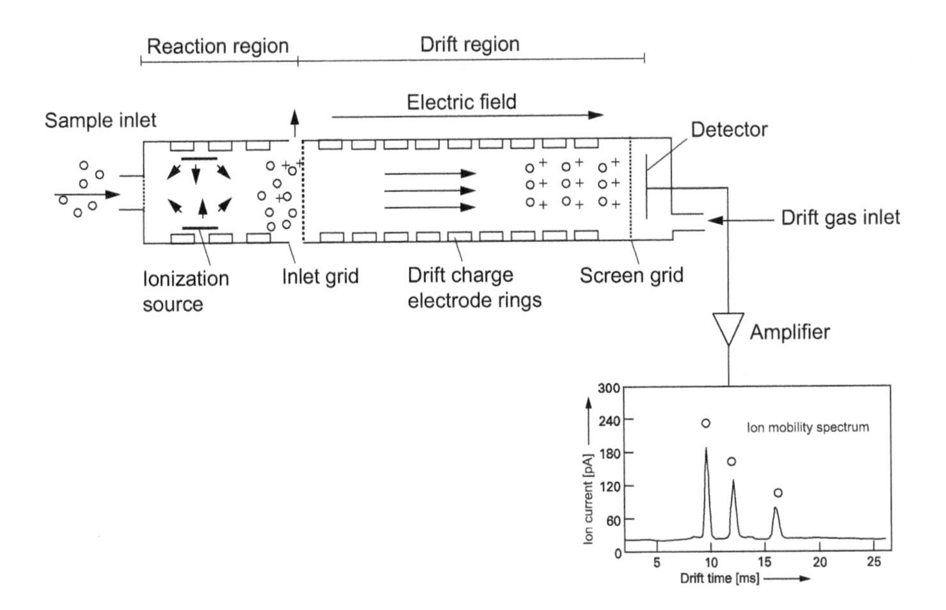

Fig. 19.4: Schematic diagram of an ion mobility spectrometer.

detector. This now registers the ions arriving one after the other as a function of their ion mobility. The separations that can be achieved are comparatively low, but the applicable geometry and the absence of a high-vacuum device allow the construction of compact IMS devices only a few centimeters in size. When appropriate drift conditions and evaluation algorithms are used, selected gas species can be separated and identified. Thus, IMS has been used in rapid warfare gas sensing since the 1970s.

19.4 Two-dimensional mass spectrometric separations

19.4.1 Separation of ion species in the magnetic sector field – single focusing mass spectrometers

If a monoenergetic ion beam passes through a magnetic sector field, ions with different *m/e ratios* are deflected to different degrees. At the same time, a so-called *directional focusing* occurs, in that a somewhat divergent ion bundle emerging from a slit is reunited behind the sector field in a point (or, when viewed spatially, in a line corresponding to the slit) (Fig. 19.5). Using terminology taken from optics, the process can be described as *imaging of the slit a at the points* b_1 *and* b_2.

The sector angle of the magnetic field can be chosen differently, e.g. instruments with 30°, 60° or 180° magnetic fields are in use. Figure 19.6 shows a schematic diagram of a mass spectrometer with a 180° sector, ion source and collector; the mass spectrum is obtained by continuously changing the strength of the magnetic field so that the individual ion types enter the collector one after the other.

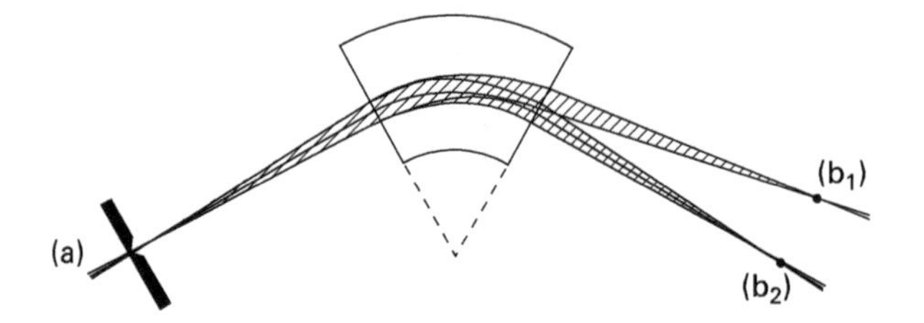

Fig. 19.5: Behavior of a monoenergetic ion beam with two ion types in the magnetic sector field. (a) Entrance slit; (b_1, b_2) focusing sites of the two ion species.

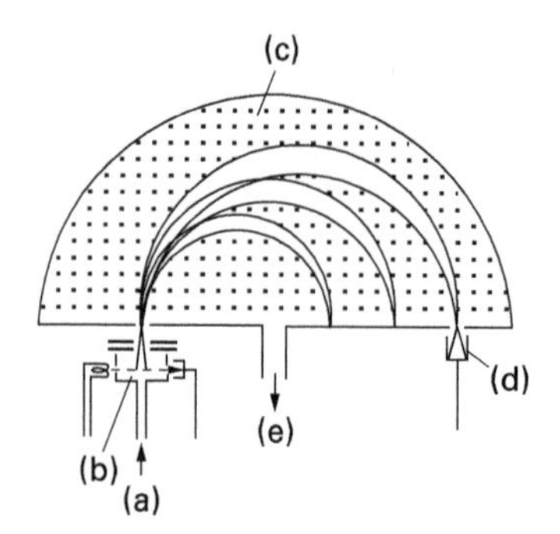

Fig. 19.6: Mass spectrometer with 180° magnetic field.
a) Gas inlet;
b) ion source;
c) magnetic field (field lines perpendicular to the paper plane);
d) collector;
e) pump connection.

19.4.2 Behavior of ion species in the electric sector field – double focusing mass spectrometers

If the ion beam entering the mass spectrometer from the ion source is not monoenergetic, the single focusing instruments with magnetic sector field fail; the slit is then not sharply imaged. However, by additionally using an electric sector field, focusing can be achieved even in such cases.

The behavior of a somewhat divergent ion beam containing ions of different energy when passing through an electric sector field is shown in Fig. 19.7: Ions of the same energy are deflected by the same amounts, even if their masses are different; on the other hand, ions of different energy are deflected differently (the larger the energy, the smaller the deflection). At the same time, a directional focusing of the divergent ion beam is obtained here as well. Ions that get a little closer to the positive plate of the capacitor are slowed down by the field and then deflected more strongly, and ions that get closer to the negative plate are accelerated and deflected more weakly.

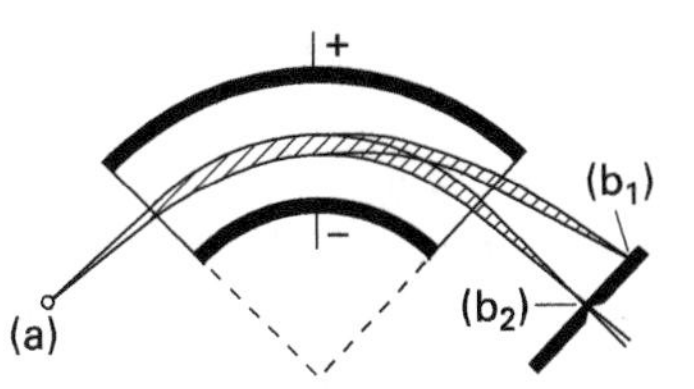

Fig. 19.7: Behavior of an ion beam with ions of different energy in the electric sector field.
(a) Entrance slit; (b_1, b_2) focusing sites of ions of different energy.

Thus, with the help of an *electric sector field*, one could first separate ions according to their energy, blank out a monoenergetic beam behind the field, and finally separate this beam according to masses (or *m/e ratios*) in a *magnetic sector field*. However, such a device cannot be realized for intensity reasons; after a considerable part of the ion stream is already lost at the entrance slit, a necessarily narrow aperture between the two fields again eliminates most of the ions, so that the fraction finally reaching the catcher is too small for evaluation.

Only Mattauch and Herzog succeeded in achieving an increase in intensity by a trick. If the sector angles of the two fields are adjusted to each other in a certain way, a relatively wide intermediate aperture can be used, so that the ion loss becomes bearable. For a magnetic sector of 90°, the sector angle of the electric field is 31.49° (cf. Fig. 19.8).

Mass spectrometers of this type are *double-focusing* in the sense that they provide directional focusing of the divergent ion beam on the one hand and focusing of ions with the same *m/e ratio* but different velocities on the other (energy focusing).

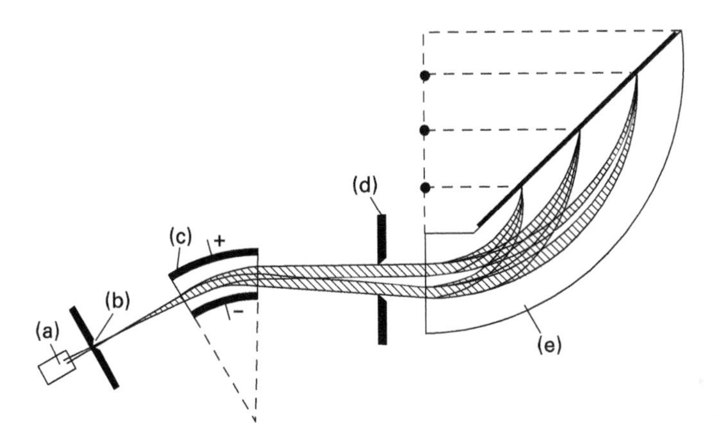

Fig. 19.8: Double-focusing mass spectrograph.
(a) Ion source; (b) entrance slit; (c) electric sector field; (d) intermediate aperture; (e) magnetic sector field.

19.4.3 Quadrupole mass spectrometer

In the so-called *quadrupole mass spectrometers*, the separation of ions is achieved by oscillations perpendicular to the direction of motion. Four round rod electrodes are arranged parallel to each other and a DC voltage of the same sign is applied to two opposite rods (Fig. 19.9). Furthermore, a high-frequency alternating voltage is superimposed on the direct voltage.

If one shoots a fine ion beam parallel to the longitudinal direction into the interior space between the rods, the ions are excited by the effect of the high-frequency field into oscillations whose amplitudes increase more and more as they travel through the system until the ions strike one of the rods and are thus eliminated from the beam. Only for ions of a certain mass does the amplitude of oscillation remain small enough to travel through the entire longitudinal space between the rods and be measured in a catcher at the other end. The system thus acts as a mass filter. By changing the values of the DC and AC voltages, the mass spectrum can be traversed.

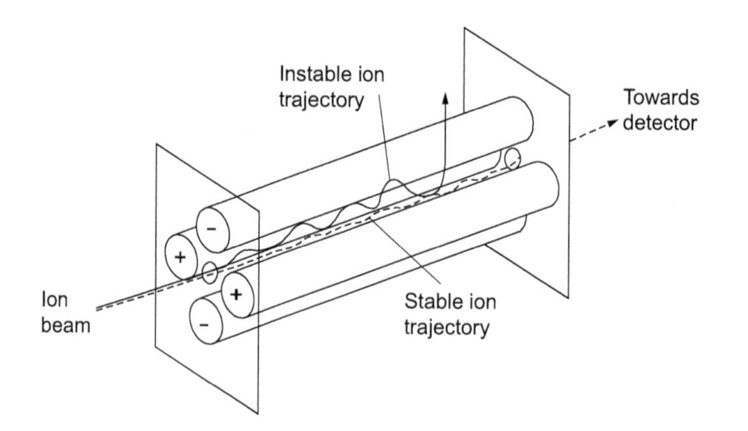

Fig. 19.9: Quadrupole mass spectrometer (principle).

19.4.4 Ion traps – mass spectrometry

Ion traps were used as early as the 1950s to study gaseous ions enclosed in electromagnetic fields as if in a cage. Wolfgang Paul received the Nobel Prize for this in 1989. In the 1980s, it was then possible to selectively release and record defined ions trapped in a trap. This paved the way for ion trap MS.

Fourier transform ion cyclotron resonance MS; FT-ICR-MS. Here, a homogeneous magnetic field is used in the ion trap, which forces the ions to orbits with a mass-dependent orbital frequency. The working equation for this can be found in the basic equations for centripetal force and Lorentz force:

$$mv^2 = e\,v\,B \tag{2}$$

expressed as the angular velocity of the accelerated analyte ion, this results in

$$\omega = v/r = eB/m \tag{3}$$

Figure 19.10 illustrates the structure.

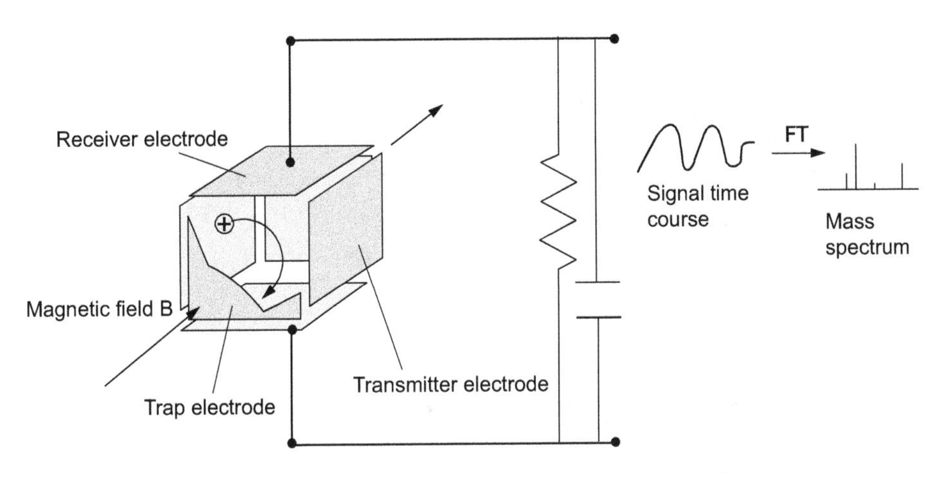

Fig. 19.10: Schematic structure of a cubic ICR ion trap.

The analyte ions are first brought into phase with an excitation pulse. By means of an alternating electric field perpendicular to the magnetic field, a *cyclotron resonance* is generated. If the frequency of the irradiated alternating field and the cyclotron angular frequency of the accelerated ion mass coincide, resonance occurs and the cyclotron radius of the accelerated analyte ion increases by absorbing energy from the alternating field. These changes induce signals at the detector plates of the ion trap. To detect ions with different masses, the irradiated alternating field is varied and the measured signal is Fourier transformed. *FT-ICR-MS* instruments thus achieve enormous mass resolutions. The resolution of FT-ICR-MS also increases with the intensity of the magnetic field (up to 15 Tesla), which is nowadays generated by means of superconducting magnets. The resolution can be up to R = 2,000,000.

Orbitrap – Mass Spectrometry. The *Orbitrap* configuration goes back to Makarov (1999) and is shown in Fig. 19.11.

The principle can be traced back to the work of Kingdon (1923). The gaseous analyte ions to be separated are introduced into a rotationally symmetrical spindle-shaped cell. The electrostatic field applied in this process forces the decentrally injected ions to move in elliptical circular motions, which are, however, electrostatically manipulated at a certain point so that they adopt the same axial frequency but experience different rotational velocities. This results in different orbital positions within the cell for ions of different m/e ratios. These induce voltage signals in the plate-like isolated wall segments that are used for quantitative analysis. The difference with the FT-ICR technique is thus the exclusive use of an electrostatic field, which is reflected in the weight of an Orbitrap system. Commercial benchtop instruments with resolutions of R in the order of 300,000 are now available.

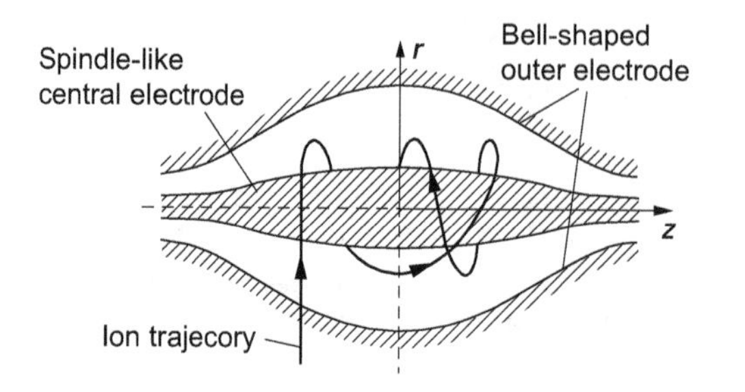

Fig. 19.11: Illustration of ion trajectories in an Orbitrap ion trap.

19.4.5 Continuous mode of operation – differential mobility analyzer

Continuous separation of charged particles is achieved by applying the orthogonal force field principle. This was first developed by Knutson and Whitby (1975) for the continuous production of monodisperse aerosols.

Differential mobility analyzer. The *differential mobility analyzer (DMA)* is used for the continuous production of monodisperse, electrically charged particles in the size range from about 2–600 nm. The production of monodisperse (= uniform particle size d_p) particles is of importance for calibration purposes of aerosol collection systems. In the meantime, however, the determination of the size distribution of polydisperse ultra-fine dusts, such as those produced in combustion processes, by means of the DMA technique has also become widespread. The structure of a DMA is shown in Fig. 19.12.

In principle, the DMA is a tubular capacitor through which polydisperse, singly charged (n = +1 or −1) aerosol particles continuously flow. The inner part of the capacitor consists of the particle-flowed annular gap, which is confined by a grounded tube outer wall and a coaxial central electrode. The flow through the vertically positioned separator is from top to bottom. The central electrode is held at a preselected electric potential so that an orthogonal electric force field E acts on the charged particles flowing downward. The horizontal particle velocity is

$$v = Z_p\, E \tag{4}$$

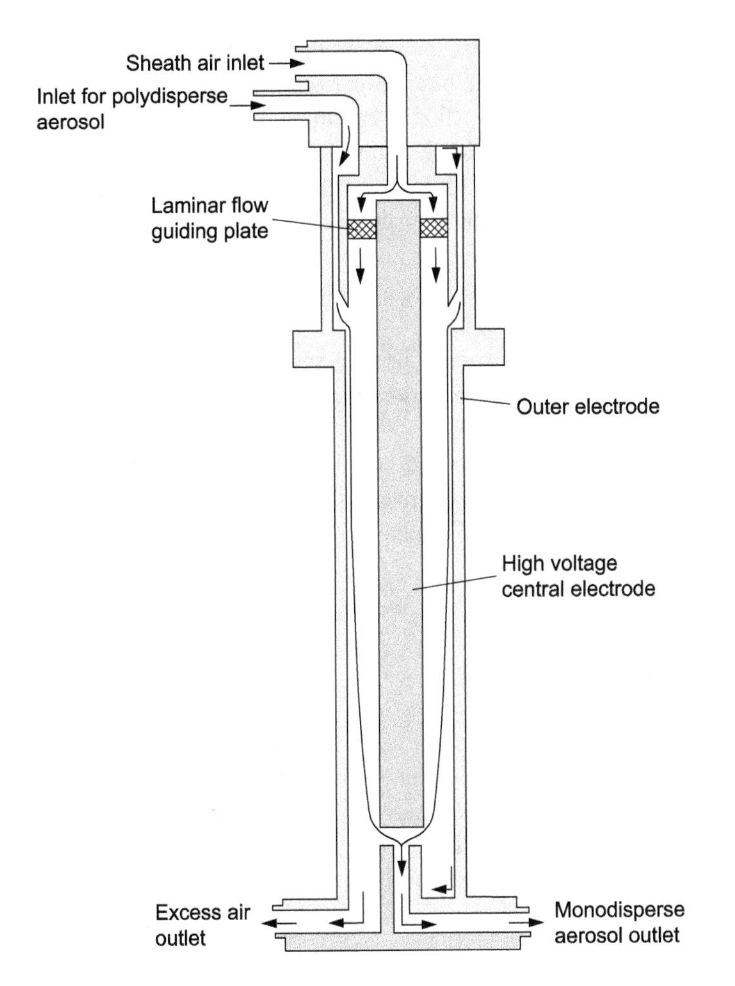

Fig. 19.12: Structure of a differential mobility analyzer.

Here Z_p is the electric mobility of the particles, and E the electric field strength between the central electrode and the tube wall. The motion of the particles is now opposed by the viscosity η of the carrier gas in which they are suspended. Furthermore, a correction factor C_c for gas friction must be taken into account. Thus, the Z_p becomes a function of the particle size d_p:

$$Z_p = neC_c(d_p)/3\pi\eta d_p \tag{5}$$

By varying the electric field, the trajectories of the particles impinging on the central electrode can thus be manipulated. The lower end of the central electrode is hollow and connected to the interior of the capacitor via a ring of orifices. As a result, a partial aerosol flow is extracted. For given potential conditions, there is now a narrow mobility fraction of particles whose trajectories lead to the position of the hole ring and can thus leave the capacitor interior through the circular orifices. Here, the finite width of the extraction holes determines the monodispersity of the particles that can continuously leave the system. The DMA acts like a bandpass filter; it *cuts out*, so to speak, a narrow fraction from a broad particle size distribution. An essential prerequisite for a defined separation is knowledge of the electric charge $n \cdot e$, since multiple charges ($n > 1$) of a particle change its mobility proportionally. In practice, the particles to be separated are separated by the addition of carrier gas ions (e.g. O_2^- or N_2^+) are electrically charged in a defined manner. The achievable resolution is low compared to mass spectrometers, at about 30 so far.

In recent years, applications for continuous protein separation or nanoparticle generation have been reported. Here, parallelization enables a higher throughput of substances.

The size distribution of unknown aerosols can be determined within a few minutes by defined, fast potential changes in connection with a continuous particle counter. A disadvantage is the need for a defined charge of the particles, since this is generally itself size-dependent.

19.4.6 Preparative mass spectrometry

While the DMA technique points the way to preparative separation via electrical mobility selection of charged particles, attempts are being made independently of this to achieve high-resolution separation of usable mass fractions by means of classical mass spectrometry. So far, various attempts have been reported for the separation of biological components, in which the detector is designed flat and the analyte can be removed after separation.

19.5 Efficacy of the method – resolving power

Resolving power. Whereas mass spectrometers with photographic registration of the spectra provide a series of images of the entrance slit at different positions on the photographic plate, the spectrum is digitally *scanned* in the instruments commonly used in analytical chemistry; the magnetic field strength (or the accelerating potential) is continuously changed so that the individual ion species migrate successively across the slit of the receiver and the dependence of the ion current on the *m/e ratio* can be registered. This results in the intensity distribution transverse to the slit images. As a result of unavoidable errors in the imaging, sharp rectangular profiles are not obtained, but bands with broadenings at the band bottom.

The selectivity of mass spectrometers is obviously conditioned by the width of the band including the band bottom. In order to arrive at a numerical statement, one determines on the abscissa the distance from the band maximum m after which the

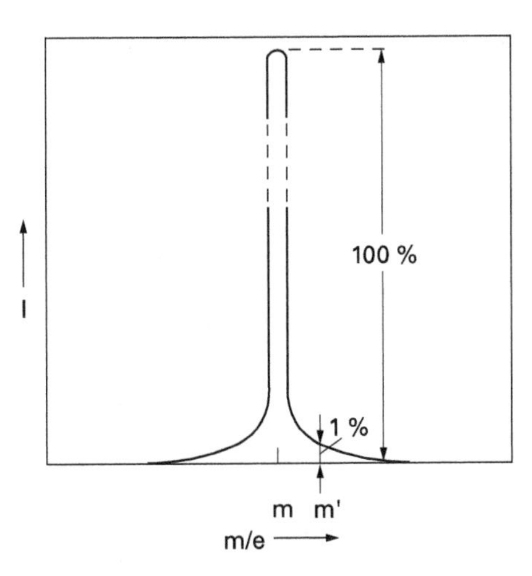

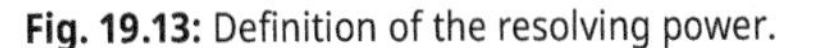

Fig. 19.13: Definition of the resolving power.

intensity I has dropped to 1% of its maximum value; this may be the case for mass m' (Fig. 19.13).

If another ion were to appear at the location m', the mass m would contribute 1% to the intensity measured there.

Since it is not indifferent at which absolute mass the 1% contribution is present, one relates the difference $m - m'$ *to* the mass m and defines (using the terminology of optics) as *resolving power A* the ratio of both values:

$$A = \frac{m}{|m - m'|} = \frac{m}{|\Delta m|}, \tag{6}$$

where Δm is always calculated as positive.

Examples. The contribution of mass 300 to mass 298 may have decayed to 1%; the resolving power then amounts to

$$\frac{300}{300 - 298} = 150.$$

Furthermore, let the mass 600 contribute 1% to the neighboring mass 601; as the resolving power results the value

$$\frac{600}{|600 - 601|} = 600;$$

on the other hand, with the same resolving power of 600, the mass 60 would already have dropped to 1% of its maximum intensity at a distance of 1/10 mass unit.

The arbitrariness in this definition of resolving power lies in the choice of the 1% contribution, and other values (e.g. 0.1%) have also been used. Therefore, one must always indicate the calculation method, i.e. the chosen percentage contribution to the neighboring mass, when giving figures.

Another definition of the resolving power is based on the concept of the *valley* between neighboring masses. Two neighboring masses of the same intensity are called resolved if the minimum between them is only 2% of the intensity of the bands in the minimum. The choice of the percentage for the minimum is arbitrary, and other values (e.g. 1% or 10%) have been suggested.

The resolving power of mass spectrometers with magnetic sector fields is independent of the size of the separated masses. Single focusing instruments give values of about 500–1000 (1% contribution), which is sufficient for solving many analytical problems. For precision mass determinations, which are important in nuclear physics and in certain tasks of organic analysis, one needs a resolving power of > 20 000, which can be achieved with double focusing mass spectrometers.

For time-of-flight mass spectrometers, the resolving power depends on the mass. For masses of about 500, values of about 1000 are achieved with a separation distance of 1 m.

In terms of selectivity and size of the application range, mass spectrometry is not even approached by any other separation method. For example, a resolving power of 600 (1% contribution) means that masses 600 and 601 are separated with a separation factor of 99/1: 1/99 ≃ 10 000 when they occur with the same intensity. The separation of two chemically very similar hydrocarbons such as n-decane (M = 142) and n-undecane (M = 156) would be practically quantitative (separation factor > 10^6) even with a much poorer resolving power.

Despite this extraordinary performance in selectivity and sensitivity, mass spectrometry has not yet played a role as a preparative separation method in analytical chemistry, since the quantities of substances separated are extremely small (in terms of size, about 10^{-8} –10^{-15} g), so that no further detection or determination methods can often be used afterwards. Furthermore, a more or less extensive fragmentation usually occurs during the separation of organic molecules. Therefore, one refrains from isolating the separated ions and is satisfied with determining the *m/e ratios* as well as the intensities of the ionic currents. From these data, versatile conclusions about the qualitative and quantitative composition of the samples can be drawn for trace analysis. For trace analysis of organic compounds, especially in combination with chromatographic separations (GC, LC), there is currently no superior analytical instrumentation available.

General literature

J. Becker, Inorganic Mass Spectrometry: Principles and Applications, John Wiley & Sons Ltd., West Sussex, UK (2007).

R. Cole, Electrospray and MALDI Mass Spectrometry: Fundamentals, Instrumentation, Practicalities, And Biological Applications, John Wiley & Sons, Inc, Hoboken, N.J. (2010).
J. Gross, Mass Spectrometry: A Textbook, Springer, Heidelberg (2004).
B. Kanawati & P. Schmitt-Kopplin, Fundamentals and Applications of Fourier Transform Mass Spectrometry, Elsevier, Amsterdam (2019).
J. Lang, Handbook on Mass Spectrometry: Instrumentation, Data and Analysis, and Applications, Nova Science Publishers, Inc, Hauppauge (2009).
J. Laskin & C. Lifshitz, Principles of Mass Spectrometry Applied to Biomolecules, John Wiley & Sons, Inc, Hoboken (2006).
M. Lee, Mass Spectrometry Handbook, John Wiley & Sons, Inc, Hoboken, (2012).
P. Liu, M. Lu, Q. Zheng, Y. Zhang, H. Dewald, & H. Chen, Recent advances of electrochemical mass spectrometry, Analyst 138, 5519–5539 (2013).
H. Lu, Mass Spectrometry–Based Glycoproteomics and Its Clinic Application, CRC Press, Boca Raton (2021).
A. McAnoy, Novel Applications of Mass Spectrometry to Research in Chemical Defence, DSTO Fishermans Bend, Australia (2010).
B. Shrestha, Introduction to Spatial Mapping of Biomolecules by Imaging Mass Spectrometry, Elsevier, Amsterdam (2021).

Literature to the text

Differential Mobility Analyzer

G. Allmaier, A. Maißer, C. Laschober, P. Messner & W. Szymanski, Parallel differential mobility analysis for electrostatic characterization and manipulation of nanoparticles and viruses, Tr AC – Trends in Analytical Chemistry *30*, 123–132 (2011).
R. Flagan, Continuous-flow differential mobility analysis of nanoparticles and biomolecules, Annual Review of Chemical and Biomolecular Engineering *5*, 255–279 (2014).
P. Intra & N. Tippayawong, An overview of differential mobility analyzers for size classification of nanometer-sized aerosol particles Songklanakarin Journal of Science and Technology *30*, 243–256(2008).
L. Pease, J. Elliott, D.-H. Tsai, M. Zachariah, and M. Tarlov, Determination of protein aggregation with differential mobility analysis: application to IgG antibody, Biotechnology and Bioengineering *101*, 1214–1222 (2008).
S.-H. Zhang, Y. Akutsu, L. Russell, R. Flagan, & J. Seinfeld, Radial differential mobility analyzer, Aerosol Science & Technology *23*, 357–372 (1995).

Fast Atom Bombardment Mass Spectrometry

M. Claeys & J. Claereboudt, Fast atom bombardment ionization in mass spectrometry (Book Chapter), in: Encyclopedia of Spectroscopy and Spectrometry, pp. 581–587 (2017).
D. Harvey, Mass spectrometry | Ionization methods overview (Book Chapter), in: Encyclopedia of Analytical Science, pp. 398–410 (2019).
M. Kay & J. Goodman, Mapping of senescent cell antigen on brain anion exchanger protein (AE) isoforms using HPLC and fast atom bombardment ionization mass spectrometry (FAB-MS), Journal of Molecular Recognition *17*, 33–40 (2004).

S. Ogawa & M. Tanaka, Study of chemical states of lanthanoid nitrate in solution by electrospray ionization and fast atom bombardment mass spectrometry, Journal of Molecular Liquids *291*, 111309 (2019).

Ion Mobility Mass Spectrometry

M. Belov, M. Buschbach, D. Prior, K. Tang & R. Smith, Multiplexed ion mobility spectrometry-orthogonal time-of-flight mass spectrometry, Analytical Chemistry *79*, 2451–2462 (2007).

G. Eiceman, Z. Karpas, and H. Hill, Ion Mobility Spectrometry, Taylor & Francis, Boca Raton, (2013).

M. Ewing, M. Glover & D. Clemmer, Hybrid ion mobility and mass spectrometry as a separation tool Journal of Chromatography A *1439*, 3–25 (2016).

A. Kanu, P. Dwivedi, M. Tam, L. Matz, & H. Hill Jr., Ion mobility-mass spectrometry, Journal of Mass Spectrometry *43*, 1–22 (2008).

C. Wilkins & S. Trimpin, Ion Mobility Spectrometry – Mass Spectrometry: Theory And Applications, CRC Press, Boca Raton (2011).

Ion cyclotron resonance mass spectrometry

S. Brown, G. Kruppa, and J.-L. Dasseux, Metabolomics applications of FT-ICR mass spectrometry, Mass Spectrometry Reviews *24*, 223–231 (2005).

S. Forcisi, M. Moritz, B. Kanawati, D. Tziotis, R. Lehmann, and P. Schmitt-Kopplin, Liquid chromatography-mass spectrometry in metabolomics research: mass analyzers in ultra high pressure liquid chromatography coupling, Journal of Chromatography A *1292*, 51–65 (2013).

C. Gosset-Erard, F. Aubriet, F. Leize-Wagner, Y.-N. François & P. Chaimbault, Hyphenation of Fourier transform ion cyclotron resonance mass spectrometry (FT-ICR MS) with separation methods: The art of compromises and the possible – A review, Talanta *257*, 124324 (2023).

J. Laskin & J. Futrell, Collisional activation of peptide ions in FT-ICR mass spectrometry, Mass Spectrometry Reviews *22*, 158–181 (2003).

Y. Park & C. Lebrilla, Application of Fourier transform ion cyclotron resonance mass spectrometry to oligosaccharides, Mass Spectrometry Reviews *24*, 232–264 (2005).

R. Sleighter & P. Hatcher, The application of electrospray ionization coupled to ultrahigh resolution mass spectrometry for the molecular characterization of natural organic matter Journal of Mass Spectrometry *42*, 559–574 (2007).

Orbitrap mass spectrometry

H. Dong, Y. Xu, H. Ye, (. . .), W. Bai & D. Luo, Advances in Analysis of Contaminants in Foodstuffs on the Basis of Orbitrap Mass Spectrometry: a Review Food Analytical Methods *15*, 803–819 (2022).

Q. Hu, R. Noll, H. Li, A. Makarov, M. Hardman, and G. Cooks, The Orbitrap: A new mass spectrometer, Journal of Mass Spectrometry *40*, 430–443 (2005).

R. March & J. Todd, Practical Aspects of Trapped Ion Mass Spectrometry, Vol. IV & V, CRC Press, Boca Raton (2016).

J. Olsen, L. de Godoy, G. Li, (. . .), S. Horning & M. Mann, Parts per million mass accuracy on an orbitrap mass spectrometer via lock mass injection into a C-trap, Molecular and Cellular Proteomics *4*, 2010–2021 (2005).

Photoionization mass spectrometry

U. Bösl, Laser mass spectrometry for environmental and industrial chemical trace analysis, Journal of Mass Spectrometry *35*, 289–304 (2000).

L. Hanley & R. Zimmermann, Light and molecular ions: The emergence of sacuum UV Single-photon ionization in MS, Analytical Chemistry *81*, 4174–4182 (2009).

R. Huang, Q. Yu, L. Li, L. Lin, W. Hang, J. He, & B. Huang, High irradiance laser ionization orthogonal time-of-flight mass spectrometry: A versatile tool for solid analysis, Mass Spectrometry Reviews *30*, 1256–1268 (2011).

Y.Li & F. Qi, Recent applications of synchrotron VUV photoionization mass spectrometry: Insight into Combustion Chemistry, Accounts of Chemical Research *43*, 68–78 (2010).

A. Raffaelli & A. Saba, Atmospheric pressure photoionization mass spectrometry, Mass Spectrometry Reviews *22*, 318–331 (2003).

T. Streibel & R. Zimmermann, Resonance-enhanced multiphoton ionization mass spectrometry (REMPI-MS): Applications for process analysis, Annual Review of Analytical Chemistry *7*, 361–381 (2014).

Z. Zhou, H. Guo, and F. Qi, Recent developments in synchrotron vacuum ultraviolet photoionization coupled to mass spectrometry, TrAC, Trends in Analytical Chemistry *30*, 1400–1409 (2011).

Preparative mass spectrometry

T.A. Blake, Instrumentation for Preparative Mass Spectrometry, PhD thesis, Purdue Univ., West Lafayette, IN, USA (2006).

J. Benesch, B. Ruotolo, D. Simmons, N. Barrera, N. Morgner, L. Wang, H. Saibil, & C. Robinson,Separating and visualizing protein assemblies by means of preparative mass spectrometry and microscopy, Journal of Structural Biology *172*, 161–168 (2010).

S. Rauschenbach, M. Ternes, L. Harnau & K. Kern, Mass Spectrometry as a Preparative Tool for the Surface Science of Large Molecules Annual Review of Analytical Chemistry *9*, 473–498 (2016).

P. Su, M. Espenship & J. Laskin, Principles of Operation of a Rotating Wall Mass Analyzer for Preparative Mass Spectrometry, Journal of the American Society for Mass Spectrometry *31*, 1875–1884 (2020).

G. Verbeck, W. Hoffmann, and B. Walton, Soft-landing preparative mass spectrometry, Analyst *137*, 4393–4407 (2012).

Quadrupole Mass Spectrometer

M. Balogh, A mass spectrometry primer, part III, LCGC North America *26*, 1176, 1178, 1180, 1182, 1184–1189 (2008).

R. March, Quadrupole ion trap mass spectrometry. A view at the turn of the century, International Journal of Mass Spectrometry *200*, 285–312 (2000).

R. March & J. Todd, Practical Aspects of Trapped Ion Mass Spectrometry, Vol. IV & V, Springer, Heidelberg (2016).

M. Petrovic & D. Barcelo, Application of liquid chromatography/quadrupole time-of-flight mass spectrometry (LC-QqTOF-MS) in environmental analysis, Journal of Mass Spectrometry *41*, 1259–1267 (2006).

20 Migration of dissolved charge carriers in an electric field (electrophoresis; electrodialysis)

20.1 Historical development

The fundamental studies on the migration of dissolved ions in the electric field were made by Kohlrausch (1897). Separations based on differences in migration rates were first described by Kendall (1923), but the method came to greater prominence only through Tiselius (1930), who was able to achieve separations of high molecular weight biological substances with a different type of apparatus. The method was later simplified in terms of apparatus by the introduction of thin-layer procedures (paper electrophoresis according to v. Klobusitzky; 1939) and extended to continuous separations by the two-dimensional mode of operation (Philpot; 1940).

20.2 General

20.2.1 Definitions

The migration of charged particles in solutions when an electric field is applied is referred to as *electrophoresis*. If the charged particles are small ions, this is also referred to as *ionophoresis*. For methods in which the liquid phase is absorbed in a porous carrier material, the terms *electropherography* and *carrier electrophoresis* are used; if – as is usually the case – the analysis sample is applied in the form of a thin strip, the method is *zone electrophoresis*. In *electrodialysis* there is a semi-permeable wall in the separating section which is permeable only to small particles.

20.2.2 Appearances during current passage through electrolytes (formation of interfaces – heat of current)

While in section 12.2.5 of the second part (*Electrolysis*) the chemical reactions at the electrodes were in the foreground, now the phenomena in the electrolyte itself will be considered.

When an electric voltage is applied to two electrodes located in an electrolyte solution, the cations move to the cathode under the influence of the field, and the anions move to the anode; both types of ions migrate independently, and the rates of migration are usually different.

If the solution is homogeneous, no concentration changes are observed in the electrolyte – at least at the beginning of the experiment (these are limited to the parts immediately adjacent to the electrodes, i.e. they take place at the solid – liquid interfaces).

https://doi.org/10.1515/9783111181417-020

However, the picture changes if the solution contains inhomogeneities, e.g. if it consists of two salt solutions separated by a sharp liquid boundary.

In a vertical tube, a specifically lighter sodium acetate solution may be layered over a heavier sodium chloride solution so that the interface between the two is not destroyed by convection (Fig. 20.1). When a voltage is applied in the sense indicated, the original interface remains stationary, but the formation of an additional interface is observed to slowly move downward. This separates acetate and chloride ions, the stationary one existing between two Na-acetate solutions of different concentrations.

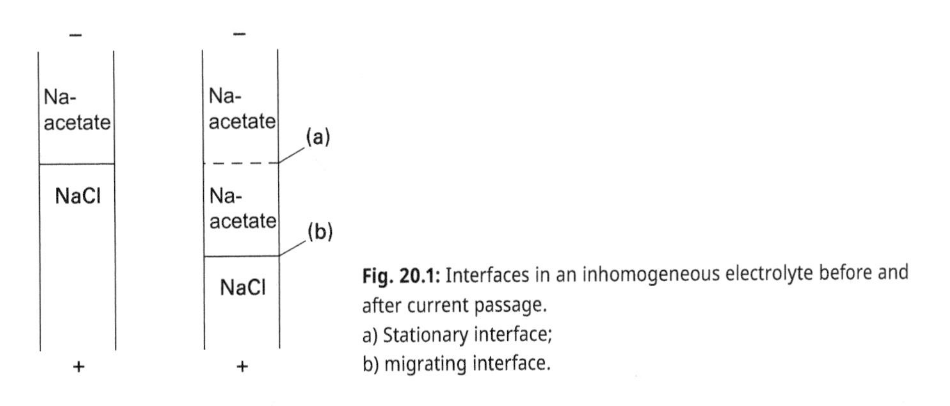

Fig. 20.1: Interfaces in an inhomogeneous electrolyte before and after current passage.
a) Stationary interface;
b) migrating interface.

The number of interfaces is given by the number of ions on either side of the original inhomogeneity; in the presence of n ions, $n - 1$ interfaces are formed, one of which is stationary.

As will be explained in more detail below, the concentration ratios on either side of a migrating interface are not arbitrarily selectable, but are fixed by law.

A side effect that is troublesome for electrophoretic separations is the appearance of heat (*heat of current*) as the current passes through solutions. The evolved heat H (in cal/s) amounts to

$$H = \frac{V \cdot I}{4,18} = \frac{R \cdot I^2}{4,18} \tag{1}$$

(with the voltage V, the current I and the resistance R). Accordingly, the heat of current increases with the square of the current intensity, which is particularly troublesome when electrophoreses are to be carried out at high field strengths. Although separations are hardly affected by heating the electrolyte, the interfaces can be disturbed by convection currents and the solutions can be altered by increased evaporation.

20.2.3 Ion migration in liquid phase – ion mobilities

When particles move in a liquid, frictional forces occur; charged particles are initially accelerated after a field is applied, but after a very short time a constant velocity v is established, which is equal to the quotient of the acting force K and the friction constant W; the force K in turn is equal to the product of the field strength Q and the charge Z of the particle:

$$v = \frac{K}{W} = \frac{Q \cdot Z}{W}. \tag{2}$$

If the particle is spherical, Stokes' law applies to the friction constant:

$$W = 6\pi \cdot \eta \cdot r$$
$$(\eta = \text{viscosity of liquid};\ r = \text{particle radius}) \tag{3}$$

Thus in this case

$$v = \frac{Z \cdot Q}{6\pi \cdot \eta \cdot r}. \tag{4}$$

In order to compare the migration velocities (dimension cm/s) of different ions, they are referred to a uniform field strength of 1 V/cm; the values obtained are called *ion mobilities* (dimension $\frac{\text{cm/s}}{\text{V/cm}} = \text{cm}^2 \cdot \text{s}^{-1} \cdot \text{V}^{-1}$). They mean the migration distances in cm that the ions travel per s in a field of 1 V/cm. The mobilities of cations are usually calculated positively, those of anions negatively. Some ionic mobilities are given in Tab. 20.1; the values are only intended to give an approximate guide, they depend on temperature, pH and, above all, on the ionic strength of the solution. The relatively high mobilities of the H^+ and OH^- ions are remarkable.

Equation (4), irrespective of its validity in principle, only unsatisfactorily reflects the actual conditions, since it does not cover several factors which are important for the ion migration velocity. First of all, it is not the ion radius determined from crystal structure determinations that is to be used in eq. (4), but the radius of the solvated ions in solution, which is difficult to determine. Furthermore, the ions – especially those of organic compounds – are often not spherical, and the migration speed or mobility still depends on the ionic strength of the solution and on the degree of dissociation. The latter, in turn, may be determined by the pH of the solution. And finally, peculiarities of the electric field (inconstancy and presence of AC components) also influence the mobilities. Since not all of these influences can be determined by calculation, the migration velocities must be determined experimentally.

In eq. (4) the mass of the particles does not appear; a refined theory must also take this into account, as has been shown to some extent by successful experiments on electrophoretic isotope separation. For all practically occurring chemical separations, however, the influence of the ion masses can be neglected.

Tab. 20.1: Ion mobilities in $cm^2 \cdot s^{-1} \cdot V^{-1}$ (examples).

Ion	Mobility	Ion	Mobility
H^+	+0,00326	OH^-	−0,00180
Na^+	+0,00025	Cl^-	−0,00032
K^+	+0,00033	$H_2PO_4^-$	−0,00016
Diethanolamine	+0,00016	Acetate	−0,00019
		Nucleic acid	−0,00014
Triethanolamine	+0,00014	Albumin	−0,00006
		Globulin	−0,00001

In the following, the most important variables on which the ionic mobilities depend are discussed in more detail.

Influence of pH on the mobilities. The ionic mobilities of strong electrolytes are, to a first approximation, pH-independent, since their dissociation is always practically complete. In the case of weak acids and bases and ampholytes, on the other hand, a pronounced pH influence is evident; although the mobilities of ions should not be directly affected by the acid concentration of the solution, interfaces formed by such compounds migrate more or less rapidly depending on the pH of the solution. The explanation will be given for the anion of a weak acid on the basis of Fig. 20.2.

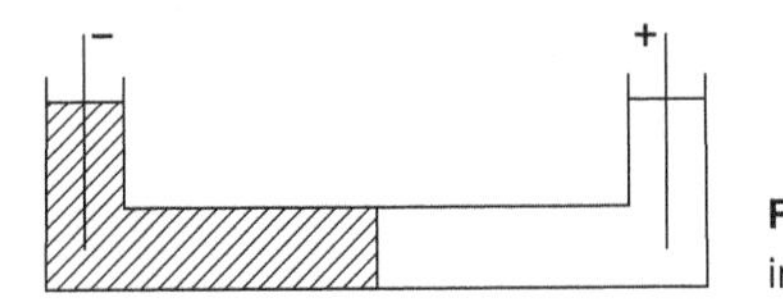

Fig. 20.2: Influence of pH on the migration velocity of an interface.

Let a tube be filled with a solution of a strong acid, and in one section of this tube there is also a weak acid (drawn hatched). In the electric field, the anion of this weak acid moves to the right with a certain velocity; when it enters the part of the tube containing only the strong acid via the interface, the migration velocity will not change, but since the (far predominant) undissociated portion has not migrated with it, it will be almost completely converted into the non-migrating undissociated form by reaction with the H^+ ions of the strong acid present. Only when a sufficient concentration of undissociated weak acid has built up immediately to the right of the interface are noticeable amounts of anions again available to migrate further.

The (reduced) net ionic mobilities u_n of an incompletely dissociated compound can be calculated from the degree of dissociation α and the (real) mobilities *u of* the ions as follows: The degree of dissociation α can be viewed as the fraction of time t_1 that the relevant fraction of the molecule spends in the dissociated form; in the nonionized form, it spends the fraction of time t_2, and the total time period considered is $t_1 + t_2$, i.e.

$$\alpha = \frac{t_1}{t_1 + t_2}. \quad (5)$$

The migration distance d, which is observed experimentally, is in the period $t_1 + t_2$

$$d = u_n(t_1 + t_2). \quad (6)$$

However, this migration distance is now also given by the mobility u of the dissociated fraction and the time t_1 in which it is dissociated:

$$d = u \cdot t_1. \quad (7)$$

This results in

$$u_\text{n} = \frac{d}{t_1 + t_2} = \frac{u \cdot t_1}{t_1 + t_2} = \alpha \cdot u. \quad (8)$$

The net mobilities u_n are therefore equal to the product of the true mobilities of the ions and the degree of dissociation of the compound supplying the ions; the pH dependence of the net mobilities must therefore parallel that of the degree of dissociation. This in turn can be calculated according to the law of mass action.

For a weak acid, for example, (A^- = anion):

$$\frac{[H^+][A^-]}{[HA]} = K. \quad (9)$$

By logarithmizing and transforming we get

$$\text{pH} = \text{p}K + \log\frac{[A^-]}{[HA]} = \text{p}K + \log\frac{\alpha}{1-\alpha}, \quad (10)$$

where *pK* is the negative logarithm of the dissociation constant *K*. Plotting the degree of dissociation (or, more graphically, the percentage dissociation) against pH yields the familiar s-shaped curves whose position is given by the *pK* value (Fig. 20.3).

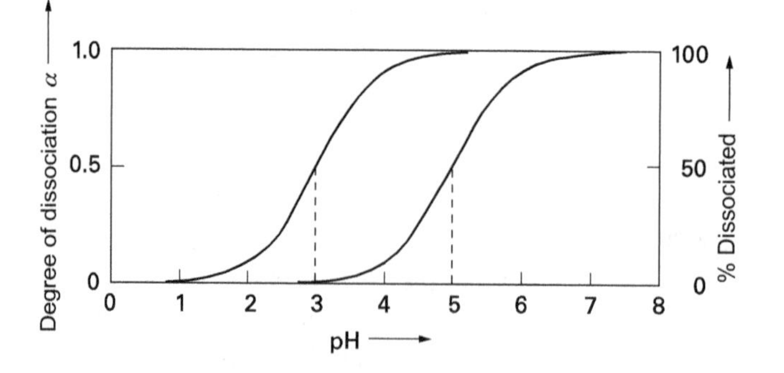

Fig. 20.3: Dependence of the degree of dissociation on pH for two acids with *pK* values of 3 and 5.

For the pH influence on the net ion mobility, corresponding curves are obtained according to eq. (8), the height of which is given by the true ion mobility u. Figure 20.4 shows some such curves of cations (positive values for u_n) and anions (negative values for u_n) with different mobilities u, where in each case the dissociation constants should have the same value.

A change in the dissociation constants at the same mobilities u is expressed by a parallel shift of the curves; it is only necessary to replace the degree of dissociation by the net mobility on the ordinate of Fig. 20.3.

The change in ionic mobilities (like that of the degree of dissociation) takes place essentially within a pH range of about ±2 units from the pH corresponding to the pK value; within this range, the percent dissociation changes from about 1% to about 99% and the net mobility u_n correspondingly changes from about 1% to about 99% of the value of u.

Ampholytes are present in the form of anions or cations, depending on the pH of the solution; their direction of migration in the electric field is positive in acidic solution; as the pH increases, it decreases until it becomes zero when the isoelectric point is reached. After exceeding this point, the migration velocity increases again with a change in direction (cf. Fig. 20.5).

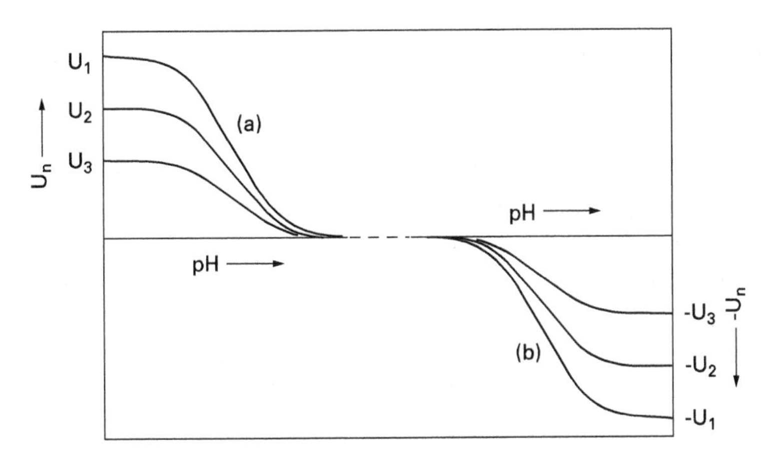

Fig. 20.4: pH dependence of net mobilities for ions with the same dissociation constants but different mobilities *u*.
a) Cations; b) anions.

The separation of ions with pH-dependent migration rates is most convenient to perform at the pH at which the difference in net mobilities is a maximum. This can be calculated from the ionic mobilities and the dissociation constant, but the graphical determination after plotting the mobility curves (as in Fig. 20.4) is preferable, especially since the ratios can then be better overlooked in the presence of more than 2 ions.

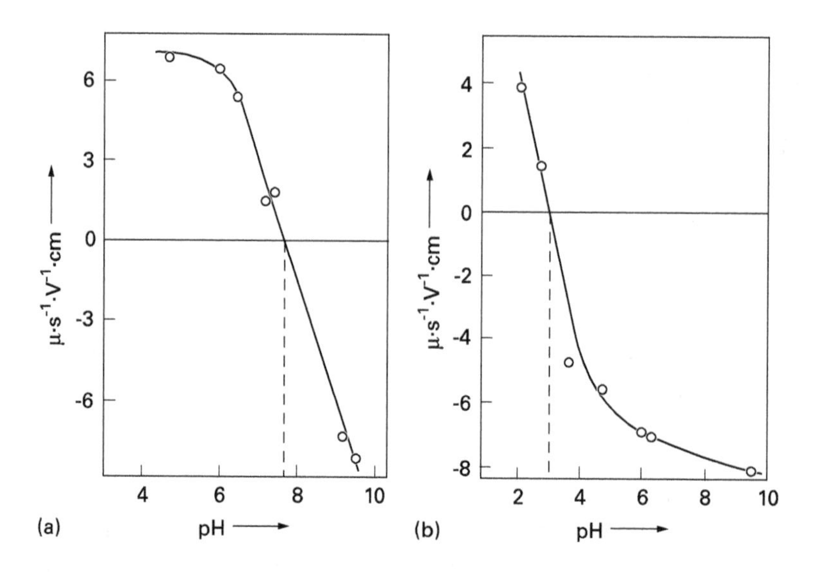

Fig. 20.5: Electrophoretic migration rates of ampholytes as a function of pH.
(a) Histidine; (b) glutamic acid.

As a consequence of this consideration, it further follows that ions of the same mobilities u can also be separated electrophoretically if only the dissociation constants are different.

Influence of complexing agents on the mobilities of ions. Complexation reactions cause a change in the mobilities through their influence on the ionic radii. Of much greater importance, however, is that in many cases the charge and possibly even the sign of the charge also change, reversing the direction of migration. One can compare the influence of complexing agents with that of the pH value. The concentration of the complexing agent corresponds to that of the H^+ ions, the complex formation constant to the dissociation constant. If, at a given concentration of the complexing agent, the concentrations of negatively and positively charged particles are equal, the component in question does not migrate, a behavior similar to that of ampholytes at the isoelectric point.

Influence of ionic strength on mobility. The greater the ionic strength of a solution, the more the ions hinder each other in their migration, since the forces of attraction increase with the approach of oppositely charged particles (Fig. 20.6).

For relatively large charged particles, it is approximately true that the migration velocity is proportional to $\frac{1}{\sqrt{\mu}}$ (μ = ionic strength). On the whole, the influence of ionic strength is complex, since it also changes the field strength (via conductivity) and the viscosity of the solution. In general, one tries to perform electrophoretic separations at the lowest possible ionic strength; values of about 0.01–0.1 mol/l are common. On the other hand, a sufficient buffer capacity must normally be present.

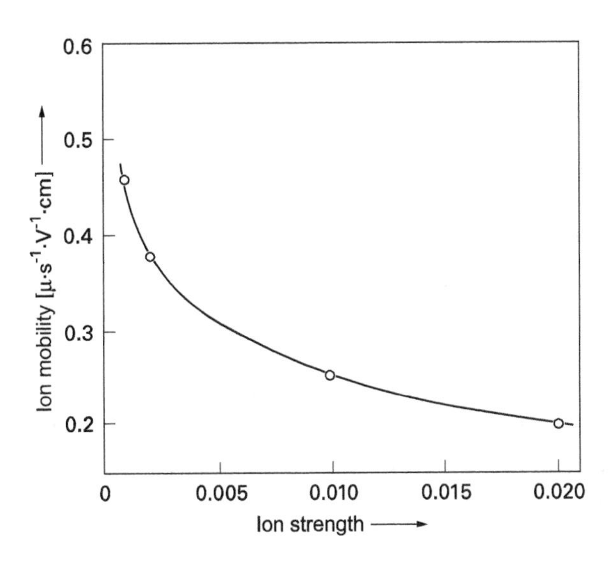

Fig. 20.6: Influence of ionic strength on the mobility of leucine.

Influence of temperature on mobility. The temperature influence on the ion mobility is essentially given by the change in the viscosity of the solution. If the mobility u, the viscosity η at a certain temperature and the temperature function of the viscosity are known, the mobility at other temperatures can be calculated roughly according to the following relationship:

$$\eta \cdot u = \text{constant}. \tag{11}$$

20.3 One-dimensional electrophoretic separations without carriers (Tiselius method)

The principle of the method worked out by Tiselius (1955) is described with reference to Fig. 20.7. In the lower part of a U-tube there is a buffer solution which also contains the ions to be separated. In both legs, pure buffer is layered on top so that sharp interfaces exist between the solutions. Buffer and compounds to be separated should have the anion in common, thus only the cations should be separated. In this arrangement, the lower solution (buffer & cations) is specifically heavier than the protruding pure buffer solution.

If a voltage is applied to the electrodes, the cations migrate towards the cathode at different speeds, and new interfaces form in the legs of the vessel.

When carrying out such separations by means of equipment, the electrode compartments must be separated from the legs of the separating vessel so that the solution cannot be contaminated by electrolysis products. Furthermore, one performs the separations at + 4 °C, the maximum density of water, to eliminate disturbances due to

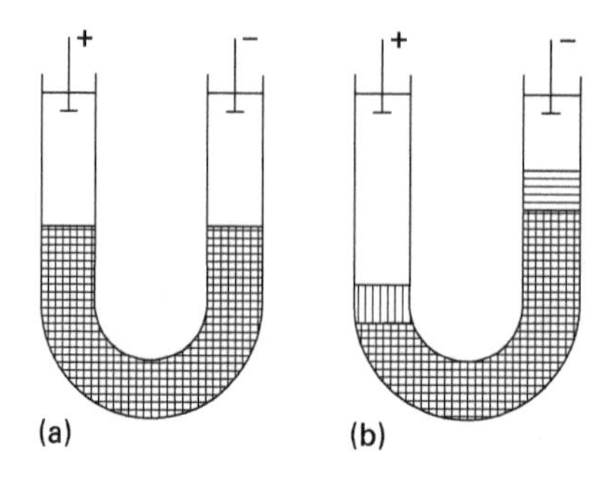

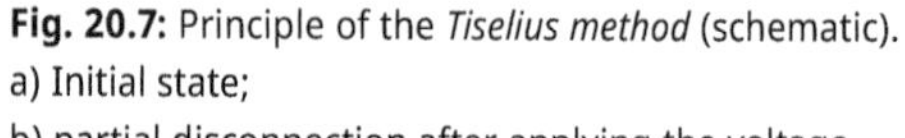

Fig. 20.7: Principle of the *Tiselius method* (schematic).
a) Initial state;
b) partial disconnection after applying the voltage.

heat convection as much as possible. The field strength is usually about 5–10 V/cm, it is limited by the strongly increased heat generation with increasing current strength. The substance quantities are in the order of a few hundred milligrams, but quantities of about 1–5 mg protein can also be separated by reducing the size of the cuvettes.

The migration of colored ions can be followed directly by eye; colorless ions are visualized by optical streak methods, conveniently working in electrophoresis vessels of rectangular cross-section.

After a sufficient experimental time, ions migrating at different speeds should be completely separated from each other. However, this cannot be achieved for several reasons: on the one hand, the interfaces cannot be kept sharp for a sufficiently long time because of convection, which cannot be completely eliminated after all, and because of diffusion of the solute (especially when relatively rapidly diffusing small ions are present); on the other hand, in the case of quantitative separation, a zone with pure buffer solution would lie between two zones with buffer + separated substance. Since the buffer is the specifically lightest phase in the system, it would be overlaid by a heavier phase, which would have to lead to the destruction of the interface.

The method therefore only allows the zones to be pulled apart somewhat, so that only part of the fastest-moving component in one limb and part of the slowest-moving in the other limb can be kept pure. In this respect, it is similar to the frontal analysis discussed earlier (see Part 2, Section 6.7.5). Although the difficulties can be eliminated by working in capillaries or by applying a density gradient in the solution, the method is now only rarely used, especially since simpler methods in terms of apparatus and more efficient in terms of separation have been devised. However, the Tiselius method still performs well today in the determination of ion migration velocities.

20.4 One-dimensional electrophoretic separations with carrier (carrier electrophoresis)

20.4.1 Carrier materials – special features of the use of carriers (adsorption – electroosmosis – suction flow – change in ion migration velocities – relative mobilities)

The flow phenomena which considerably disturb the Tiselius method can be prevented by absorbing the solutions into porous carrier materials. As such, various substances which are as inert as possible are used in powder, gel or membrane form (cf. Tab. 20.2). Electrophoresis in gels in particular has led to effective separations.

In addition to the elimination of convection currents, the use of carriers brings another significant change compared with the *free* method: the substance mixture can be applied to the carrier material as a narrow zone in the form of a line, so that separation of the individual components into zones separated by areas of pure buffer solution takes place after relatively short migration distances (*zone electrophoresis*). On the other hand, an essential feature of the Tiselius method is retained; as in the latter, the substances to be separated migrate in a uniform buffer solution, and the concentration ratios are chosen so that current conduction is mainly performed by the ions of the buffer. The phenomena of adsorption on the carrier, electroosmosis and suction flow are disturbing.

Tab. 20.2: Carrier substances for electrophoretic separations (examples).

Powder form	Gel form	Membrane shape
SiO_2	Silica gel	Cellulose (paper)
Glass	Agar gel	Cellulose acetate
Starch	Starch gel	PVC film
PVC powder	Polyacrylamide gel	Glass fiber paper
Al_2O_3		Polyester films

Adsorption. The dissolved ions are adsorbed to a greater or lesser extent on the carrier, so that a chromatography effect (and in some gels also a sieve effect) is superimposed on the electrophoretic migration. Since adsorption usually leads to an undesirable zone broadening, it is avoided as far as possible by selecting suitable carrier materials. However, there are also processes in which the separation is based more on differences in adsorption than on electrophoresis; this is then referred to as *electro-chromatography*.

Electroosmosis. When carriers are used, a directed current is observed in the solution, which is called *electroosmosis*. This is the reverse of electrophoresis.

In a sense, the carrier mimicks the role of the dissolved ions; since it cannot move, the moving force is transferred to the liquid.

The strength of the flow depends, among other things, on the type of carrier material; in glass powder, for example, it is relatively strong, while in some organic gels, e.g. polyacrylamide, it is much lower.

Electroosmosis has a particularly disturbing effect in the determination of ionic mobilities. This effect is eliminated by dissolving a neutral, non-migrating compound, e.g. sugar, with the ion to be investigated and subtracting its displacement from that of the ion. Separations are hardly influenced by electroosmosis.

Suction flow. In electrophoresis instruments, where the heat of the current can cause some of the liquid to evaporate from the separating section, suction flow occurs in addition to electroosmosis. This results from the after-flow of solution from the electrode vessels due to the capillary forces of the carrier's pores.

When determining migration velocities, suction flow must also be taken into account; it also occasionally gives rise to disturbances, since the buffer concentration and thus the conductivity of the solution are changed.

Change of migration velocities in carrier substances – relative mobilities. In carrier substances, the migration velocities of ions are apparently reduced compared to those in free electrophoresis; the explanation is essentially to be found in the increased travel distance due to the turns of the pores. The effect depends on the type of carrier material and is also different for different types of paper, for example.

Since it is difficult to convert the ionic mobilities found in carriers to the values of carrier-free electrophoresis, one often makes do with the determination of relative mobilities, i.e. one relates the values found to those of a standard substance which has migrated under the same conditions. The ratio of the migration distance of the investigated substance to that of the standard (taking into account the electroosmotic flow) gives the relative mobility u_r. These values are readily reproducible and easy to obtain, especially since even minor temperature fluctuations need not be taken into account, since they generally have the same effect on both ions.

20.4.2 Column electrophoresis

For the separation of larger amounts of substance (approx. 1–10 g), columnar arrangements of the carrier are used. The electrodes are located at the top and bottom of the column, and furthermore solution leaving the column by electroosmotic flow must be returned to it (cf. Fig. 20.8). If the diameter of the column is not too large, an outer cooling jacket is sufficient, otherwise a second cooling tube must be placed inside so that the support is in an annular gap between the two cooling tubes.

After completion of the separation, the solution can be allowed to flow down and collected fractionally as in the chromatographic methods. In another method, the lower part of the column is continuously rinsed with buffer solution so that the substances migrating out of the support layer are removed from the apparatus one by one (cf. Fig. 20.8).

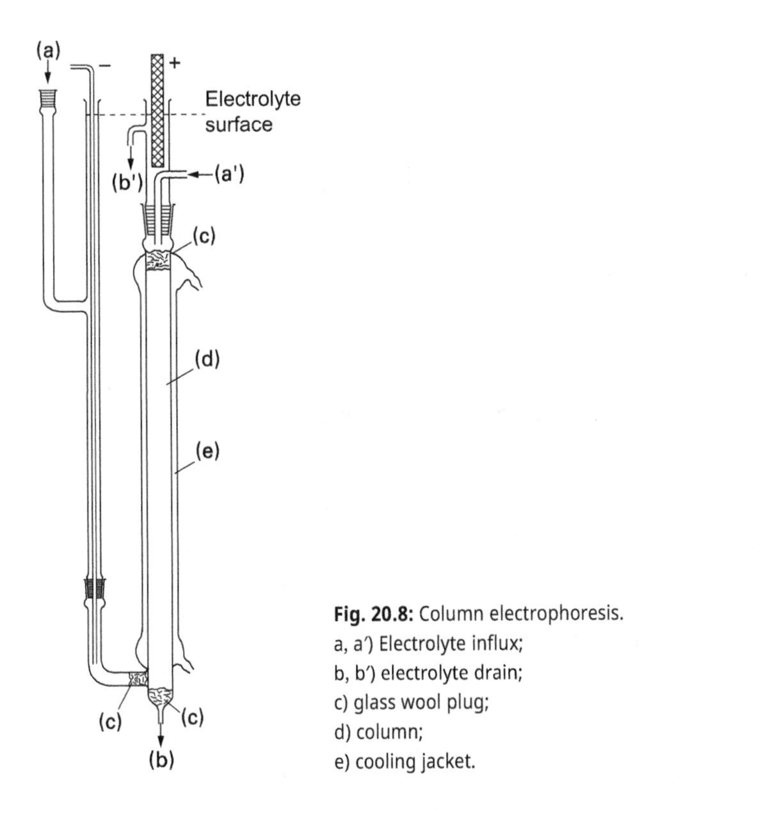

Fig. 20.8: Column electrophoresis.
a, a′) Electrolyte influx;
b, b′) electrolyte drain;
c) glass wool plug;
d) column;
e) cooling jacket.

20.4.3 Thin-layer methods – paper electrophoresis – high-voltage electrophoresis – micro- and ultramicro methods

Carrier electrophoresis can be simplified considerably by using the thin-film technique. The carrier impregnated with buffer solution is placed on a non-conductive plate and both ends are connected by wicks to buffer reservoirs containing the electrodes. The mixture of substances to be separated is applied in the form of a thin strip on one side or – if both cations and anions are to be separated – in the middle of the thin film plate. After the voltage is applied, the ions of the analysis sample migrate at different speeds so that the original application site shifts and separates into a series of zones; these gradually widen due to diffusion.

Of the thin-layer methods, *paper electrophoresis* has probably found the most widespread application, since the apparatus used for this method is particularly simple. A long strip of paper is placed on a support, e.g. on a glass plate, or attached to two supports in a freely suspended manner (cf. Fig. 20.9). The ends are immersed in the buffer reservoirs; the electrodes are shielded by porous separating plates to prevent the electrolysis products from changing the composition of the solution on the paper strip. The whole apparatus is placed in a closed chamber to avoid evaporation losses.

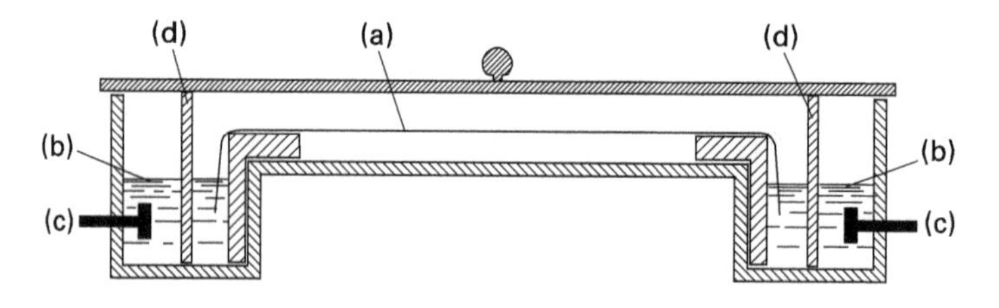

Fig. 20.9: Paper electrophoresis (principle).
(a) Paper strips; (b) buffer solution; (c) electrodes; (d) porous separating plate.

Instead of paper, various other substances in film form are also used as substrate materials. Some plastic films (e.g. cellulose acetate or polyacrylamide) have proved to be particularly favorable, as they can be obtained with relatively uniform pore sizes under defined manufacturing conditions.

Colorless zones are usually made visible after separation by staining with suitable reagents, as in the chromatographic methods. The separated compounds are then determined photometrically directly on the layer or after dissolution.

High-voltage electrophoresis. The zone broadening that occurs as a result of diffusion can be considerably reduced by increasing the migration velocities while shortening the test times. This is achieved by higher field strengths (approx. 20–100 V/cm instead of the usual 5–10 V/cm). Such procedures are referred to as *high-voltage electrophoresis.*

The main difficulty here is the dissipation of the heat of the current, which – as mentioned above – increases with the square of the current intensity and thus also with the square of the voltage. It is therefore necessary to use particularly effective cooling devices. In so-called *liquid cooling,* the paper strip is suspended in a trough filled with an organic liquid, but this must neither dissolve nor chemically alter the substances to be separated. More generally applicable are methods in which the paper strip lies on a cooled glass plate or, better still, between two cooled plates (*sandwich technique*). This completely prevents evaporation of the electrolyte.

Micro and ultramicro methods. In the case of paper electrophoresis, the weights are usually in the range of about 0.1–1 mg, i.e. it is a distinctly micro process. On very narrow paper strips and especially on cellulose acetate films, samples of a few μg down to about 0.01 μg can still be separated. An extremely sensitive method with microscopic observation is the so-called *capillary electrophoresis* for substance amounts of < pg (= 10^{-12} g). Laser fluorescence, for example, is used to detect the separated analytes.

20.5 Special effects with inhomogeneous separating zones

In the electrophoretic methods discussed so far, the separation zone was uniformly filled with buffer solution of constant composition, and the ions to be separated were present in relatively low concentration. However, special effects can be achieved when solutions of different compositions abut against each other in the separation zone or when gradients are present; under certain conditions, the interfaces can be stabilized against the influence of diffusion, the zones containing the mixture of substances to be separated can be narrowed, and the ions can be focused at certain points in the separation section.

20.5.1 Interface stabilization according to Kendall

The effect of interfacial stabilization is explained using the example of a sodium acetate solution and a sodium chloride solution which abut against each other in a tube forming a sharp interface (Fig. 20.10). The electric field should be directed in such a way that the acetate ion with the smaller mobility u_{Ac} – follows the chloride ion with the larger mobility u_{Cl} – (in Fig. 20.10, both anions thus migrate to the right).

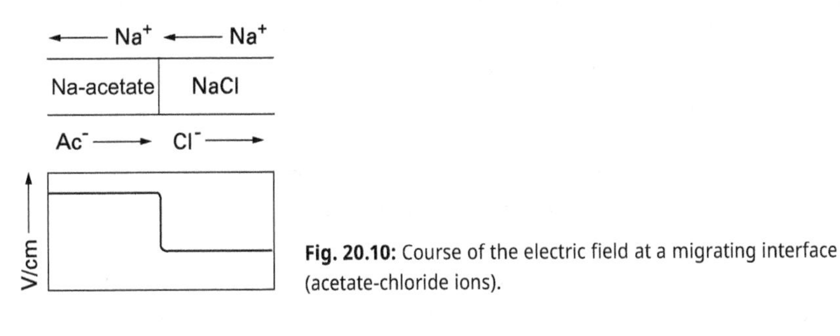

Fig. 20.10: Course of the electric field at a migrating interface (acetate-chloride ions).

After some time, because of the greater mobility of the Cl^- ion, there should be a gap between the two anions, but this is ruled out by the electroneutrality condition (the gap could contain only Na^+ ions). This means that the two anions must migrate at the same rate; therefore, the following applies

$$v_{Ac^-} = v_{Cl^-}\,. \tag{12}$$

However, equal migration velocities v can only exist if the field strengths in the two sections of the migration path are different, because the migration velocities are given by the mobility u and the field strength Q according to what was said earlier,

$$v_{Ac^-} = u_{Ac^-} \cdot Q_{Ac} \tag{13}$$

$$v_{Cl^-} = u_{Cl^-} \cdot Q_{Ch}\,. \tag{13a}$$

(Q_{Ac} and Q_{Ch} are the field strengths in the acetate and chloride sections, respectively.)

It follows:

$$u_{Ac^-} \cdot Q_{Ac} = u_{Cl^-} \cdot Q_{Ch} \tag{14}$$

and

$$\frac{Q_{\text{Ch}}}{Q_{\text{Ac}}} = \frac{u_{\text{Ac}^-}}{u_{\text{Cl}^-}}. \tag{15}$$

The field strengths in the two sections of the apparatus thus behave inversely to the migration velocities of the anions. The field strength decreases abruptly at the interface from left to right (in Fig. 20.10).

This abrupt change in field strength now causes a stabilization of the interface, in which the influence of diffusion is eliminated. If an acetate ion travels ahead by diffusion, it reaches a region of lower field strength in the chloride section, so that it migrates more slowly here and stays behind again. Conversely, if a Cl^- ion is left behind the interface by the leftward component of diffusion, it will enter a region of higher field strength in the acetate section, so that it will migrate faster until it arrives back in the chloride section.

The separation process given by Kendall (1923) exploits this stabilization effect several times. For example, in an experiment to separate chloride and thiocyanate ions, the separation zone, which consists of a horizontally lying glass tube, contains the following solutions in succession: Na-acetate/Na-chloride + Na-thiocyanate/Na-hydroxide. To avoid currents, agar gel is used as a support (Fig. 20.11).

After the voltage is applied, the anions move to the right, and a new interface forms between SCN^- and Cl^- ions. The three interfaces continue to move at equal rate without becoming fuzzy, since in each case an ion with lower mobility follows an ion with greater mobility.

Since the length of the separation distance can be chosen arbitrarily, very effective separations can be achieved in this way; for example, in (albeit unsuccessful) experiments for the separation of isotopes, migration distances of 30 m were achieved, with the interfaces remaining completely sharp. A disadvantage of this method is that the zones of the separated ions are directly adjacent to each other, so that complete separation is not possible for purely mechanical reasons (this difficulty can, however, be eliminated by interposing auxiliary substances).

20.5.2 Zone sharpening – disc electrophoresis

A high separation efficiency can only be achieved in zone electrophoresis if the individual zones are kept as narrow as possible. This is achieved by applying the substance mixture in the form of a thin line across the direction of travel. In addition, however, there are methods by which the width of a migrating zone can be reduced during the separation process; this effect of *zone sharpening* is achieved by

- Exploiting the *regulatory function* according to Kohlrausch (see below),
- by slowing down the zone front or
- by accelerating the back of the zone.

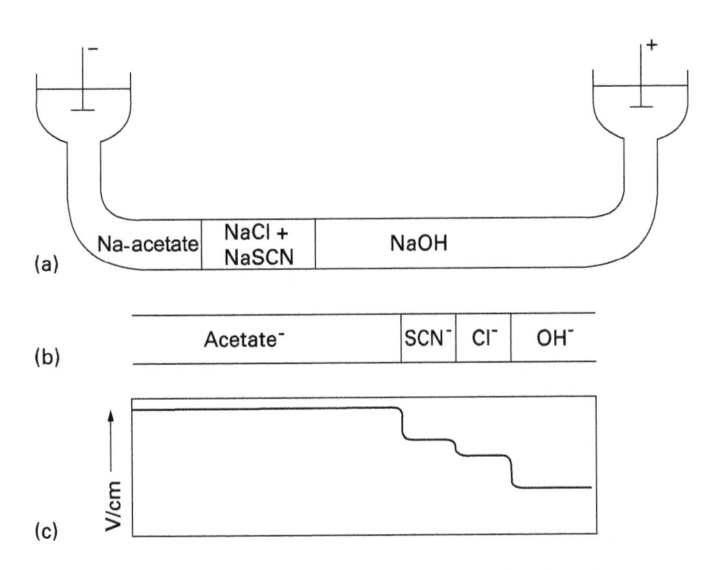

Fig. 20.11: Electrophoretic separations according Kendall (schematic).
a) Initial state; b) zones after separation; c) field strength along the separation path.

Regulating function according to Kohlrausch. In the previous section it was shown that the field strength ratio on both sides of a moving interface is given by the mobilities of the ions. As Kohlrausch (1897) found, a law-like relationship also applies to the concentration ratios.

Consider again the arrangement shown in Fig. 20.10. During electrophoresis, the current strength must obviously be the same in both sections of the tube; this is equal to the product of the specific conductance k and the field strength Q. If we again denote the chloride side by the subscript Ch and the acetate side by the subscript Ac, then:

$$k_{\mathrm{Ch}} \cdot Q_{\mathrm{Ch}} = k_{\mathrm{Ac}} \cdot Q_{\mathrm{Ac}} \tag{16}$$

respectively

$$\frac{k_{\mathrm{Ch}}}{k_{\mathrm{Ac}}} = \frac{Q_{\mathrm{Ac}}}{Q_{\mathrm{Ch}}}. \tag{16a}$$

Now the specific conductivity of an electrolyte is given by the concentrations C and the mobilities u of the ions it contains. The following therefore applies to the chloride section

$$k_{\mathrm{Ch}} = C_{\mathrm{Na^+}} \cdot u_{\mathrm{Na^+}} + C_{\mathrm{Cl^-}} \cdot u_{\mathrm{Cl^-}}. \tag{17}$$

Accordingly, in the acetate section

$$k_{\mathrm{Ac}} = C_{\mathrm{Na^+}} \cdot u_{\mathrm{Na^+}} + C_{\mathrm{Ac^-}} \cdot u_{\mathrm{Ac^-}}. \tag{18}$$

Since the molar concentrations of cation and anion must be the same at any point in the solution because of the electroneutrality condition, eqs. (17) and (18) can be transformed:

$$k_{Ch} = C_{Cl^-} \cdot u_{Na^+} + C_{Na^+} \cdot u_{Cl^-} \tag{19}$$

and

$$k_{Ac} = C_{Ac^-} \cdot u_{Na^+} + C_{Na^+} \cdot u_{Ac^-}. \tag{20}$$

From eqs. (19), (20), and (16) and the mobility eq. (15), we get:

$$\frac{k_{Ch}}{k_{Ac}} = \frac{C_{Cl^-} \cdot u_{Na^+} + C_{Na^+} \cdot u_{Cl^-} \text{ (in Ch section)}}{C_{Ac^-} \cdot u_{Na^+} + C_{Na^+} \cdot u_{Ac^-} \text{ (in Ac section)}} = \frac{Q_{Ac}}{Q_{Ch}} = \frac{u_{Cl^-}}{u_{Ac^-}}. \tag{21}$$

If you reshape and divide both sides by u_{Na^+} we get

$$\begin{aligned} &\frac{C_{Cl^-} \cdot u_{Na^+} + C_{Na^+} \cdot u_{Cl^-}}{u_{Cl^-} \cdot u_{Na^+}} \text{(in Ch section)} = \\ &\frac{C_{Ac^-} \cdot u_{Na^+} + C_{Na^+} \cdot u_{Ac^-}}{u_{Cl^-} \cdot u_{Na^+}} \text{(in Ac section)} \end{aligned} \tag{22}$$

or:

$$\frac{C_{Cl^-}}{u_{Cl^-}} + \frac{C_{Na^+}}{u_{Na^+}} \text{(in Ch section)} = \frac{C_{Ac^-}}{u_{Ac^-}} + \frac{C_{Na^+}}{u_{Na^+}} \text{(in Ac section)}. \tag{23}$$

eq. (23) states in general that the sum of the concentrations of each type of ion, divided by the corresponding mobilities, on one side of the migrating interface is equal to the sum of the concentrations, also divided by the mobilities, on the other side. This equation is known as the *Kohlrausch persistent function* or *regulating function*. It means that on both sides of a migrating interface, the concentration ratios are also fixed.

Let us now consider again an arrangement as already used by Kendall: A fast migrating anion (Cl^-) is followed by a zone with a slower migrating protein anion (P^-) in a glycinate buffer, which in turn is adjacent to a zone with pure glycinate buffer; the migration speed of the glycinate ion is even lower than that of the protein ion (Fig. 20.12).

If we assume for simplicity that the mobilities of the ions in front of and behind the interface have similar values, then in the above example the protein concentration after formation of the protein zone must be set so that it corresponds to that of the chloride ions in terms of magnitude. This above all determines the protein concentration behind the migrating interface.

The length of the protein zone depends on the concentration, the absolute amount of protein in the system and the diameter of the tube of the electrophoresis apparatus. If the amount of protein is low and the concentration is relatively high, a narrow zone will form. Its width is independent of the width of the initial zone before electrophoresis and thus also independent of the initial concentration of the protein solution.

The concentrations of anion and cation behind the protein zone are again determined by the Kohlrausch function. They are set in such a way that a stationary interface is created through which electrolyte flows when the protein zone advances.

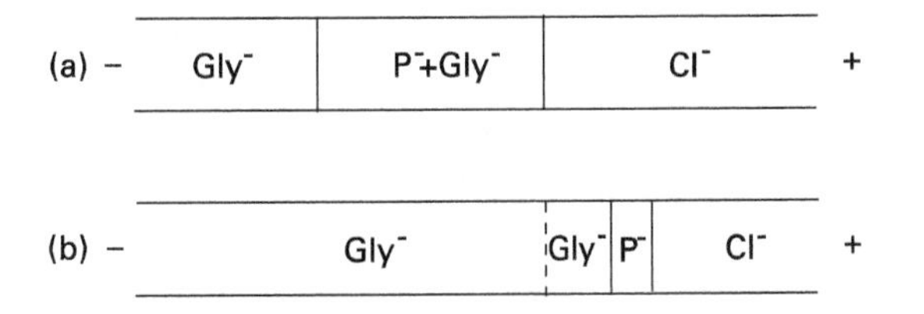

Fig. 20.12: Zone sharpening by applying the *Kohlrausch regulating function*.
a) Initial state; b) state after separation of the anions.

Zone sharpening by slowing down the zone front. If a zone enters from a region of higher migration velocity into a region of lower migration velocity, it is compressed in the ratio of the two migration velocities. This effect can be achieved in several ways.

The migration speed decreases if the field strength on the path of the zone decreases by increasing the buffer concentration (and thus the electrical conductivity). Another possibility is to influence ion mobility by changing the pH, and finally, the migration rate slows down when the zone enters a gel from a clear solution or when it migrates in a gel whose pore size decreases. For example, ion mobility in 5% polyacrylamide gels is smaller than in 3% gels.

The reduction in migration rate can occur either by abruptly changing the conductivity, pH, or pore size of gels at a particular point along the separation path, or by gradual changes in a gradient along the separation path.

Zone sharpening by accelerating the backside of the zone. The effect of zone sharpening can further be achieved by accelerating the migration rate of the backside of a zone. This acceleration occurs when the zone is overtaken by a migrating interface behind which the field strength increases.

Unlike zone sharpening using the Kohlrausch function, where the width of the zone does not depend on its original dimension, the other sharpening methods cause it to be compressed to a specific fraction of its original extent.

Disc electrophoresis. In *disc electrophoresis* (after disc = discontinous and disc = disk), as described by Ornstein (1964) and Davis (1964), several of the described sharpening effects are exploited together. One application of this is *isotachophoresis*.

A vertical tube (Fig. 20.13, left) contains the cathode electrolyte (glycinate buffer solution of pH 8.3) at the top, a zone with the mixture of substances to be separated below, and a zone with buffer solution again below. These two zones are protected against interference by convection by a wide-pore polyacrylamide gel, the electrolyte contained therein has a pH of 6.7; both contain Cl^- anions.

During electrophoresis, the fast-migrating Cl^- ions first separate from the slower protein ions in the substance zone. The glycinate ions of the cathode zone enter the area with pH 6.7, where their migration speed is very small, since this pH is close to that of their isoelectric point. Thus, there is an arrangement as in the Kendall method.

The protein ions arrange themselves according to their individual migration velocities in contiguous layers moving downward at equal velocities. Since they are followed by an anion with smaller mobility and preceded by one with greater mobility, interfacial stabilization occurs. At the same time, a strong sharpening effect according to the Kohlrausch function is obtained by suitable adjustment of the ion concentration in the third layer; the thickness of the individual protein layers after concentration is of the order of about 10 μm (Fig. 20.13, center).

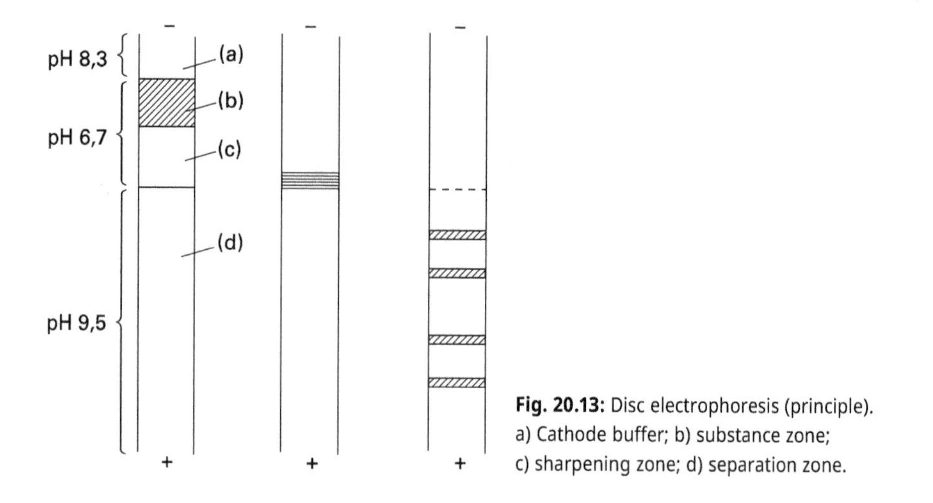

Fig. 20.13: Disc electrophoresis (principle). a) Cathode buffer; b) substance zone; c) sharpening zone; d) separation zone.

Advancing further, the protein zone now enters a fourth section of the separation path, where both the pH and the pore size of the gel change. The buffer solution is adjusted to pH 9.5, and the polyacrylamide gel is more narrow-pored due to higher concentration during polymerization. Accordingly, the protein zone migrates more slowly in this gel, resulting in zone sharpening. The pH change increases the migration speed of the subsequent glycinate ions to such an extent that they overtake the protein zone and compress it again by accelerating the zone back. Subsequently, the proteins migrate in a region of uniform glycinate buffer, where they are pulled apart into separate zones according to the principle of ordinary carrier electrophoresis (Fig. 20.13, right). A stationary interface forms at the boundary of the two gels, across which the pH difference is maintained.

Disc electrophoresis is usually performed as a micro-method in small tubes about 7 cm long and 5 mm in diameter. The efficiency of the method is demonstrated, among other things, by the fact that human serum protein, which is split into only 5 fractions in paper electrophoresis, yields over 30 zones in the disc method. Furthermore, by selecting different buffer systems, the method can be adapted to a wide variety of separation problems.

20.5.3 Ion focusing

As already mentioned, the migration rate of ampholytes assumes the value zero at the isoelectric point. If an ampholyte is allowed to migrate along a separation path which has a pH gradient in the direction of the electric current, the cationic form in the acidic range moves to the left (Fig. 20.14) and the anionic form in the basic range moves to the right until the isoelectric point is reached in each case.

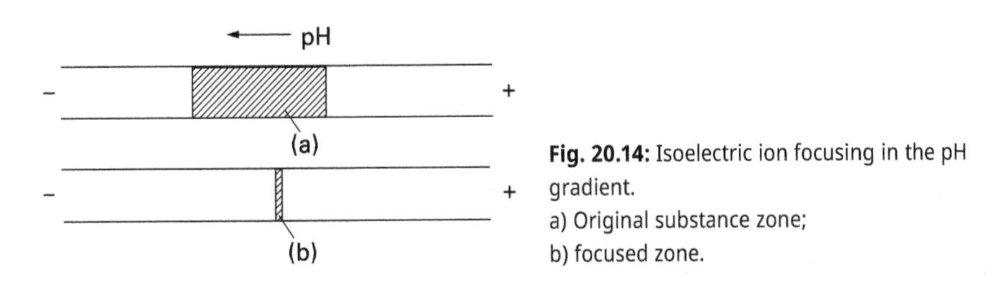

Fig. 20.14: Isoelectric ion focusing in the pH gradient.
a) Original substance zone;
b) focused zone.

The ampholyte is thus focused at the pH of this point, regardless of where the original application site was on the pH gradient and how wide the initial zone was.

The sharpness of the focusing depends on the steepness of the pH gradient and the amount of ampholyte present. The concentration distribution in the focusing zone corresponds to a Gaussian curve. If a mixture of substances contains several ampholytes with different isoelectric points, these accumulate at different points of the pH gradient and form a so-called *isoelectric spectrum*.

The main difficulty with this method is to keep the pH gradient constant over time during electrophoresis, since the H^+ ions also migrate in the electric field. According to a suggestion by Svensson, a mixture of polyaminopolycarboxylic acids is added to the solution. These compounds, which are commercially available under the name Ampholine®, are also ampholytes and therefore arrange themselves according to their isoelectric points along the separation path, forming a constant pH gradient.

Another focusing method, used mainly for inorganic ions, was given by Schumacher. A strong acid is added to the anode vessel of an apparatus for paper electrophoresis and a complexing agent, e.g., the diammonium salt of nitrilotriacetic acid, is added to the cathode compartment (Fig. 20.15). Both solutions diffuse into the paper strip; a zone is formed in the center containing a gradient of acid concentration increasing from left to right and a gradient of complexing agent concentration increasing from right to left. Since the free nitrilotriacetic acid is very little dissociated, the concentration of the nitrilotriacetate ions, which are decisive for complex formation, is very low in the acidic range, but increases sharply with increasing pH, i.e., with the approach to the cathode.

If a mixture of cations capable of forming negatively charged complexes with the nitrilotriacetate ions (e.g. ions of the trivalent rare earths) is applied in a wide zone to the center of the paper, anionic complexes are formed on the cathode side which

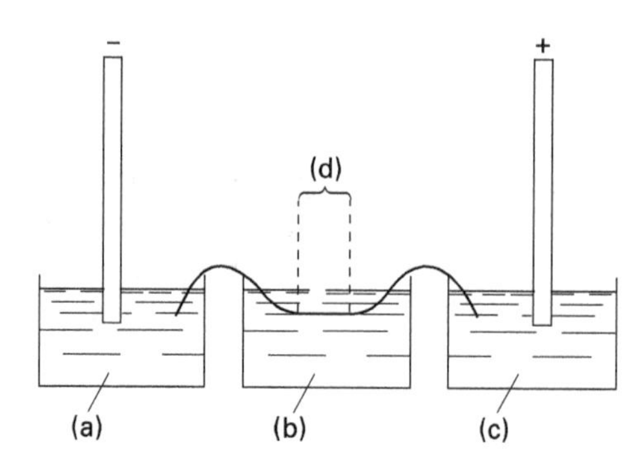

Fig. 20.15: Electrophoretic ion focusing in the complexing agent gradient.
a) Cathode compartment with complexing agent;
b) coolant (CCl_4);
c) anode compartment with acid;
d) substance zone (on paper strips).

migrate to the right toward the anode in the electric field. On the anode side of the paper strip, there are hardly any nitrilotriacetate ions because of the high acid concentration; the cations of the added rare earths therefore remain as such in the electrolyte and migrate to the left toward the cathode. At a certain point, given by the concentration of complexing agent and the dissociation constant of the complex, cations and anions of a certain rare earth are present in equal concentrations; here the net mobility is zero, and at this point the substance in question is focused to a narrow non-migrating zone.

Ions with different complexation constants result in focusing zones at different locations on the complexation gradient. The separations achievable in this way are only due to differences in the complexation constants, not to differences in the ionic mobilities.

20.6 One-dimensional separations with semipermeable membrane (electrodialysis)

The method of electrodialysis causes separations with the help of membranes through which ions with small diameter can pass, while ions with larger dimensions and colloids are retained.

The membranes are usually made of plastic films; cellulose and nitrocellulose are most commonly used. Various pore widths can be achieved by controlled manufacturing conditions (see section 4, Dialysis). Membranes made of ion exchangers occupy a special position with regard to selectivity: cation exchanger membranes are practically permeable only for cations, anion exchanger membranes only for anions.

For laboratory purposes, apparatuses with three separate chambers are often used; a central chamber containing the substance mixture is bounded by two membranes, and the electrodes are located in the two side chambers (Fig. 20.16). Apparatus with ion-exchange membranes usually contain only a single membrane separating the anode and cathode compartments.

Electrodialysis is mostly used on a preparative scale, mainly to remove low molecular weight components from solutions of high molecular weight compounds (e.g. protein).

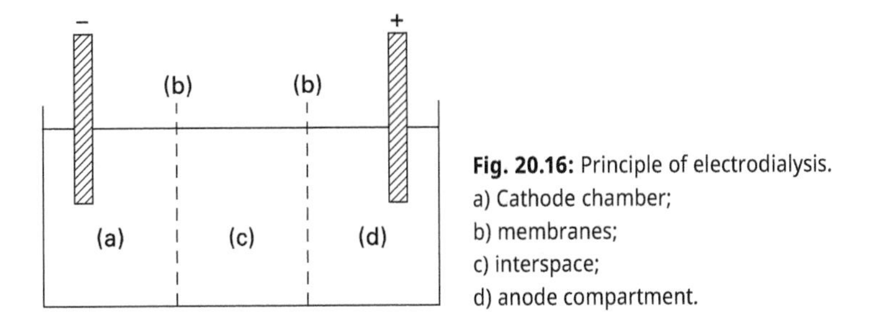

Fig. 20.16: Principle of electrodialysis.
a) Cathode chamber;
b) membranes;
c) interspace;
d) anode compartment.

As an example of the analytical application of ion exchange membranes, consider the separation of boron traces from sodium metal according to Logie. The metal is dissolved in water and the NaOH solution containing sodium borate is poured into the anode compartment of an electrodialysis cell divided with a cation exchange membrane; the cathode compartment contains 0.5 N NaOH solution. During electrodialysis, Na^+ ions migrate through the membrane, while borate anions (along with aluminate, chromate, and others) remain in the anode compartment. Electroneutrality is maintained by discharging hydroxyl ions at the anode to form water. Boron contents down to about 1 ppm have been determined using this method.

20.7 Countercurrent electrophoresis

If ions with only slight differences in migration rates are to be separated from one another by electrophoresis, long runs, i.e. relatively large apparatus, are required. However, it is possible to shorten the running distances by allowing the electrolyte to flow against the direction of migration of the ions. Experimentally, this can be done by electrophoresis in a tube provided with inlet and outlet, or by unwinding a paper tape (as a carrier) which is pulled through the cathode and anode compartments. Such procedures are somewhat cumbersome, however; they have been devised primarily for isotope separation experiments.

20.8 Two-dimensional mode of operation

20.8.1 Overview

Continuous electrophoretic separations can be effected by the two-dimensional method of operation, in which the electrolyte flows perpendicular to the direction of the electric field. Several methods have been devised for this purpose, the most common of which is that using uniform buffer solution (Fig. 20.17, top). Another arrangement is derived from the Kendall method: The electrolyte consists of three partial streams with graded ionic mobilities, namely the cathode buffer, the anode buffer, and the substance zone in between (Fig. 20.17, middle). Finally, the focusing method can also be performed continuously by maintaining a pH gradient in the electrolyte flowing perpendicular to the field (Fig. 20.17, bottom).

Of these methods, only the first mentioned will be discussed in more detail, as the others have not gained any importance so far. In addition, a method called *immunoelectrophoresis* should be mentioned, which is to be addressed as a two-dimensional but discontinuous method.

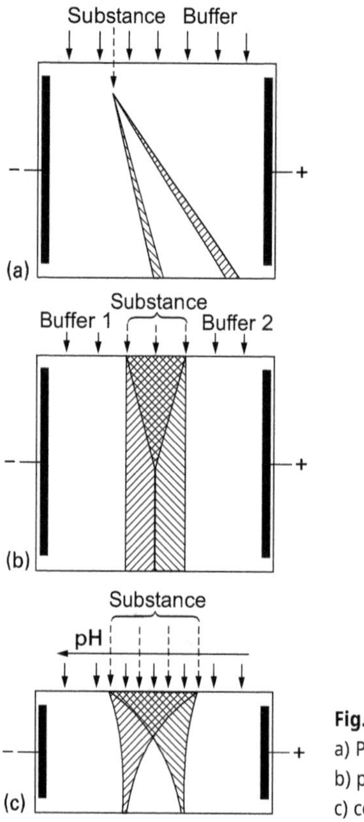

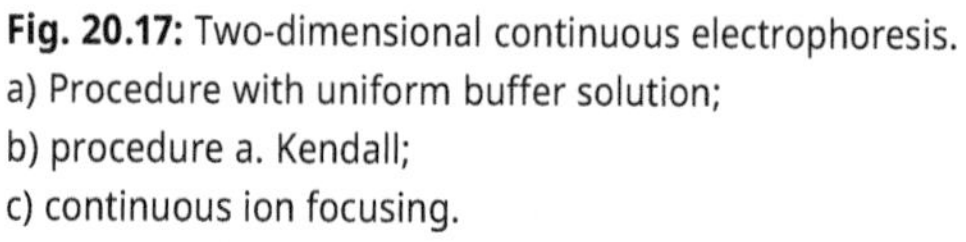
Fig. 20.17: Two-dimensional continuous electrophoresis.
a) Procedure with uniform buffer solution;
b) procedure a. Kendall;
c) continuous ion focusing.

20.8.2 Procedure with uniform buffer solution

Continuous electrophoretic separations with uniform buffer solution can be carried out carrier-free or in a carrier. In the carrier-free mode of operation (*Free-flow electrophoresis; FFE*), the buffer solution flows in a narrow gap about 0.5 mm wide between two cooled glass plates which are placed at a slight angle. The substance solution is introduced into the buffer solution flowing into the upper side at a specific point. The separated components are continuously collected at the lower edge with the aid of a larger number of thin discharge tubes.

Since this method is rather complicated in terms of equipment, the carrier method is usually preferred. A vertical paper curtain is immersed in a trough with buffer solution attached to the upper end (Fig. 20.18).

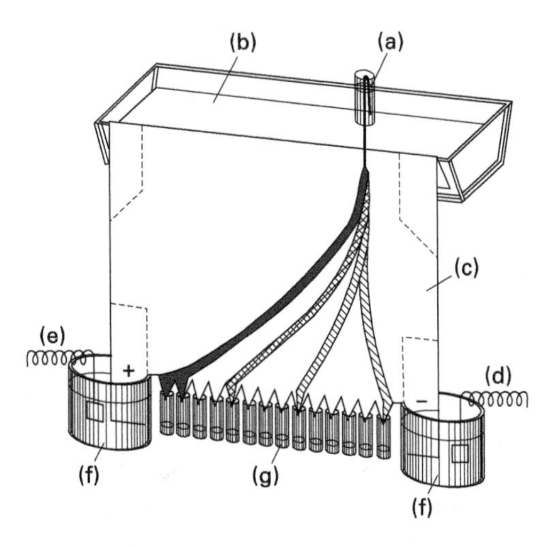

Fig. 20.18: Continuous two-dimensional carrier electrophoresis.
a) Substance container with wick;
b) trough with buffer solution;
c) paper curtain;
(d, e) electrodes;
f) electrode container;
g) collecting vessels.

The capillary forces draw in the liquid, which then flows down the paper. The substance is supplied via a wick; the mixture is pulled apart by the effect of the electric field, and the individual components can be collected separately at the bottom.

20.8.3 Immuno-electrophoresis

In the *immuno-electrophoresis* indicated by Grabar, an immunochemical precipitation reaction is performed in addition to the electrophoretic separation to identify proteins. First, the protein mixture is separated into zones by agar gel electrophoresis. Along the separation zone there is a groove in the agar plate, which is filled with an antigen solution (Fig. 20.19). This diffuses into the gel perpendicular to the longitudinal direction of the channel, while the proteins from the electrophoretically separated narrow zones in turn diffuse radially in the opposite direction. If the protein in

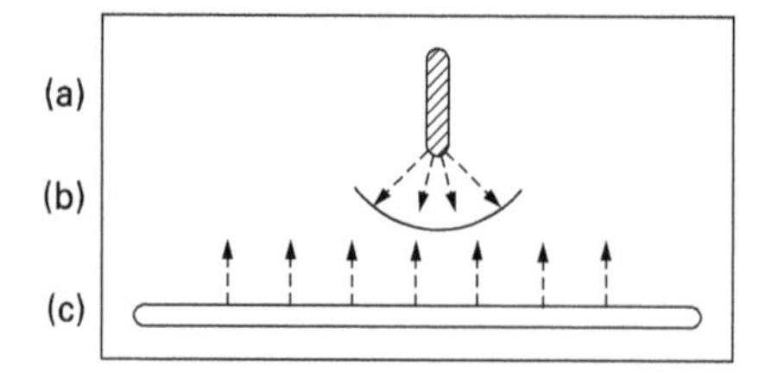

Fig. 20.19: Immuno-electrophoresis (principle).
a) Agar gel with protein zone;
b) reaction zone;
c) trough with antigen solution.

question reacts with the antigen, precipitation occurs at sites with favorable concentration ratios (cf. Part 2, section 7.2.3). The precipitation zones are ellipse sections because of the different diffusion directions.

20.9 Efficacy and scope of the method

By electrophoresis, separations of ions are possible with relatively little experimental effort even in the case of great chemical similarity; the amounts of substance used can vary within wide limits.

In the field of inorganic chemistry, the method is used for the separation of stable complexes. Furthermore, mixtures of radioactive ions from nuclear fission experiments have occasionally been investigated using the focusing method. The advantage of the method is that it is particularly suitable for extremely small quantities of substances.

However, the main field of application of electrophoresis is in the field of organic analysis, and there above all in the field of biochemistry. With the aid of this analytical method, low-molecular compounds such as amino acids (cf. Fig. 20.20), amines, alkaloids, acids, phenols, sugars and polysaccharides (as boric acid complexes), dyes, hormones, vitamins and many other substances can be separated; in addition, the method can be used for high-molecular substances, and – also because of the extremely gentle conditions under which it works – it has attained a unique position among all analytical methods, especially for the separation of sensitive protein substances. An important branch of application is clinical diagnosis, which makes use of electrophoresis of blood serum for the diagnosis of kidney, liver and blood diseases (cf. Fig. 20.21). Furthermore, larger particles such as viruses, cell components and bacteria can also be distinguished and in some cases separated on the basis of their migration speeds.

A disadvantage is the considerable time required (if one does not work on an ultramicro scale with very small migration distances), furthermore losses due to adsorption are possible with carrier electrophoresis.

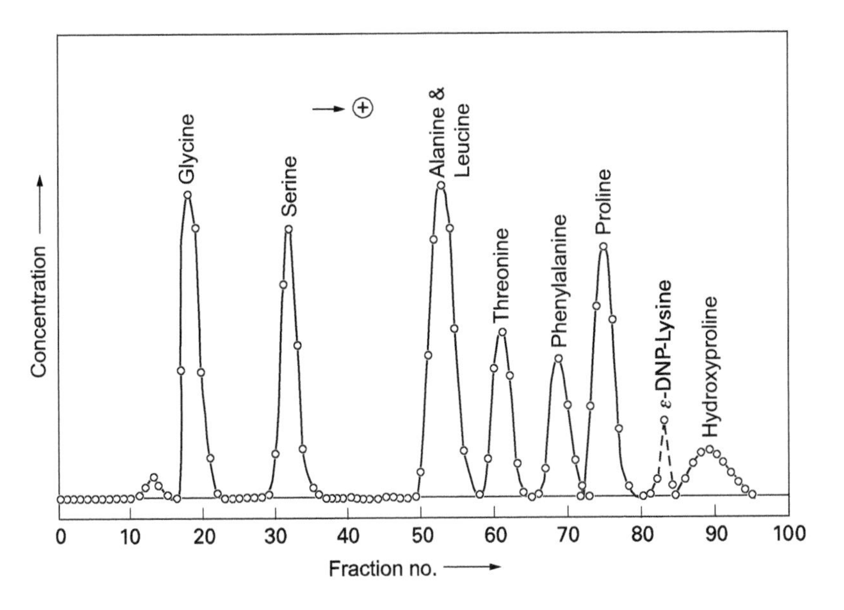

Fig. 20.20: Separation of amino acids by column electrophoresis.

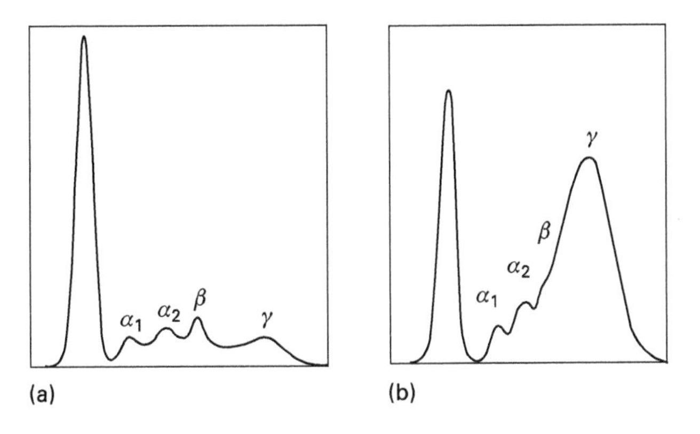

Fig. 20.21: Electrophoresis of human serum.
Main band = albumin; α_1, α_2, β, γ = globulins.
a) Normal serum;
b) patient with liver cirrhosis.

General literature

R. Frazier, J. Ames, & H. Nursten, Capillary Electrophoresis for Food Analysis: Method Development, Royal Society of Columbus, Cambridge, UK (2000).

B. Gas, Fundamentals of Electrophoresis, Wiley-VCH Verlag, Weinheim (2008).

C. Garcia, K. Chumbimuni-Torres, and E. Carrilho, Capillary Electrophoresis and Microchip Capillary Electrophoresis: Principles, Applications and Limitations, John Wiley & Sons, Inc., Hoboken, N.J. (2013).

S. Magdeldin, Gel Electrophoresis Principles and Basics, InTech, Rijeka, Croatia (2012).

P. Righetti, Capillary electrophoresis, in: Handbook of Proteins, edited by M. Cox & G. Phillips, Vol. 2, 855–858. John Wiley & Sons Ltd, Chichester, UK (2007).

R. Waetzig, Quantitation and Validation in Electrophoresis, Wiley-VCH Verlag, Weinheim (2005).

Literature to the text

Disc electrophoresis

Y. Michikawa, A. Umetsu, and T. Imai, Application of preparative disc gel electrophoresis system to proteomics research, Bunseki Kagaku *12*, 667–671 (2005).

Electrodialysis

C. Huang, T. Xu, Y. Zhang, Y. Xue, and G. Chen, Application of electrodialysis to the production of organic acids: State-of-the-art and recent developments, Journal of Membrane Science *288*, 1–12 (2007).

M. Kariduraganavar, R. Nagarale, A. Kittur & S. Kulkarni, Ion-exchange membranes: preparative methods for electrodialysis and fuel cell applications, Desalination *197*, 225–246 (2006).

T. Xu & C. Huang, Electrodialysis-based separation technologies: a critical review, AIChE Journal *54*, 3147–3159 (2008).

Focus

P. Righetti, E. Fasoli & S. Righetti, Conventional isoelectric focusing in gel slabs and capillaries and immobilized pH gradients, Methods of Biochemical Analysis *54* (Protein Purification), 379–409 (2011).

Immuno-Electrophoresis

L. Amundsen & H. Siren, Immunoaffinity CE in clinical analysis of body fluids and tissues, Electrophoresis *28*, 99–113 (2007).

D. Schmalzing, S. Buonocore, and C. Piggee, Capillary electrophoresis-based immunoassays, Electrophoresis *21*, 3919–3930 (2000).

Isotachophoresis

P. Kosobucki & B. Buszewski, Isotachophoresis, Springer Series in Chemical Physics *105* (Electromigration Techniques), 93–117 (2013).
S. Shrivastava, A. Jain, S. Mhaske, and S. Nayak, Isotachophoresis: a technique of electrophoresis for the separation of charged particles, Research Journal of Pharmacy and Technology *2*, 642–647 (2009).
P. Smejkal, D. Bottenus, M. Breadmore, R. Guijt, C. Ivory, F. Foret, and M. Macka, Microfluidic isotachophoresis: a review, Electrophoresis *34*, 1493–1509 (2013).

Capillary electrophoresis techniques

V. Dolnik, S. Liu & S. Jovanovich, Capillary electrophoresis on microchip, Electrophoresis *21*, 41–54 (2000).
V. Kasicka, Recent developments in capillary electrophoresis and capillary electrochromatography of peptides, Electrophoresis *27*, 142–175 (2006).
X. Subirats, D. Blaas & E. Kenndler, Recent developments in capillary and chip electrophoresis of bioparticles: Viruses, organelles, and cells, Electrophoresis *32*, 1579–1590 (2011).
W. Thormann, Capillary electrophoretic separations, Methods of Biochemical Analysis *54* (Protein Purification), 451–485 (2011).

Carrier-free two-dimensional electrophoresis

M. Bowser & R. Turgeon, Micro free-flow electrophoresis: theory and applications, Analytical and Bioanalytical Chemistry *394*, 187–119 (2009).
D. Kohlheyer, J. Eijkel, and A. van den Berg, Miniaturizing free-flow electrophoresis – a critical review, Electrophoresis *29*, 977–993 (2008).
G. Weber & R. Wildgruber, A versatile free-flow electrophoresis system for proteomics applications, American Biotechnology Laboratory *22*, 26, 28 (2004).

21 Migration of particles in the concentration gradient (diffusion)

21.1 Historical development

The formulation of the laws governing diffusion is due to Fick. Separations by diffusion using semipermeable membranes were first carried out by Graham (1861); the decisive advance in the production of membranes with reproducible pore sizes was achieved by Elford (1930).

21.2 General

21.2.1 Definitions

If two different gases or two miscible solutions of different compositions are adjacent to each other, mutual mixing, i.e., concentration equalization, occurs over time as a result of the random Brownian molecular motion until the composition of the system is uniform at all points. This process is called *diffusion*.

Separations by diffusion in the liquid phase with the aid of a membrane which is permeable only for a part of the dissolved substances are called *dialysis*. The initial solution containing the compounds to be dialyzed is the *dialysate*, the liquid into which the fractions to be separated diffuse is the *diffusate*.

21.2.2 Fick's law

The amount dS of a substance diffusing in time dt through a section of thickness dx with cross-section q (e.g. of a pipe) depends, among other things, on the concentration gradient $\frac{\mathrm{d}c}{\mathrm{d}x}$ depends:

$$\frac{\mathrm{d}S}{\mathrm{d}t} = -D \cdot q \cdot \frac{\mathrm{d}c}{\mathrm{d}x} \tag{1}$$

or

$$\mathrm{d}S = -D \cdot q \cdot \frac{\mathrm{d}c}{\mathrm{d}x} \cdot \mathrm{d}t. \tag{1a}$$

The expression on the right side of the equation has a negative sign because the solute diffuses in the direction of the smaller concentration.

The constant D is the so-called *diffusion coefficient*, which is a quantity characteristic of the diffusing substance. D gives the number of moles of substance diffusing in the

https://doi.org/10.1515/9783111181417-021

unit of time through a cross-section of 1 cm^2 at a concentration gradient of 1 mol/l. The diffusion coefficient of undissociated compounds is independent of concentration to a first approximation, but changes strongly with a change in temperature or solvent.

D is related to the size of the diffusing particle or molecule in a medium as follows:

$$D = K \cdot T \cdot B \tag{2}$$

K = Boltzmann constant
T = Temperature
B = Particle mobility

where

$$B = [1 + A \cdot l/r_p + Q \cdot l/r_p \cdot \exp(-b\, r_p/l)] 6\pi\, \nu\, r_p \tag{3}$$

b, Q, A = Empirical constants
l = Mean free path of a gas molecule
r_p = Particle radius
ν = Gas viscosity

is. Thus, a particle or molecule size can be expressed by its diffusion coefficient *D*.

Gases diffuse relatively quickly, so that in vessels that are not too large, differences in concentration can be expected to equalize in periods of minutes to hours. In solutions, however, the process takes much longer; here, days, weeks or even months are required until complete mixing takes place. Particles in the nanometer and micrometer range, such as those that exist in the atmosphere from combustion processes, require much longer periods for complete mixing.

21.2.3 Membranes

Since diffusion is an undirected movement running in all directions, in which all components of a system participate, it cannot easily be exploited for effective separations. The separation effect is improved either by the two-dimensional method of operation (see below) or by the use of special partition walls which are permeable to only some of the substances to be separated but retain the others. These partition walls are usually used in the form of thin membranes.

When such membranes are used in separations in aqueous solutions, it is customary to stir intensively both the solution to be dialyzed on one side and the liquid on the other side of the membrane in order to accelerate the separation. Then the diffusion process no longer takes place in the entire volume of the system, but exclusively in the membrane, and Fick's law can be transformed in the following way:

$$dS = K \cdot A \cdot \frac{C_1 - C_2}{dx} \cdot dt. \tag{4}$$

The difference of the concentrations $c_1 - c_2$ on both sides of the membrane has taken the place of the expression *dc*; choosing as c_1 the smaller concentration in the diffusate eliminates the minus sign of the original equation. *A* is the membrane area, and *K* is called the *permeability constant*. This depends (in addition to temperature) on the nature of the diffusing substance and the permeability of the membrane. If the permeability constants of different membranes with the same diffusing substance are determined, comparisons can be made about their permeabilities.

There are several types of such partitions:
- Porous membranes,
- ion exchange membranes and
- *Solution Membranes.*

Porous membranes for the separation of gases consist of sintered glass or metal powder with pore widths of about 1–5 µm. The relatively narrow pores allow atoms or molecules of light gases, such as H_2 or He, to diffuse through much faster than larger organic molecules. However, one does not achieve quantitative separations with such partitions, but only more or less good enrichments.

For the separation of dissolved compounds, membranes made of animal skins (pig bladders, intestines), which were used in the past, have now been replaced by films made of plastics; these can be obtained with reproducible pore widths by controlled manufacturing conditions. Cellulose, acetyl cellulose, nitrocellulose, PTFE and polyacrylates are among the materials used.

Nitrocellulose films with specific pore sizes are obtained, for example, by pouring a solution of nitrocellulose in ether-alcohol mixtures onto glass plates and evaporating the solvent. The porosity can be varied by the mixing ratio of the two solvent components; the higher the ether content, the denser the membrane.

The average pore widths of commercially available membranes are usually between about 3 µm and 500–700 µm, but the range can be extended somewhat; for example, nitrocellulose membranes with pore widths of about 1–3000 µm are available.

Ion exchange membranes are occasionally used in separations by electrodialysis, but will be discussed here in context.

The production of sufficiently mechanically stable homogeneous ion exchange membranes is difficult; therefore, mixtures of ion exchange powder with an inert plastic binder material (polyethylene, polystyrene, rubber, phenolic resin, etc.) are usually used. A high percentage of the exchanger must be present in the finished membrane so that the individual particles are not completely enveloped by the inert material, but can form continuous areas.

The essential feature of such membranes is the high electrical charge density, which is due to the large concentration of dissociating functional groups (cellulose and collodion membranes also carry electrical charges, but in such low density that their properties are relatively little affected by this). As a result, ion-exchange membranes put a very high resistance to the passage of ions carrying the same charge sign as the exchanger framework, while ions with opposite charge sign can diffuse through. Cation-exchange membranes are thus permeable only to cations, anion-exchange membranes only to anions. This phenomenon is referred to as *permselectivity*.

In the third type of membranes, which will be called *solution membranes* here, a solution mechanism is to be regarded as decisive for diffusion. The materials used are films of hydrophobic compounds, such as silicone polymers, acetyl cellulose, etc., through which nonpolar organic substances can pass, while ions and hydrophilic substances are retained. Separation processes based on this principle were termed *diasolysis* by Brintzinger.

An extension of the method can be derived from separations in which substances are shaken out from aqueous solutions with organic solvents (Part 2, Chap. 7). For example, iron(III) chloride can be extracted from strongly saline solutions with β,β'-dichlorodiethyl ether as the solvated compound $HFeCl_4$. If a membrane of polyvinyl chloride is prepared with a portion of this ether, it will allow Fe^{3+} to pass through, while other chlorides, e.g., of Al^{3+}, are quantitatively retained.

Accordingly, uranyl nitrate can be removed from aqueous nitrate-containing solutions by PVC membranes with tributyl phosphate (including phosphoric acid esters) as plasticizer. The method, which can undoubtedly still be extended, works very selectively.

Finally, the diffusion of hydrogen through palladium membranes should be mentioned here, based on the solubility of the gas in the metal.

21.3 One-dimensional separations by diffusion

21.3.1 Separations in the gas phase

Diffusion separation requires the presence of a *diffusion sink*. This can be realized by rapid, continuous removal of an analyte in the gas stream or by irreversible fixation of the analyte by adsorption or chemical bonding. Above the surface of a diffusion sink, the analyte concentration is approximately zero, so that, according to Fick's laws, the substance flows to the sink, where it attempts to compensate for the concentration imbalance.

Gas permeation. The most effective of all separations by diffusion in the gas phase is based on the diffusion of hydrogen through palladium membranes at temperatures from about 200 °C. The process is mainly used for the preparative recovery of extremely pure hydrogen, but – especially in the two-dimensional mode of operation – also has analytical significance (see below).

Other diffusion processes of gases, e.g. diffusion of H_2 and He through quartz membranes at elevated temperature or of He through membranes of fluorinated plastics, are less used.

However, *gas permeation* is of certain importance for the production of defined calibration gas concentrations (e.g. volatile organic halogenated hydrocarbons, H_2S, SO_2, NO_2, HF, etc.). A constant partial pressure of the gas is set at a constant temperature. The gas molecules then permeate through the separation membrane at a similarly constant rate. As on the other side of the membrane the gas is removed by a constantly flowing dilution gas (e.g. air or nitrogen). This allows trace gas concentrations in the pptv range to be generated continuously.

Conversely, when selected membranes are used, gaseous analytes can be separated selectively from the media surrounding the membrane surface (external atmosphere, water). This has been realized, for example, for inlet systems of mobile mass spectrometers (polysiloxane membrane) for monitoring toxic gases in the air.

21.3.2 Separations in liquid phase (dialysis)

Dissolved substances are separated by diffusion exclusively with the aid of membranes. Since the pore sizes are not uniform but vary with a certain scattering around the mean value, effective separations can only be achieved if the molecular sizes of the compounds to be separated show considerable differences. The method is therefore mainly used to remove low molecular weight compounds from solutions of high polymers; however, membrane approaches with graded pore sizes can also be used to separate higher molecular weight compounds to a limited extent.

Since the diffusion velocities in solutions are very low, it is advisable to accelerate the separations by means of equipment. The evaluation of eq. (2) shows that the largest possible membrane area in relation to the solution volume should be aimed for. Furthermore, small diffusion distances, i.e. thin membranes, and large concentration gradients are favorable. The latter are achieved by a large volume of diffusate and frequent renewal of the solvent. Furthermore, stirring should be used to ensure that the diffusing compounds are continuously fed to or away from the membrane so that additional diffusion processes do not have to take place in the solutions. Finally, the process can be accelerated considerably by increasing the temperature.

In the experimental performance of dialysis, membranes are often used in the form of hollow fibers, tubular or bag-shaped containers that contain the dialysate and are suspended in a larger vessel containing water. Other devices contain flat membranes that divide the container into chambers. An example of the numerous designs is given in Fig. 21.1.

If the dialyzed substances are to be recovered from the diffusate, a device in which the liquid is continuously evaporated and returned to the cell is recommended; in this way, the evaporation of large quantities of liquid after dialysis can be avoided (Fig. 21.2).

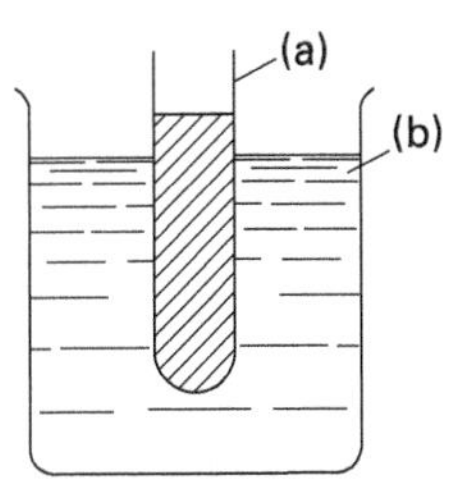

Fig. 21.1: Dialysis (principle).
a) Bag-shaped membrane containing the dialysate;
b) Diffusate.

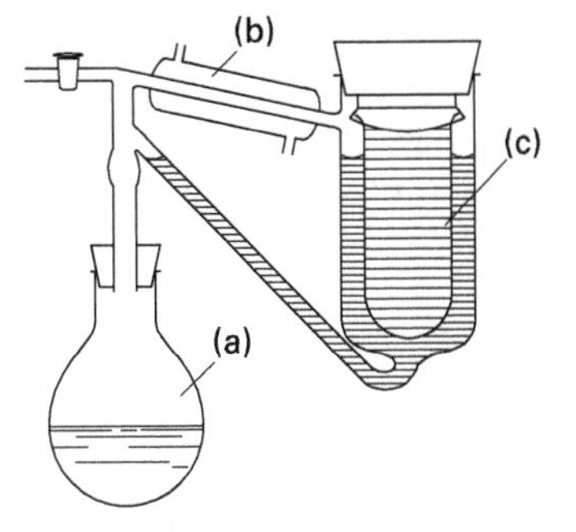

Fig. 21.2: Dialyzer with continuous evaporation of the diffusate.
a) Flask for evaporation of the diffusate;
b) cooler;
c) dialysate.

In the case of the so-called *high-flux dialyzers*, the main aim is to increase the membrane surface area. For example, a device described by Craig (1955) has a membrane of 45 cm^2 surface area with only 0.6 ml solution volume.

Reverse osmosis. *Reverse osmosis* is another method of accelerating separations by diffusion; the solution to be dialyzed is forced through the semipermeable membrane at an overpressure of several atmospheres. This causes the reversal of the otherwise voluntary process of concentration equilibration. For large-scale seawater desalination, pressures of up to 80 bar are applied. The membrane structure is optimized in such a way that only the solvent water can pass through the membrane, but not the substances dissolved in it. Other applications include the concentration of fruit juices in the food sector. Hollow fiber and nanofiltration membranes are mainly being used.

21.4 Pervaporation

The term *pervaporation* derives from the two underlying processes: Permeation of the solute through the membrane and evaporation into the gas phase. In this process, the liquid mixture to be separated is present on one side of the membrane under slight overpressure, while on the downstream side a vacuum brings the permeate into the gas phase. In some cases, a purge gas can also be used to remove the permeate and thus create the necessary diffusion sink. The driving force here is the difference in the partial pressures of the analytes involved. The separation is based on the different solubility of the analytes in the membrane. This can be described by a solution-diffusion model. The separation process can be controlled by specific hydrophilization or hydrophobization.

21.5 Countercurrent method

The simplest way to implement countercurrent dialysis is to lay a dialysis tube inside a tube through which water flows and through which the dialysate flows in the opposite direction. For larger amounts of substance, machines have been developed with several chambers through which dialysate and diffusate flow alternately.

Countercurrent dialysis is used primarily for preparative purposes; the process is probably of no significance analytically. Especially in the form of hollow fiber modules, this technique is used for the production of pure water.

21.6 Two-dimensional separations by diffusion

The two-dimensional mode of operation is present in the so-called *molecular separators*, of which the separating nozzle according to Becker (1955), the frit separator according to Watson and Biemann (1964), the palladium separator according to Lovelock (1969) and the gap separator according to Brunnée (1969) should be mentioned. In these devices, the component of diffusion running perpendicular to the direction of flow of a gas stream is exploited.

Separation nozzle according to Becker and Ryhage. The gas mixture to be separated emerges from a narrow nozzle, flows freely for a short distance and then reenters a tube (Fig. 21.3). On the free path, diffusion causes the gas jet to widen, with the most rapidly diffusing components reaching the farthest outside. On re-entry into the tube, the outer part of the gas stream is peeled off and removed from the outer space by suction. The slower diffusing components are thus enriched in the remaining inner part. The efficiency of the process can be increased by connecting several such *separation nozzles* in series. A further improvement in separability is achieved by using a crossed laser beam.

Frit separators consist of a porous tube made of glass or metal with pore widths of about 1–5 µm through which the gas mixture flows (Fig. 21.4). Since small molecules diffuse through the pores of the tube faster than larger ones, partial separation occurs with enrichment of the higher molecular weight compounds in the remaining gas stream.

Similar in principle is a separator specified by Lovelock, in which the gas mixture flows through a palladium tube heated to about 250 °C. The device is used to separate hydrogen, which is the only gas that can diffuse through the metal and is completely removed if the tube is long enough. The gas is exhausted from the outside space or burned with oxygen on the wall of the palladium tube.

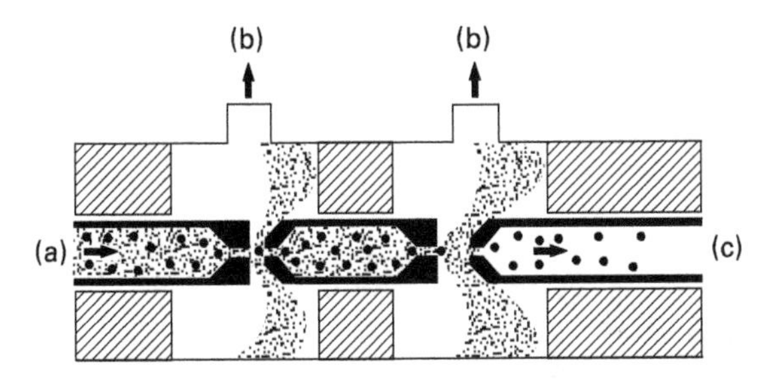

Fig. 21.3: Separating nozzles in two-stage arrangement.
a) Gas ingress;
b) pump connections;
c) gas leakage.

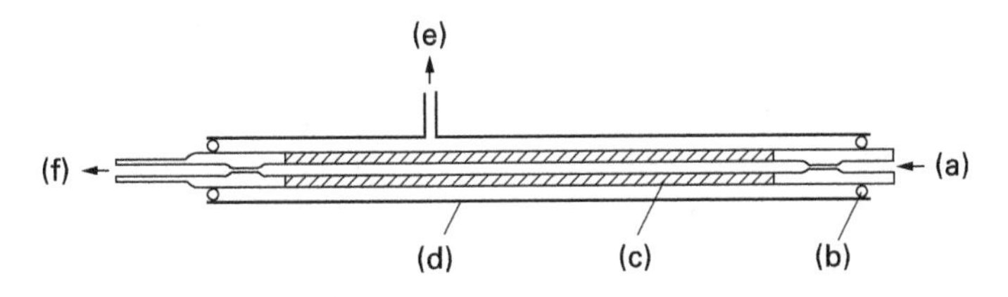

Fig. 21.4: Molecule separator.
(a) Gas inlet; (b) sealing ring; (c) porous tube; (d) evacuable outer tube; (e) pump connection; (f) gas leakage.

Since pure palladium becomes brittle after some time due to hydrogen action, palladium-silver alloys with about 25% Ag are used. The effectiveness of the tubes is destroyed by mercury, iodine and sulfur compounds; unsaturated compounds in the gas stream can be hydrogenated by the catalytic action of the palladium.

Finally, Brunnée specified a so-called *gap separator* in which the gas mixture flows in a circular channel between two concentric ribs. The channel is bounded by a lid which does not close completely at the top, so that gas can diffuse outward as well as inward through the annular gap thus formed (Fig. 21.5). The part of the gas flow leaving the channel on the side opposite the inlet opening is enriched with heavy components. The main advantage of the system is that the gap width can be set as desired between 0 and 50 μm, so that the quantity ratio of entering to enriched gas can be varied.

Diffusion separator for gas/particle separation. *Diffusion separators* (*denuder*) have become established for the enriching sampling of trace gases from the atmospheric aerosol. In numerous air monitoring problems, the trace component sought must be separated from the aerosol (particles or droplets; suspended in a carrier gas), since the gas species sought cannot always be distinguished from particle components in the subsequent analytical determination procedure.

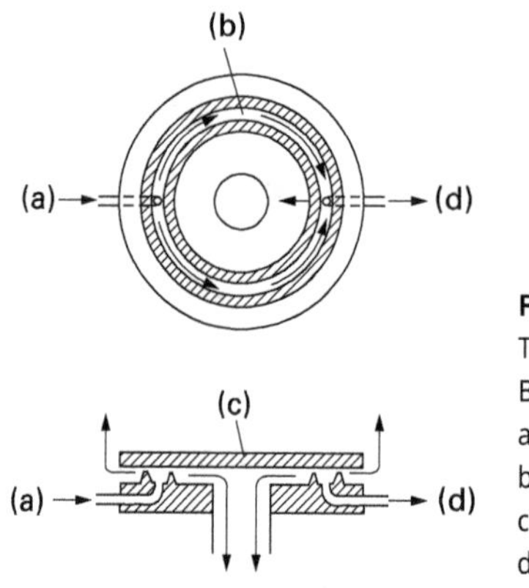

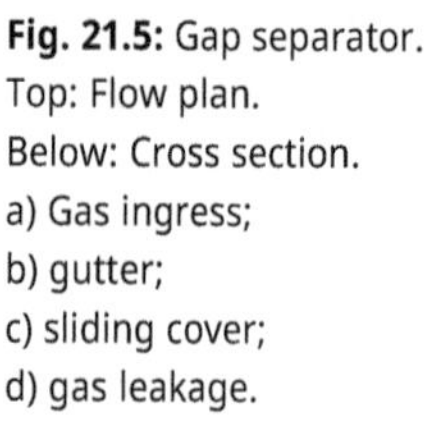

Fig. 21.5: Gap separator.
Top: Flow plan.
Below: Cross section.
a) Gas ingress;
b) gutter;
c) sliding cover;
d) gas leakage.

Tubular separators are frequently used for this purpose (see principle of operation in Fig. 21.6), the inner walls of which represent the diffusion sink. Since gases have a diffusion coefficient several logarithmic orders of magnitude higher than that of the interfering atmospheric particle constituents (3 nm < d_p < 10 μm), gas molecules often reach the inner wall during passage through a tube due to Brownian molecular motion, but not the aerosol particles. If irreversible binding of the gas molecules occurs there, further gas molecules are also deposited during passage. The aerosol is therefore freed from this separated trace gas component, which is enriched on the inner wall.

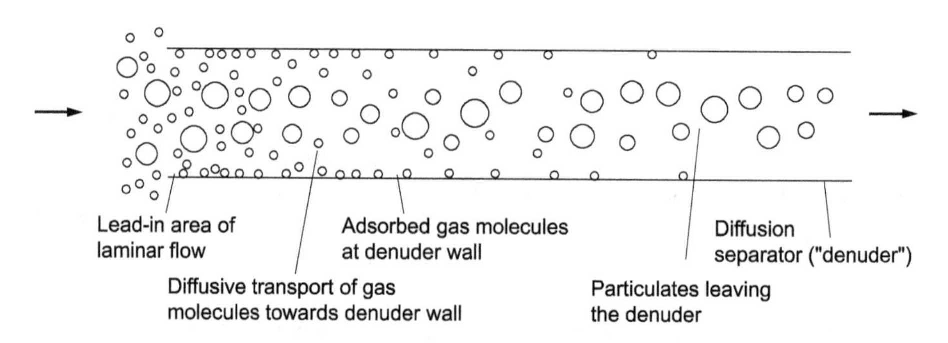

Fig. 21.6: Operating principle of a diffusion separator (denuder) for gas/particle separation.

Figure 21.7 shows a sequentially arranged denuder system for the separation of various gaseous and thermally labile sulfate aerosol constituents.

In this process, the aerosol under investigation is drawn through four sequentially arranged denuders at a constant flow rate:

a) K_2CO_3 or PbO_2 denuder to bind H_2S and SO_2 (T = RT);
b) Activated carbon denuder for the separation of organic volatile sulfur components (T = RT);

c) Cu/CuO denuder for deposition of thermally volatilizable H_2SO_4 (T = 120 °C);
d) Cu/CuO denuder for the separation of thermally volatilizable ammonium sulfate aerosols (T = 220 °C).

After an enrichment period, the sulfate components deposited in the Cu/CuO denuders can be individually converted to SO_2 under a switched-on hydrogen atmosphere after rapid bakeout and continuously quantitatively measured with a flame photometric detector. During the bakeout phase, samples continue to be taken in a parallel arrangement. In this way, the automated quasi-continuous separate measurement of atmospheric ammonium sulfate and sulfuric acid is possible. Analogous measuring methods have been developed for the determination of nitric acid and ammonium nitrate in ambient air.

Diffusion battery for nanoparticle separation. Townsend (1900) already recognized the possibility of inferring the particle size of gas ions by measuring their diffusion losses during penetration through a tube. Gormley & Kennedy (1949) summarized the penetration losses P based exclusively on diffusion separation as follows:

$$P = N/N_0 = 0.819\exp(-3.657\Delta) + 0.097\exp(-22.3\Delta) + 0.032\exp(-57\Delta) + \ldots \quad (5)$$

N = Particle number concentration at channel end;
D = diffusion coefficient;
N_0 = particle number concentration at channel inlet.

Whereby

$$\Delta = Dx/R^2\, v \quad (6)$$

x = Channel length;
R = channel diameter;
v = mean flow velocity.

is.

It is interesting to note that diffusion separation is independent of the material as long as the energy transfer of the colliding carrier gas molecules to the particles suspended in it is exclusively elastic. The question of the selectivity of particle separation based on diffusion can be answered by considering the dependence of the diffusion coefficient D on the particle size $2r_p$ (see Fig. 21.8).

It can be seen that the smaller the particle size, the separation efficiency by diffusion separation increases logarithmically. If particles in the size range below a few hundred nanometers are to be separated, very large channel lengths are required for this purpose according to eq. (5). This can be circumvented by using so-called *diffusion batteries*, a parallel arrangement of plate-shaped channels, tube bundles or stacked screens. A diffusion battery of stacked screens, whose open mesh size is in the size range of about 10–30 μm, has become established. Figure 21.9 shows the dependence of the penetration P *of* monodisperse aerosols through 55 stacked sieves.

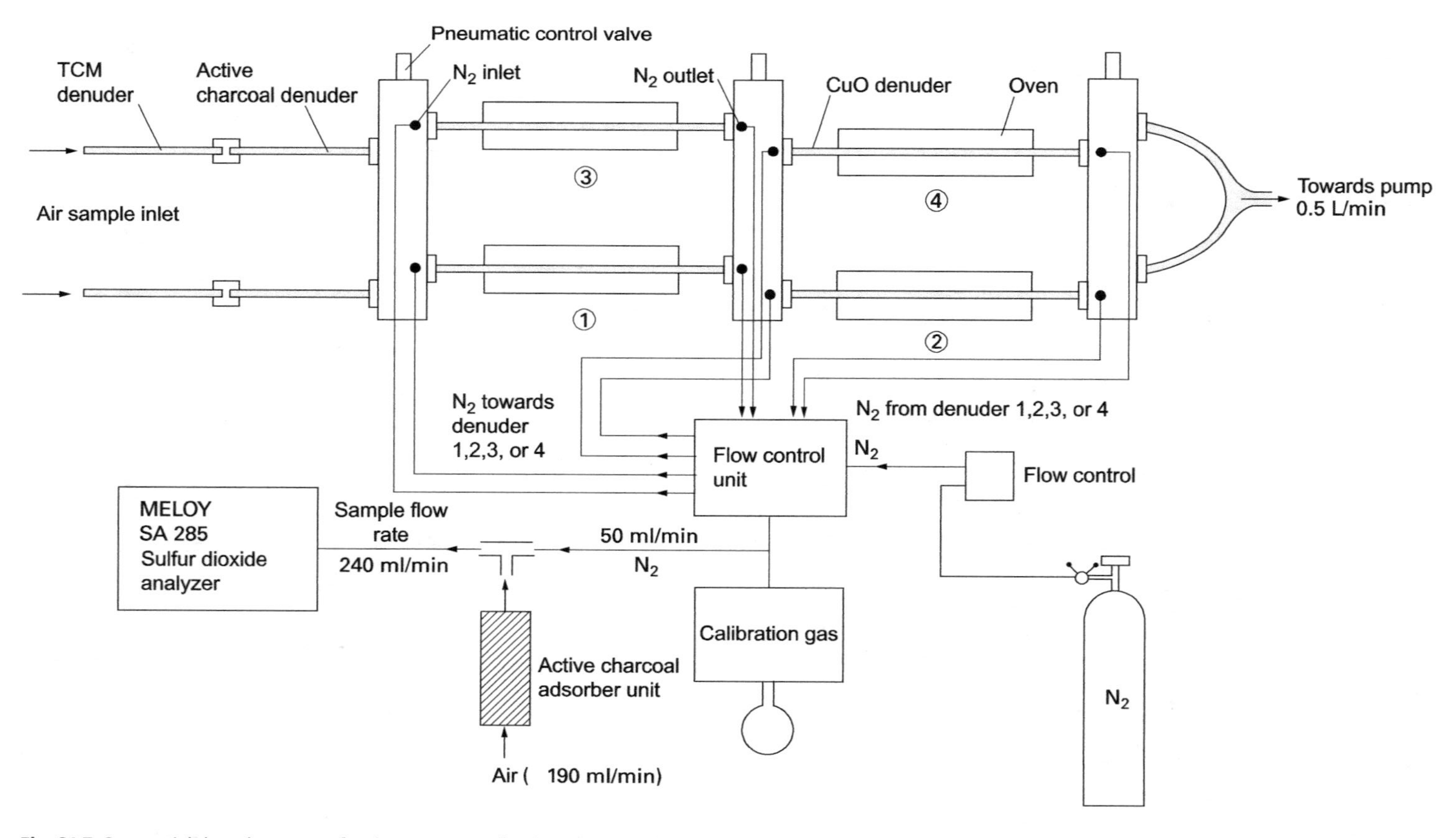

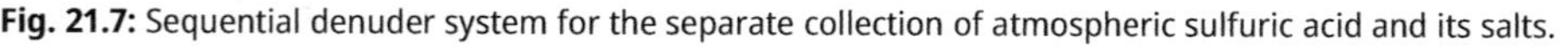
Fig. 21.7: Sequential denuder system for the separate collection of atmospheric sulfuric acid and its salts.

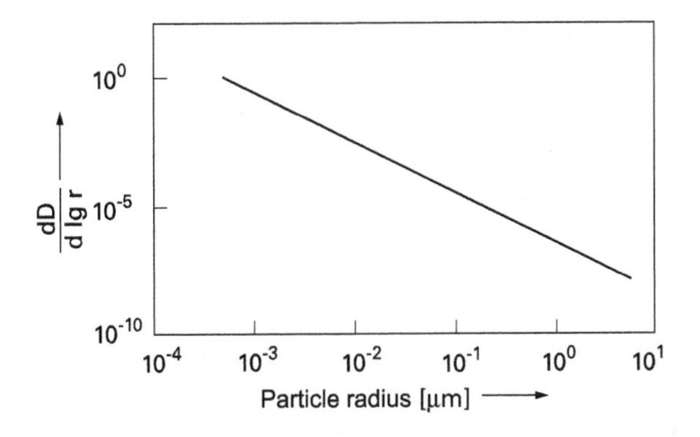

Fig. 21.8: Relative differences in diffusion coefficient as a function of particle size.

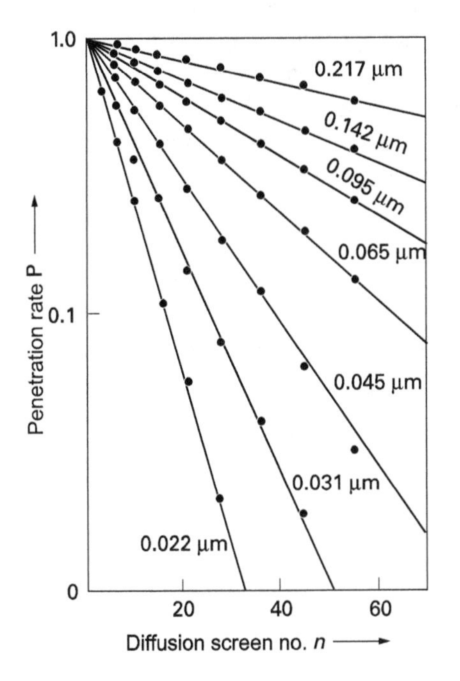

Fig. 21.9: Penetration of nanometer aerosol particles through a screen diffusion battery (SS 400 – sieves).

It can be seen that in the range smaller than 30 nm a resolution of ±1 nm is achieved. In the range below 20 nm particle diameter size resolutions in the Ångstrom range become possible.

21.7 Efficacy and scope of the method

Molecular separators are used in the coupling of gas chromatographs with mass spectrometers as an intermediate element to separate part of the carrier gas before it enters the mass spectrometer and to increase the concentration of the separated compounds contained in the eluate of the gas chromatograph. Since the greater the difference in the molecular weights of the carrier gas and substances, the more effective the process, in this case the gas chromatographs are operated exclusively with hydrogen or helium.

Palladium separators permit practically complete separation of the hydrogen at 100 per cent yield of analysis substance; in order that this can still be transported into the mass spectrometer, either part of the hydrogen must remain in the gas stream or a small quantity of another gas must be added; e.g. with an addition of 1% nitrogen and complete H_2 removal, an enrichment of the analysis substances by a factor of 100 is achieved. The disadvantage of these separators is the relatively high operating temperature.

Separation nozzles and frit separators are much less effective; one can expect enrichment factors of about 50 to 100 with substance losses of 50–80%. The advantage of these devices is that they can also use helium as a carrier gas. The operation of frit separators has proved to be simpler than that of separation nozzles, since in the latter the pressure ratios between the separated and passing parts of the gas stream must be carefully matched.

Diffusion-based separators are of great importance in the nuclear industry for isotope separation and enrichment.

Although the field of application of the dialysis of dissolved substances is essentially in the area of preparative chemistry and health care, the method is also occasionally used for (bio)analytical tasks. It can be used to separate low-molecular compounds from high-molecular ones, which can be important in the study of natural products (e.g. antibodies), among other things.

General literature

E. von Halle & J. Shacter, Diffusion separation methods, in Kirk-Othmer Separation Technology, John Wiley & Sons, Inc, Hoboken, N. J *1*, 816–871 (2008).

R. Krishna, Diffusion in porous crystalline materials, Chemical Society Reviews *41*, 3099–3118 (2012).

Y. Lin, I. Kumakiri, B. Nair, and H. Alsyouri, Microporous inorganic membranes, Separation and Purification Methods *31*, 229–379 (2002).

F. Richter, N. Dauphas, and F.-Z. Teng, Non-traditional fractionation of non-traditional isotopes: evaporation, chemical diffusion and soret diffusion, Chemical Geology *258*, 92–103 (2009).

Literature to the text

Diasolysis

J. Luo, C. Wu, T. Xu, & Y. Wu, Diffusion dialysis-concept, principle and applications, Journal of Membrane Science *366*, 1–16 (2011).

Diffusion separator

Y.-S. Cheng, Instruments and samplers based on diffusional separation, in P. Kulkarni, P. Baron, & K. Willeke, Aerosol Measurement, John Wiley & Sons, Inc, Hoboken, N. J, 365–379 (2011).
J.-C. Wolf & R. Niessner, High-capacity NO_2 denuder systems operated at various temperatures (298–473 K), Analytical and Bioanalytical Chemistry *404*, 2901–2907 (2012).

Diffusion battery

P. Feldpausch, M. Fiebig, L. Fritzsche, and A. Petzold, Measurement of ultrafine aerosol size distributions by a combination of diffusion screen separators and condensation particle counters, Journal of Aerosol Science *37*, 577–597 (2006).
M. Fierz, S. Weimer & H. Burtscher, Design and performance of an optimized electrical diffusion battery, Journal of Aerosol Science *40*, 152–163 (2009).
V. Gomez, F.Alguacil and M. Alonso, Deposition of aerosol particles below 10 nm on a mixed screen-type diffusion battery, Aerosol and Air Quality Research *12*, 457–462 (2012).
Y. Julanov, A. Lushnikov, and V. Zagaynov, Diffusion aerosol spectrometer, Atmospheric Research *62*, 295–302 (2002).

Diaphragms

S. El-Safty, Designs for size-exclusion separation of macromolecules by densely-engineered mesofilters, TrAC, Trends in Analytical Chemistry *30*, 447–458 (2011).
M.-B. Haegg, Membranes in gas separation, in: A. Pabby, S. Rizvi & A. Sastre, in: Handbook of Membrane Separations, CRC Press, Boca Raton, 65–105 (2008).
A. Kumar, Fundamentals of membrane processes, in: T. Zhang et al., Membrane Technology and Environmental Applications, ASCE, Reston 75–95 (2012).
C. Lau, P. Li, F. Li, T.-S. Chung, and D. Paul, Reverse-selective polymeric membranes for gas separations, Progress in Polymer Science *38*, 740–766 (2013).
M. Luque de Castro, Membrane-based separation techniques: dialysis, gas diffusion and pervaporation, in: S. Kolev &I. McKelvie, Comprehensive Analytical Chemistry, Elsevier, Amsterdam *54*, 203–234 (2008).
A. Pabby & Sastre, Hollow fiber membrane-based separation technology: performance and design perspectives, in: M. Aguilar u. J. Cortina, Solvent Extraction and Liquid Membranes, CRC Press, Boca Raton, 91–140 (2008).

P. Pinacci & A. Basile, Palladium-based composite membranes for hydrogen separation in membrane reactors, Woodhead Publishing Series in Energy *55* (Handbook of Membrane Reactors, Volume 1), 149–182 (2013).

D. Sanders, Z. Smith, R. Guo, L. Robeson, J. McGrath, D. Paul, & B. Freeman, Energy-efficient polymeric gas separation membranes for a sustainable future: A review, Polymer *54*, 4729–4761 (2013).

Molecular separators

J. Eerkens & J. Kim, Isotope separation by selective laser-assisted repression of condensation in supersonic free jets, AIChE Journal *56*, 2331–2337 (2010).

Palladium separators

A. Santucci, S.Tosti & A. Basile, Alternatives to palladium in membranes for hydrogen separation: nickel, niobium and vanadium alloys, ceramic supports for metal alloys and porous glass membranes, in: A. Basile, Handbook of Membrane Reactors, Woodhead Publishing Ltd, Cambridge *1*, 183–217 (2013).

Y. Ma, I. Mardilovich, and E. Engwall, Thin composite palladium and palladium/alloy membranes for hydrogen separation, Annals of the New York Academy of Sciences *984* (Advanced Membrane Technology), 346–360 (2003).

Y. Ma & F. Guazzone, Metallic membranes for the separation of hydrogen at high temperatures, Annales de Chimie (Cachan, France) *32*, 179–195 (2007).

S. Yun & S. Oyama, Correlations in palladium membranes for hydrogen separation: A review, Journal of Membrane Science *375*, 28–45 (2011).

Pervaporation

J. Fontalvo Alzate, Design and performance of two-phase flow pervaporation and hybrid distillation prodess, Technische Universiteit Eindhoven, The Netherlands: JWL Boekproducties (2006).

P. Shao & R. Huang, Polymeric membrane pervaporation, Journal of Membrane Science *287*, 162–179(2007).

22 Migration of particles in the gravitational field (sedimentation – flotation)

22.1 General – definitions – centrifugal force

The migration of particles in the gravitational field is called *sedimentation* if the movement is against the gravitational field, it is called *buoyancy* or *flotation*. Sedimentation can occur either in the Earth's force field or by the centrifugal force of a centrifuge. Sedimentation or flotation occurs only when there is a density difference between the particle in question and the surrounding medium.

While sedimentation of coarser particles suspended in liquids generally does not cause any difficulties, Brownian molecular motion prevails in the case of dissolved molecules, causing diffusion, so that normally no sedimentation occurs. Only in extremely high gravitational fields, which are achieved in very fast running so-called *ultracentrifuges* can a sedimentation of molecules be forced, but only if the molecular weights are not too small.

In a circular motion, the radial acceleration b *is* equal to the product of the radius r and the square of the angular velocity ω:[1]

$$b = r \cdot \omega^2. \tag{1}$$

The centrifugal force Z acting on a particle of mass m *is* thus given by

$$Z = m \cdot r \cdot \omega^2. \tag{2}$$

As eq. (2) shows, the centrifugal force grows proportional to the radius, but with the square of the angular velocity. To achieve high gravitational fields, it is therefore most favorable to aim for large angular velocities.

The acceleration experienced by a particle in a *centrifuge* is often given in multiples of the acceleration due to gravity $g = 9.81\ \mathrm{m \cdot s^{-2}}$. Commercially available ultracentrifuges operate at rotational speeds of about 50 000–60 000 per min and centrifugal forces of about 150 000–200 000 g, but much higher values have been achieved.

---

1 Circular motions are described by two quantities. The path velocity v (also called linear velocity) with the dimension $\mathrm{m \cdot s^{-1}}$, which indicates the distance covered by a point on the circumference of the circle in the unit of time, and the angular velocity ω (dimension $\mathrm{s^{-1}}$), which indicates the angle covered by a radius of the circle in the unit of time. The angle α is usually not given in degrees, but in radians (as arc α). The circumference of the circle is $2\pi r$, so for $r = 1$ it is 2π. The angle 1 measured in radians then corresponds to $360°/2\pi = 57.3°$; this quantity is also called the *radian*. The angular velocity $\omega = 1$ therefore means that the radius of the circle covers an angle of 57.3° per second, which corresponds to a number of revolutions of $60/2\pi \approx 9.5$ rev/min.

https://doi.org/10.1515/9783111181417-022

22.2 Separations using heavy liquids

If a mineral mixture is suspended in a liquid whose density is between that of different components of the mixture, the specifically heavier components sink to the bottom and the lighter ones float. Prerequisites for such separations are insolubility of the compounds concerned and exposure of the individual mineral grains.

Aqueous solutions of K_2HgI_4 (Thoulet's solution), thallium malonate + thallium formate solution (Clerici's solution), bromoform and methylene iodide are used as heavy liquids in this method. Intermediate density values can be obtained by adding benzene or ethanol in various proportions to the organic liquids. The separation process can be accelerated by centrifugation. Some examples of mineral separations are given in Tab. 22.1.

Tab. 22.1: Mineral separations by heavy liquids (examples).

Calcite (*d* 2.72)	–	Aragonite (*d* 2.95)
Gypsum (*d* 2.3–2.4)	–	Anhydrite (*d* 2.9–3.0)
Barite (*d* 4.48)	–	Fluorspar (*d* 3.1–3.2)
Sillimanite (*d* 3,2)	–	Disthene (*d* 3.5–3.7)
Quartz (*d* 2.6)	–	Almandine (*d* approx. 4.2)
Feldspar (*d* 2.6–2.7)	–	Pyroxene (*d* 3.1–3.5

22.3 Separations by sedimentation in the gravitational field of the earth

When *settling* suspensions, separations can be achieved in this way, since both coarser and specifically heavier particles sink faster in a liquid than smaller or specifically lighter ones. This process plays a major role in technology, but has little significance for analytical chemistry.

Also worthy of mention is the flotation process, much used in technology (e.g. ore separation), in which fine air bubbles are attached to individual components of mineral mixtures, causing the components concerned to float selectively out of suspension.

22.4 Separations with the aid of the ultracentrifuge

The separation of precipitates by centrifugation (instead of filtration) is to be regarded only as an aid (cf. Part 2, Section 12, Solubility) and is not to be dealt with further here. In contrast, separations of dissolved molecules by ultracentrifugation are to be treated as a separate separation process.

In extremely strong gravitational fields of ultracentrifuges, the sedimentation velocity of dissolved molecules can become greater than the diffusion velocity, so that settling can be achieved. If several types of molecules with different sedimentation velocities are present in the solution, separation effects occur; however, it is not readily possible to obtain useful separations, but in principle only a portion of the slowest sedimenting component in the upper part of the liquid can be obtained pure. The method is similar in this respect to the frontal analysis discussed earlier; it is not used in this form for separations; however, it performs excellent service in determining the molecular weight of macromolecules and in determining the uniformity of isolated high molecular weight natural products and of fractions of synthetic polymers.

Good separations, on the other hand, can be obtained – as first shown by Pickels (1942) – by centrifuging solutions with a density gradient. A solution with an upwardly decreasing density is placed in the centrifuge tube and overlaid with a narrow zone of the substance mixture to be separated. During centrifugation, molecular species with different sedimentation rates migrate downwards at different rates and form separate zones after some time. If centrifugation is carried out for too long, all components will eventually be at the bottom of the vessel. The concentration gradient need not be very steep; it is only necessary so that the individual zones do not flow together again when the centrifuge is switched off as a result of the overlapping of less dense zones by denser ones.

Another method also works with a density gradient, but according to a different principle: The mixture of substances to be separated is homogeneously distributed throughout the solution, which is additionally given a density gradient by the addition of a dissolved inert material. During centrifugation, each substance type migrates to the point of the gradient corresponding to its own density and remains stationary here. The width of the zones formed in this way depends on the steepness of the density gradient and the strength of the gravitational field. In contrast to the method described first, where differences in sedimentation rates were decisive for the separations, here the density differences are the decisive factor.

The density gradients are prepared by mixing pure solvent with solutions of low molecular weight substances, usually sugars, in the usual way. It is simpler to dissolve a compound of relatively high molecular weight, e.g. Cs-Br, together with the analytical sample; during centrifugation, the cesium bromide forms the density gradient by itself. The various procedures are shown in Fig. 22.1.

22.5 Efficacy and scope of the method

Separations by heavy liquids belong to the field of phase analysis (cf. Part 2, Chap. 13, Extraction), since individual crystal species can be isolated here. The procedure is a valuable addition to the few other methods available in this field, but its application is limited because no liquids with densities above about 3.3 g / ml are available. Currently the separation of microplastic particles from sea sediment is widely in use. It is

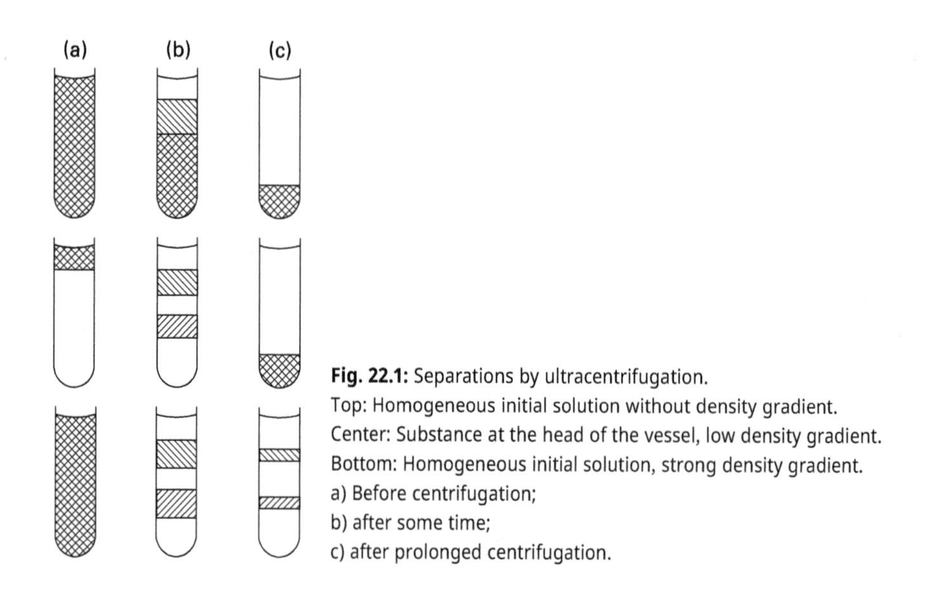

Fig. 22.1: Separations by ultracentrifugation.
Top: Homogeneous initial solution without density gradient.
Center: Substance at the head of the vessel, low density gradient.
Bottom: Homogeneous initial solution, strong density gradient.
a) Before centrifugation;
b) after some time;
c) after prolonged centrifugation.

difficult to make general statements about the effectiveness of the separations, since the intergrowth and degree of comminution of the minerals are of decisive influence.

The ultracentrifuge – insofar as it is used for separations – is primarily employed in the field of high-molecular biochemically significant compounds. Larger particles, e.g. bacteriophages and virus species, have also been separated. The method is experimentally quite complex, but has a wide range of applications.

General literature

N. Boujtita & Noles, Ultracentrifugation – application possibilities and obstacles, LaborPraxis *34*, 46–48 (2010).

J. Lebowitz, M. Lewis, & P. Schuck, Modern analytical ultracentrifugation in protein science: a tutorial review, Protein Science *11*, 2067–2079 (2002).

A. Letki & N. Corner-Walker, Centrifugal separation, in: A. Seidel, Kirk-Othmer Encyclopedia of Chemical Technology (5th Edition) John Wiley & Sons, Inc., Hoboken *5*, 505–551 (2004).

A. Lynch, G. Harbort & M. Nelson, History of flotation. (Australasian Institute of Mining and Metallurgy Spectrum Series, Number 18), Australasian Institute of Mining and Metallurgy, Victoria, Australia (2010).

H. Nirschl, Effect of physical chemistry on separation of nanoparticles from liquids, Chemie Ingenieur Technik *79*, 1797–1807(2007).

L. Svarovsky, Sedimentation, in A. Seidel, Kirk-Othmer Encyclopedia of Chemical Technology (5th Edition), John Wiley & Sons, Inc, Hoboken *22*, 50–71 (2006).

Literature to the text

Flotation

S. Farrokhpay, The significance of froth stability in mineral flotation – A review, Advances in Colloid and Interface Science *166*, 1–7 (2011).

J. Kohmuench, M. Mankosa, E. Yan, H. Wyslouzil, L. Christodoulou, & G. Luttrell, Advances in coarse particle recovery – fluidized-bed flotation, Publications of the Australasian Institute of Mining and Metallurgy *7/2010*, 2065–2076 (2010).

S. Lu, Z. Song & C. Sun, Review on some research methods of mineral crystal chemistry and computer simulation on flotation, Publications of the Australasian Institute of Mining and Metallurgy *7/2010*, 3269–3275 (2010).

C. Luo, H. Li, L. Qiao, & X. Liu, Development of surface tension-driven microboats and microflotillas, Microsystem Technologies *18*, 1525–1541 (2012).

J. Rubio, M. Souza, and R. Smith, Overview of flotation as a wastewater treatment technique, Minerals Engineering *15*, 139–155 (2002).

L. Wang, M.-H. Wang, N. Shammas & M. Krofta, Innovative and cost-effective flotation technologies for municipal and industrial wastes treatment, in: Y.-T. Hung, L. Wang u. N. Shammas, Handbook of Environment and Waste Management, World Scientific Publishing Co. Pte. Ltd, Singapore, Singapore 1151–1176 (2012).

Separation by sedimentation

H. Imhof, J. Schmid, R. Niessner, N. Ivleva, and C. Laforsch, A novel, highly efficient method for the separation and quantification of plastic particles in sediments of aquatic environments, Limnology and Oceanography: Methods *10*, 524–537 (2012).

A. Majumder, Settling velocities of particulate systems – a critical review of some useful models, Minerals & Metallurgical Processing *24*, 237–242 (2007).

S. Vesaratchanon, A. Nikolov, and D. Wasan, Sedimentation in nano-colloidal dispersions: effects of collective interactions and particle charge, Advances in Colloid and Interface Science *134–135*, 268–278 (2007).

S. Yu, Particle sedimentation, in: H. Masuda, K. Higashitani, Y. Ko & H. Yoshida, Powder Technology Handbook, CRC Press, Boca Raton, 133–141 (2006).

Ultracentrifuge

T. Arakawa, T. Niikura, Y. Kita, and F. Arisaka, Structure analysis of short peptides by analytical ultracentrifugation: review, Drug Discoveries & Therapeutics *3*, 208–214 (2009).

T. Arakawa, T. Niikura, Y. Kita, and F. Arisaka, Structure analysis of short peptides by analytical ultracentrifugation: review, Drug Discoveries & Therapeutics *3*, 208–214 (2009).

H. Cölfen & W. Wohlleben, Analytical ultracentrifugation of latexes, in: L. Gugliotta &J. Vega, Measurement of Particle Size Distribution of Polymer Latexes, Research Signpost, Trivandrum, India, 183–222 (2010).

J. Cole, J. Correia, & W. Stafford, The use of analytical sedimentation velocity to extract thermodynamic linkage, Biophysical Chemistry *159*, 120–128 (2011).

J. Gabrielson & K. Arthur, Measuring low levels of protein aggregation by sedimentation velocity, Methods (Amsterdam, Netherlands) *54*, 83–91 (2011).

A. Ortega, D. Amoros, & J. Garcia de la Torre, Global fit and structure optimization of flexible and rigid macromolecules and nanoparticles from analytical ultracentrifugation and other dilute solution properties, Methods (Amsterdam, Netherlands) *54*, 115–123 (2011).

K. Planken & H. Cölfen, Analytical ultracentrifugation of colloids, Nanoscale *2*, 1849–1869 (2010).

23 Separation of particles in the crossed force field

23.1 General – definitions – asymmetric field-flow fractionation – applications

The application of crossed force fields allows in principle a continuous separation of particles. Examples of this have already been presented, such as continuous two-dimensional column chromatography in Chap. 10.7, the differential mobility analyzer in Chap. 19.4.5 and continuous two-dimensional electrophoresis in Chap. 20.8.2.

Common to all these methods is the use of a laminar flowing carrier stream containing the analyte molecules or particles as one acting force component, either gaseous or liquid. A second, orthogonally acting force component ensures a deflection of the analyte particles from the original flow lines and thus a spatial separation.

Giddings (1966) introduced a family of separation techniques that are becoming increasingly popular for the separation of macromolecules down to particles in the size range of several hundred μm. These field flow techniques, called *field flow fractionation: FFF* in Anglo-Saxon, represent an interesting possibility for rapid particle separation. Figure 23.1 illustrates this principle.

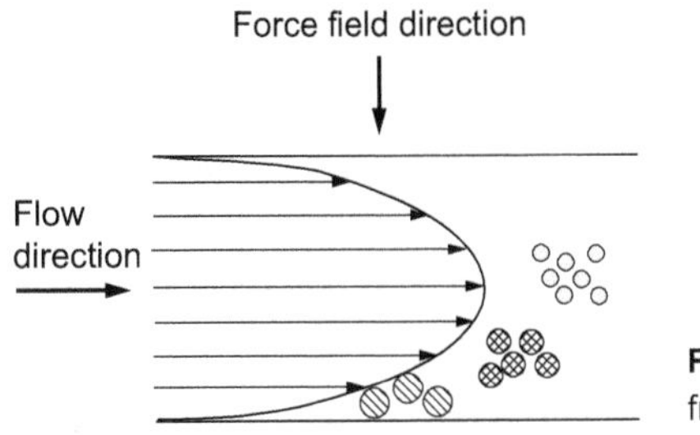

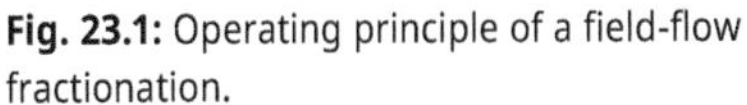

Fig. 23.1: Operating principle of a field-flow fractionation.

Particle separation takes place in a rectangular flow channel with a channel height w in the range of about 100 μm. The particles to be separated are in a laminar flow. The orthogonally acting force component F causes the deflection of the particle swarm in the direction of the lower accumulation wall. This deflecting force is in turn counteracted by the Brownian molecular motion, which causes small particles in particular to remain in the central, fast flow lines for a longer time. Thus, while larger particles move along the accumulation wall in the slower flow regions on a time-average basis, the small particles leave the separation channel early. The resulting change in retention time can be described in simplified terms as follows:

$$t_r/t_o = F\ w/6k\ T \tag{1}$$

Here, t_r is the retention time of a particle deflected by an orthogonal force component, compared to the dwell time t_o of a non-deflected particle. The time delay here follows,

https://doi.org/10.1515/9783111181417-023

as a first approximation, the applied force *F*, which should be in the range of 10–16 N for successful separation. The currently known applications of crossed force fields for particle separation are summarized in Tab. 23.1.

Tab. 23.1: Orthogonally acting force fields for particle separation in the FFF technique.

Force component 1	Force component 2
Hydrodynamic flow	Hydrodynamic flow
Hydrodynamic flow	Sedimentation/centrifugal force
Hydrodynamic flow	Dielectrophoresis
Hydrodynamic flow	Thermophoresis
Aerodynamic flow	Thermophoresis
Aerodynamic flow	Photophoresis
Hydrodynamic flow	Photophoresis
Hydrodynamic flow	Magnetic force
Aerodynamic flow	Magnetic force
Aerodynamic flow	Diffusiophoresis

Asymmetric Flow Field–Flow Fractionation (AF^4). A widely used setup of an *AF^4 separation system* is shown in Fig. 23.2.

The accumulation wall of the separation channel is made of a porous membrane. The orthogonal separation force is generated by the onset of flow division between the channel outlet and the permeation of the solvent through the membrane. In this application, the separation is not performed continuously, but sequentially intermittently, analogous to a chromatographic separation. The hydrodynamic force component is generated by a commercial HPLC pump. An ordinary scattered light or UV detector is also used for detection.

Fractograms. The separation process is usually recorded as a *fractogram*. The particles leaving the separation channel are registered and recorded by a continuous optical detector. A typical fractogram of various monodisperse polystyrene calibration latices is shown in Fig. 23.3. As can be seen, the time required for separation in the lower nanometer range is only a few minutes. Through the calibration process, the retention time here can be assigned to a hydrodynamic particle diameter. When using known biomacromolecules such as bovine serum albumin or similar, the molecular size of unknown macromolecules can be determined.

Applications. While AF^4 is already widely used in nanomaterial and polymer research for size characterization, thermophoresis is used for continuous sampling of aerosol particles from exhaust gases. Bacteria, viruses and proteins have also been successfully and gently separated. Through the combination with element- or molecule-selective continuous detection techniques, the particle size distribution of a species or element form is accessible. This enables a better understanding of sources and sinks, especially in environmental research.

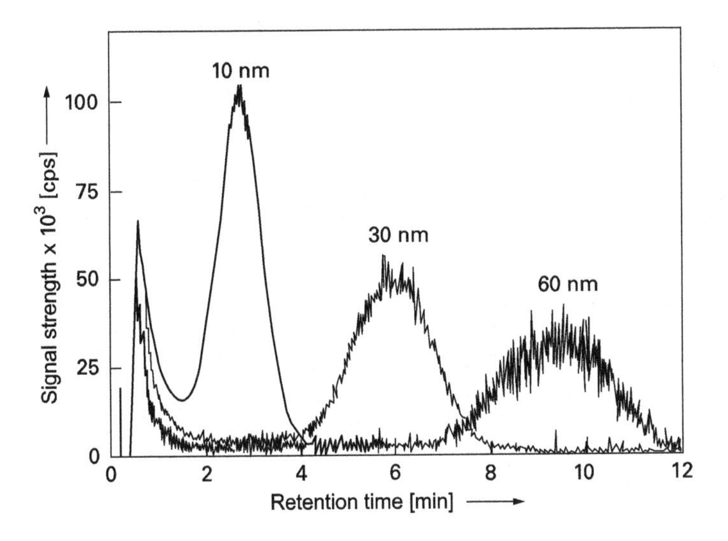

Fig. 23.2: Fractograms of various quasi-monodisperse hydrosols.

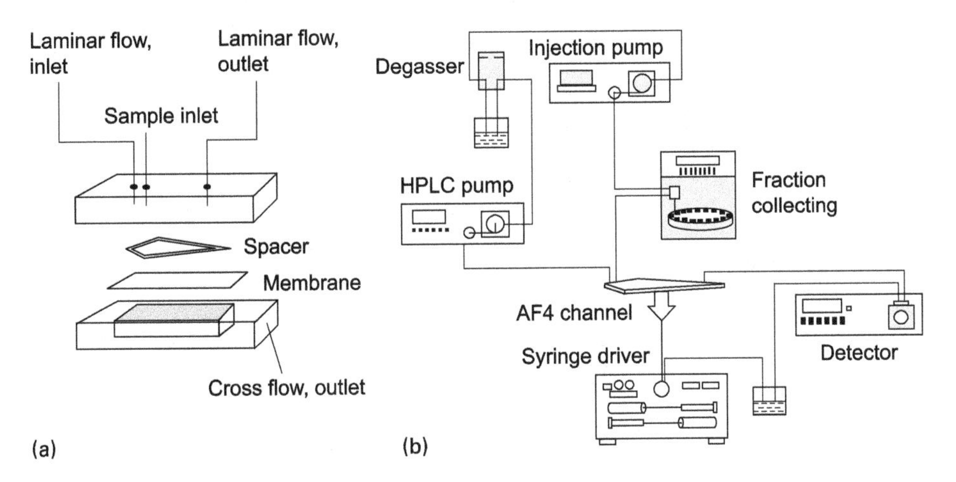

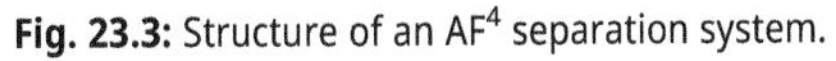
Fig. 23.3: Structure of an AF^4 separation system.

Literature to the text

J. Chmelik, Applications of field-flow fractionation in proteomics: presence and future, Proteomics textit 7, 2719–2728 (2007).

A. Exner, M. Theisen, U. Panne, and R. Niessner, Combination of asymmetric flow field-flow fractionation (AF4) and total-reflection X-ray fluorescence analysis (TXRF) for determination of heavy metals associated with colloidal humic substances, Fresenius' Journal of Analytical Chemistry *366*, 254–259 (2000).

C. Haisch, C. Kykal & R. Niessner, Photophoretic velocimetry for the characterization of aerosols, Analytical Chemistry (Washington, DC, United States) *80*, 1546–1551 (2008).

C. Helmbrecht, R. Niessner, and C. Haisch, Photophoretic velocimetry – a new way for the in situ determination of particle size distribution and refractive index of hydrocolloids, Analyst (Cambridge, United Kingdom) *136*, 1987–1994 (2011).

C. Helmbrecht, R. Niessner & C. Haisch, Photophoretic velocimetry for colloid characterization and separation in a cross-flow setup, Analytical Chemistry (Washington, DC, United States) *79*, 7097–7103 (2007).

J. Liu, J. Andya, & S. Shire, A critical review of analytical ultracentrifugation and field flow fractionation methods for measuring protein aggregation, AAPS Journal *8*, E580–E589 (2006).

S.-W. Nam, S. Kim, J.-K. Park, & S. Park, Dielectrophoresis in a slanted microchannel for separation of microparticles and bacteria, Journal of Nanoscience and Nanotechnology *13*, 7993–7997 (2013).

H. Prestel, L. Schott, R. Niessner & U. Panne, Characterization of sewage plant hydrocolloids using asymmetrical flow field-flow fractionation and ICP-mass spectrometry, Water Research *39*, 3541–3552 (2005).

P. Reschiglian & M.H. Moon, Flow field-flow fractionation: A pre-analytical method for proteomics, Journal of Proteomics *71*, 265–276 (2008).

B. Roda, A. Zattoni, P. Reschiglian, M.H. Moon, M. Mirasoli, E. Michelini, and A. Roda, Field-flow fractionation in bioanalysis: a review of recent trends, Analytica Chimica Acta *635*, 132–143 (2009).

Index

absorption 120
Activated carbon 125
adsorption 120
adsorption isotherms 120
adsorptive filtration 146
advanced oxidation process 12
AF^4 separation system 400
affinity chromatography 143
Alumina 139
amalgam exchange 68, 76
amplified distillation 307
annular gap bubble columns 304
antigen-antibody reaction 215
appearance potential 331
ascending development 158
atmospheric pressure ionization, APCI 333
azeotrope 269

batch process 145
BET isotherm 121
boiling analysis 277
boiling diagram 268
breakthrough curve 49

capillary columns 104
capillary electrophoresis 362
carrier electrophoresis 350
carrier gas method 290
cascade 36
Cellulose 140
Cellulose filters 206
cementation 235
centrifugal chromatography 160
centrifuge 393
change of state 9
chemisorption 120
chromatography 41, 137
circular chromatography 56
Clausius-Clapeyron equation 264
closed loop 296
closed loop distillation 298
cloud point 66
codistillation 292
condensation 320
continuous column chromatography 162
co-precipitation 203
corrected retention volume 48
countercurrent methods 57
Cox diagram 264
cross-flow processes 58
cross-reactivity 5
cryodiffusion 286
crystallization 253
cyclotron resonance 341

Dalton's partial pressure law 267
dead volume 48
decomposition potential 217
decontamination factor 4
denuder 385
depletion factor 4
descending method 158
developing 42
dextran gels 141
diaphragm 226
diasolysis 381
differential ionization 331
differential mobility analyzer (DMA) 342
diffusion 378
diffusion batteries 387
diffusion coefficient 378
diffusion separators 385
diffusion sink 381
dip stick techniques 159
direct isotope dilution 15
directional freezing out 255
directional solidification 255
disc electrophoresis 367
displacement method 154
displacer 50
distillation 263
distillation curve 276
distribution coefficient 14
distribution isotherm 28
double-focusing 339
drift tube 336
drying 284
drying gun 284
dynamic headspace 99

electric sector field 339
electro-chromatography 359
electrode potential 218
electrode reactions 217

https://doi.org/10.1515/9783111181417-024

electrodialysis 350
electron impact ionization 331
electroosmosis 359
electropherography 350
electrophoresis 350
electrospray ionization 333
eluate 42
eluotropic series 143
elution 42
elution curve 42
energy dispersion 335
enrichment factor 3
entrainment effect 203
entrainment isotherms 230
equilibrium diagram 268
evaporation analysis 315
exchange capacity 170
exchange isotherms 176
exchange reactions 9
extraction 61, 243
extractive distillation 305

field-flow fractionation 399
field ionization 334
filtration 206
flotation 393
fractional distillation 285
fractional precipitation 237
fractional sublimation 314
fractogram 400
fragmentation 332
freeze-drying 313
freezing out 320
Freundlich isotherm 120
frit separators 384
frontal analysis 49
FT-ICR-MS 341
functional groups 167

gap separator 385
gas chromatography 103
gas permeation 382
gas-liquid chromatography 103
gas-solid chromatography 103
Gaussian distribution 43
getter effect 282
Golay columns 104
gradient elution 44

half-power width HW 54
head space analysis 278
Henry's Law 95
HETP (= height equivalent to a theoretical plate) 50, 301
high pressure liquid chromatography 88
high-flux dialyzers 383
high-voltage electrophoresis 362
horizontal distillation 323
hot extraction 282

immunoaffinity chromatography 146
immuno-electrophoresis 372–373
IMS 336
inclusion compounds 212
inhibitors 143
internal electrolysis 226
inverse isotope dilution 16
ion exchange chromatography 186
ion exchange membranes 380
ion exchangers 141
ion mobilities 352
ion source 330
Ion traps 340
ionic liquids 106
ionophoresis 350
isoelectric spectrum 369
isotachophoresis 367
isotope dilution method 15

Kjeldahl digestion 280
Kohlrausch persistent function 366
Kohlrausch regulating function 367
Kováts retention index 114

lab-on-the-chip 87
Langmuir isotherm 121
ligand exchange 9
liquid-liquid extraction 61, 243
lyophilization 314

m/e ratio 335
magnetic sector field 339
masking 9
mass spectrometry 329
matrix 170
membrane filters 207
metal-organic frameworks (MOFs) 126

micellar extraction 64
microdiffusion 288
mobile phase 39, 108
molecular distillation 287
molecular separators 384
molecular sieves 123
molecularly imprinted polymers (MIPs) 143
monolithic columns 147
moving phase 39

Nernst distribution 28
Nernst's distribution law 61
Nernst's potential equation 219
neutral salt adsorption 182
normal hydrogen electrode 218
normal potential 220
Nylon 142

one-sided repetition 34
Orbitrap 341
overhead distillation 285
overvoltage 220

packed columns 303
paper chromatography 56, 159
paper electrophoresis 361
perforators 79
Perlon 142
permeability constant 380
permselectivity 381
pervaporation 383
phase analysis 243, 248
photoionization 334
physisorption 120
Poisson distribution 43
polyamides 142
polymer phase separation 64
porous membranes 380
post-precipitation 203, 234
potential serie 220
potentiostats 227
precipitation chromatography 238, 259
precipitation exchange 186, 235, 237
precipitation from homogeneous solution 145, 204
precipitation in a suitable medium 205
precipitation with ion exchangers 238
protein precipitation (protein crystallization 215
purge-and-trap 101
pyrohydrolysis 316

quadrupole mass spectrometers 340
quantitative separation 4

reaction gas chromatography 115
recirculation processes 34
recrystallization 258
redox buffer system 218
reflux 299
reflux ratio 301
regeneration 169
regulating function 366
relative retention time $t_{i,St}$ 48
relative retention volume $V_{i,St}$ 48
re-precipitation 236
residual current 217
resin ion exchangers 172
resolving power A 345
resolving power R 55
resonance-enhanced multiphoton ionization (REMPI) 334
resublimation 314
retention 46
retention time 47
retention volume 46
reverse osmosis 383
reversed gas chromatography 111
reversed phase 86
reversed phase chromatography 159
R_f value 46
ring-oven method 244
rotary drum columns 304
rotary evaporators 286

salt-exchange chromatography 192
saturated 196
saturation analysis 18
sedimentation 393
selectivity 5
separating funnels 76
separation factor 3–4
separation nozzles 384
separation stages 50
Sephadex 141
septum 105
short-circuit electrolysis 224
short-path distillation 287
silica gel 125, 140
SIMS 332
single-drop microextraction: SDME 78

soft ionization 332
solid phase microextraction (SPME) 99, 145
solidification 255
solid-state mass spectrometry 329
solubility product 196
solubilization chromatography 192
solution membranes 380
sorption 120
specific 5
spinning belt columns 304
splitting 105
stages 299
static headspace 99
stationary liquid phase 106
stationary phase 39
steam distillation 292
sublimand 312
sublimate 312
sublimation 312
substoichiometric reagent addition 18
substrate 143
supercritical fluid 108, 244
supercritical fluid chromatography 244
supercritical fluid chromatography: SFC 108
supersaturated 196
supported liquid membranes (SLM) 81
synergistic effect 66

tailing 45
temperature gradient 111
temperature programming 111
theoretical plate 50, 301
thin-layer chromatography 56, 158
thin-layer chromatography plates 158
time-of-flight mass spectrometer (TOF-MS) 335
Tiselius method 358
total reflux distillation 299
trace catcher 230
trays 299
triangular scheme 37

ultracentrifuges 393
unsaturated 196

vacuum melting methods 282
van Deemter equation 109, 151
volatility 275

Zechmeister's series 138
zone electrophoresis 350, 359
zone melting method 256
zone sharpening 364

Printed in the USA
CPSIA information can be obtained
at www.ICGtesting.com
LVHW080157140624
783111LV00003B/337